LIPOPROTEIN DEFICIENCY SYNDROMES

ADVANCES IN EXPERIMENTAL MEDICINE AND BIOLOGY

Recent Volumes in this Series

LIPOPROTEIN DEFICIENCY SYNDROMES

Edited by

Aubie Angel

Institute of Medical Sciences
University of Toronto
Toronto, Ontario, Canada

and

Jiri Frohlich

Shaughnessy Hospital
University of British Columbia
Vancouver, British Columbia, Canada

Springer Science+Business Media, LLC

Library of Congress Cataloging in Publication Data

International Conference on Lipoprotein Deficiency Syndromes (1985: Vancouver, B.C.)
    Lipoprotein deficiency syndromes.

    (Advances in experimental medicine and biology; v. 201)
    "Proceedings of an International Conference on Lipoprotein Deficiency Syndromes, held in conjunction with the Tenth Canadian Lipoprotein Conference, May 13–14, 1985, in Vancouver, British Columbia, Canada"—T.p. verso.
    Includes bibliographies and index.
    1. Hypolipoproteinemia—Congresses. I. Angel, Aubie. II. Canadian Lipoprotein Conference (10th: 1985: Vancouver, B.C.) III. Title. IV. Series. [DNLM: 1. Lipoproteins—deficiency—congresses. W1 AD559 v.201 / QU 85 I5917L 1985]
    RC632.H92I57   1985   616.3'99   86-16852
    ISBN 978-1-4684-1264-2

ISBN 978-1-4684-1264-2   ISBN 978-1-4684-1262-8 (eBook)
DOI 10.1007/978-1-4684-1262-8

Proceedings of an International Conference on Lipoprotein Deficiency Syndromes, held in conjunction with the tenth Canadian Lipoprotein Conference, May 13–14, 1985, in Vancouver, British Columbia, Canada

© 1986 Springer Science+Business Media, LLC, New York
Orginally published by Plenum Press
Softcover reprint of the hardcover 1st edition 1986

PREFACE

Current interest in lipoprotein deficiency states stems from the
growing realization of their importance in the etiology of premature
coronary heart disease. While hypercholesterolemia and coronary heart
disease risk are strongly correlated in their etiologic relationship, it
is becoming equally clear that deficiencies in HDL, whether congenital or
acquired, also enhance the risk for the future development of coronary
atherosclerosis. This has led to renewed attention to the lipid
hypothesis and realization of the fact that each lipoprotein class and
apoprotein species has specific functions in the transport and cellular
uptake of various lipids.

It is a truism that a biochemical correlate of disease once identified
is subsequently recognized with increasing frequency in clinical medicine.
The story of HDL was no exception. Indeed hypoalphalipoproteinemia
appears to be a disease of high prevalence approaching and perhaps even
exceeding that of familial hypercholesterolemia. Its clinical significance escaped our notice for many years largely due to a heavy emphasis on
hypercholesterolemia and to difficulties in measuring HDL reliably.

Recognizing that clinical assessment of patients suspected of
lipoprotein disorders requires understanding of newer concepts, we thought
it timely to assemble leaders in this field to highlight advances in
understanding the lipoprotein deficiency syndromes and their etiologic
mechanisms. The four chapters of this book represent areas of major
interest. We have avoided the historic context and emphasized relationships between lipoprotein deficiencies, their molecular mechanisms and
clinical correlations. The first chapter focuses on the relationship
between plasma HDL deficiency and atherosclerosis. Clinically relevant
relationships are identified, and various disease states commonly
associated with premature coronary disease are covered. Hypoalphalipoproteinemia is not a single disease but a multitude of syndromes with
varied etiologies. The second chapter explores these syndromes discussing
both the common autosomal dominant disorder and rare, mostly recessive,
diseases differing in molecular etiologies. In chapter three lecithin:
cholesterol acyl-transferase (LCAT), the enzyme responsible for plasma
cholesterol esterification is discussed in detail. The role of LCAT in
the context of cellular cholesterol efflux is also addressed. Numerous
hypertriglyceridemic syndromes are associated with deficiencies or defects
in lipoprotein lipase or hepatic lipase activity, apoprotein CII and apo
E. These are covered in chapter four. Finally, scattered throughout the
book are papers with recent insights in the molecular biology of
apoprotein synthesis and metabolism. Here the molecular etiology of
lipoprotein deficiency syndromes are explored at the genomic level.

This area will be of particular interest to those wishing to
understand the molecular biology of lipoprotein disorders.

We believe that this book will be of use to clinical lipidologists, resident trainees and research fellows as well as graduate students interested in both research and clinical aspects of lipoprotein metabolism.

A. Angel
J. Frohlich

ACKNOWLEDGMENTS

An International Conference on Lipoprotein Deficiency Syndromes was held in conjunction with the Tenth Canadian Lipoprotein Conference on May 13 to 15, 1985 at the University of British Columbia, Vancouver, B.C., Canada. This book contains the manuscripts presented by guest faculty, who are leaders in the the area of lipoprotein deficiency diseases and their cellular and molecular biology. We gratefully acknowledge the principal sponsorship of British Columbia and Canadian Heart Foundations and Bristol Laboratories of Canada.

The following Companies also supported the meeting; Abbott Laboratories, Beckman, Fisher Scientific, Herdt & Chanton, Hoechst, Hoffmann-La Roche, Kodak Company, Merck Frosst, Miles Laboratories, Parke-Davis, Pfizer, Roche Laboratories, Sandoz, Searle, CIBA Laboratories.

Finally, we are particularly indebted to Dr. Hayden Pritchard and Ms. Lori DiMonte for assisting in the development of the meeting. We also thank Ms. Lori DiMonte for organizing and preparing the manuscripts for publication.

A. Angel
J. Frohlich

CONTENTS

GENETIC HIGH DENSITY LIPOPROTEIN DEFICIENCY STATES AND ATHEROSCLEROSIS

Ernst J. Schaefer, Judith R. McNamara, Carol J. Mitri and
Jose M. Ordovas

Lipid Metabolism Laboratory, Human Nutrition Research Center
on Aging at Tufts University, 711 Washington Street
Boston, MA 02111

INTRODUCTION

High density lipoprotein (HDL) as found in human plasma have a density
of 1.063-1.21 g/ml, and are composed (weight percent) of approximately 50%
protein, 25% phospholipid, 20% cholesterol, and 5% triglyceride (1).
Fluctuations in HDL levels have been associated mainly with alterations in
HDL_2 (density, 1.063-1.125 g/ml), rather than HDL_3 (density, 1.125-1.21
g/ml (2). Apolipoproteins (apo) A-I and A-II are the major proteins of
HDL. Minor constituents include apoB, apoC-I, apoC-II, apoC-III, apoD,
apoE, apoF, apoG, apoLp(a) (3-12). Apolipoprotein not only have structural roles in HDL particles, but have other functions as well. ApoA-I and
apoC-I have both been reported to activate lecithin/cholesterol acyltransferase (LCAT), while apoA-II has been reported to enhance hepatic lipase
activity (12-15). ApoC-II activates the enzyme lipoprotein lipase, while
apoC-III has been shown to inhibit hepatic chylomicron remnant uptake
(16,17). Both apoB and apoE can bind to the apoB,E receptor (LDL receptor)
on various cell surfaces, while apoE also binds to the liver apoE receptor,
which is essential for chylomicron remnant uptake (18,19). The low density
lipoprotein (LDL) receptor appears to be crucial for regulating LDL levels
in plasma, as well as for maintaining intracellular cholesterol homeostasis.
HDL particles can bind to specific sites on various cell surfaces (20).

Constitutents of HDL can be derived by direct synthesis of HDL particles by the liver or the intestine, or as a result of the catabolism of
chylomicrons and very low density lipoproteins (VLDL) (21-31). HDL can
also pick up lipid constituents, especially free cholesterol, from
peripheral cells. Free cholesterol, apolipoproteins, and the polar head
groups of the phospholipids, are mainly on the surface of HDL particles,
while fatty acids of phospholipids, cholesterol ester and triglyceride are
found predominantly in the core.

After free cholesterol is picked up by HDL particles it is esterified
by the action of (LCAT), and is transferred to the core of the particle.
Cholesterol ester can then be transferred from HDL to other lipoproteins
by the action of a cholesterol ester transfer protein. Phospholipids can
also be exchanged to other lipoprotein particles. Triglyceride-rich
lipoproteins such as chylomicron remnants can serve as acceptors of
cholesterol ester from HDL. HDL_3 can be converted to particles in the
HDL_2 density region by picking up free cholesterol as well as phospholipid.

1

Such particles can then be cycled back into the HDL$_3$ density range by the transfer of cholesterol ester and other lipids to other lipoprotein particles, as well as by the action of hepatic lipase, an enzyme important for hydrolysis of phosopholipid and triglyceride. In addition the C and E apolipoproteins can shuttle back and forth between HDL and triglyceride-rich particles.

The plasma residence times of different HDL constituents may vary greatly, and current evidence suggests that HDL are not catabolized as a unit, but rather that individual constituents have differing catabolic rates. Studies in humans indicate that apoA-I is catabolized at a somewhat faster fractional rate than apoA-II, while animal studies have shown that the cholesterol components of HDL are catabolized more rapidly than the protein constituents (32,33). Particles containing only apoA-I within HDL may be catabolized significantly more rapidly than particles containing both apoA-I and apoA-II (34).

There has been significant recent interest in the constituents of HDL because decreased plasma concentrations of HDL cholesterol as well as apoA-I have been shown to be correlated with premature coronary artery disease (35-39). Decreased levels of HDL constituents in plasma have been associated with obesity, sedentary life style, hypertriglyceridemia, diabetes mellitis, as well as male sex (40). The plasma concentrations of HDL constituents can be increased by estrogen administration, and decreased by administration of various androgenic compounds (40,41). The increased HDL protein levels found in females have been shown to be due to increased production of apoA-I and apoA-II (32). Moreover estrogen administration has been shown to increase apoA-I production in women (41). HDL metabolism will be covered in greater detail in the capter by Dr. Krauss.

A number of genetic HDL deficiency states have been reported (42). All of these disorders are extremely rare with the exception of familial hypoalphalipoproateinemia. Other HDL deficiency states include familial LCAT deficiency, fish eye disease, the various apolipoprotein A-I variant states (apoA-I$_{Milano}$, apoA-I$_{Munster\ 1-3}$, apoAI$_{Giessen}$, and apoA-I$_{Marburg}$), HDL deficiency with planar Xanthomas, Tangier disease and the familial apolipoprotein A-I and C-III deficiency states, (variant I and variant II). The clinical and laboratory features of these disorders are summarized in tables 1 and 2.

Familial Hypoalphalipoproteinemia

This common disorder has been associated with isolated HDL deficiency, with HDL cholesterol levels less than the 10th percentile of normal values (43-46). Mean HDL choelsterol levels in affected family members are approximately 50% of normal. This disorder has been associated with premature coronary artery disease as well as stroke. No xanthomas, ocular abnormalities, oral abnormalities, or enlargement of the liver or spleen have been noted. These patients do not have evidence of neurological disease other than being prone to developing strokes. No abnormalities of apoA-I or apoA-II have been reported. Kinetic data suggest that these individuals have decreased production of HDL proteins (47). The precise defect in this disorder remains to be defined.

Recent research has indicated that familial hypoalphalipoproteinemia is associated with a specific mutation in the apoA-I, apoC-III intergenic region (48). This mutation results in a restriction fragment length polymorphism (RFLP) following Pst 1 enzyme digestion resulting in a shift of the apoA-I gene band from its normal 2.2 kb position to a unique 3.3 kb

2

TABLE 1

CLINICAL FEATURES IN GENETIC HDL DEFICIENCY STATES

DISORDER	FREQUENCY	GENETIC TRANS-MISSION	PREMATURE CAD	CORNEAL OPACIFI-CATION	XANTHOMAS
Familial Hypoalpha-lipo-proteinemia	common	autosomal dominant	+++	-	-
Familial LCAT deficiency	rare	autosomal recessive	+	+++	-
Fish eye disease	rare	autosomal codominant	-	+++	-
HDL deficiency with Planar Xanthomas	rare	autosomal codominant	++	+	+
Tangier Disease*	rare	autosomal codominant	+	+	-
Familial apolipoprotein A-I and C-III Deficiency					
Variant I	rare	autosomal codominant	+++	+	+
Variant II	rare	autosomal codominant	++	+	-

*Abnormal tonsils and hepatosplenomegaly have only been reported in Tangier Disease.

position. This polymorphism is due to a specific alteration in the DNA sequence approximately 220 base pairs adjacent to the 3' end of the apoA-I gene. The presence of this RFLP in either the heterozygous or homozygous state has been documented in 4% of the normal population, 34% of patients with premature coronary artery disease, and in 66% of familial hypoalpha-lipoproteinemia kindreds examined (48). How this mutation can affect HDL levels is not clear. Familial hypoalphalipoproteinemia may be by far the most genetic lipid abnormality observed in the general population, as well as in patients with premature coronary artery disease. Moreover our own data indicate that in a large smapling of male and female patients with premature coronary artery disease, HDL cholesterol levels of less than the 10th percaentile are found in over 50% of these individuals. Familial hypoalphalipoproteinemia will be covered in more detail in the chapters by Drs. Vergain and Glueck.

Familial Lecithin:Cholesterol Acyltransferase Deficiency

Patients with this autosomal recessive disorder often present with severe diffuse corneal opacification as well as arcus (49-51). The defect in this diesease is due to a marked decrease or an abnormality of the enzyme LCAT, which is responsible for plasma cholesterol esterification. These patients have a marked increase in the proportion of plasma chol-esterol found in the unesterified form, and as a result have a decrease in cholesterol ester which normally serves as a core constituent in a variety of plasma lipoproteins. Therefore they develop an increase in particles within the LDL and HDL density region which lack cholesterol ester in the core, and are present in plasma as circulating disks rather than spheres. LCAT-deficient patients have a marked deficiency of plasma HDL, an accumulation of abnormal, free cholesterol and phospholipid rich LDL known as Lp-X, as well as an increase in triglyceride rich lipoproteins. In addition to corneal opacification which can affect vision, these patients may also develop anemia secondary to red blood cell membrane abnormalities, proteinuria, renal insufficiency, and premature atherosclerosis. Familial LCAT deficiency is an extremely rare condition which was initally described in several Norwegian kindreds by Norum and Gjone, and has subsequently been reported in other kindreds around the world. LCAT deficiency will be covered in more detail in the chapter by Dr. Frohlich.

Fish Eye Disease

Fish eye disease has been reported by Carlson and colleagues in two Swedish kindreds with marked HDL deficiency and severe corneal opacification (52,53). Analysis, of corneas that have been removed revealed vacuoles in the stroma and Bowman's capsule containing free cholesterol. Premature cornonary artery disease has not been reported in these kindreds. In addition to HDL deficiency, these patients had hypertriglyceridemia, as well as triglyceride-rich LDL. Mean apoA-I and apoA-II concentrations in homozygotes were 2% and 12% of normal, respectively (See table 2). The disease is not expressed in heterozygotes, except that they may have mild HDL deficiency. Recent data indicates that patients with fish eye disease may have abnormalities of cholesterol esterification. The precise nature of the defect reamins to be defined.

ApoA-I Variants

A number of different mutations in apoA-I structure affecting plasma levels of HDL constituents have been reported (See table 3). The first of these mutations was documented in an Italian kindred from the Milano area. Only heterozygotes with this abnormality have been reported. These subjects had moderate hypertriglyceridemia, as well as HDL cholesterol

4

levels that were 20% of normal (54). Both LDL and HDL in affected
kindred members were triglyceride-rich. There was no increase in the ratio
of free to total cholesterol. On isoelectric focusing apoA-I bands were
present in both normal and abnormal position, and subsequent analysis
revealed that these patients had an abnormal cysteine containing apoA-I
due to a substitution of cysteine for arginine at residue position 173
(See table 3) (55). Homozygotes have not been reported. Kinetic data
indicate that apoA-I$_{Milano}$ is catabolized at a faster rate than normal
apoA-I (60). This hypercatabolism of apoA-I$_{Milano}$ presumably accounts
for the deficiency of apoA-I observed in these subjects.

Several additional apoA-I variants have been reported (56-59). These
include apoA-I$_{Marburg}$, apoA-I$_{Giessen}$, apoA-I$_{Munster\ 1-3}$. ApoA-I$_{Marburg}$
and apoA-I$_{Munster\ 2}$ were found by sequence analysis to be identical
mutations due to deletion of lysine at residue 107 in apoA-I (51). This
mutation results in a decreased ability of apoA-I to activate LCAT (57).
The inital patient with apoA-I$_{Marburg}$ was found to have mild
hypertriglyceridemia and a decreased HDL cholesterol value, and was a
heterozygote for this abnormality. Similar findings were also noted in
the apoA-I$_{2\ Munster}$ patient. Other abnormalities picked up on isoelectric
focusing and subsequently characterized by sequence analysis include
apoA-I$_{Munster\ 3A}$ due to a substitution at residue 103, and apoA-I$_{Munster\ 3B}$
and 3C and due to substitutions of arginine for proline at position 4, or
histidine for proline at position 3, respectively, in the apoA-I molecule
(57,59). All these subjects were heterozygotes, and had mild HDL deficiency.

A polymorphism in the DNA sequence in the 3' flanking region of the
human apoA-I gene has been reported following digestion with the restriction enzyme Sst (61). This polymorphism has been associated with hypertriglyceridemia. The frequency of this polymorphism (heterozygous state) was
5% in the normal population, and was noted to be present in 30% of 33
hypertriglyceridemia subjects. How this restriction fragment polymorphism
affects triglyceride and HDL metabolism is not clear. Sequencing of cDNA
of apoC-III gene in a homozygote with this abormality was within normal
limits. Further work on polymorphisms of DNA in the intergenic region
between the flanking apoA-I and apoC-III genes are indicated. Whether
this area is a region where enhancers are present which can affect apoA-I
polymorphisms in this area could be linked to other mutations which affect
apoA-I or apoC-III production.

HDL Deficiency with Planar Xanthomas

A kindred from Sweden was reported with marked HDL deficiency. The
proband for this kindred was a 48 year old female with a history of wide-
spread yellow skin discoloration around the eyes and in the groin since
early childhood (62,63). She also had a history of angina. The yellowish
discoloration was also in the neck, upper chest, axillae, antecubital
fossa, inguinal regions, vulva, perineum, perianal region, and in the area
of the soft palate mucosa of the floor of the mouth. Both upper eyelids
were thickened and infiltrated with small firm nodules, especially at the
margins. Histochemical tests revealed intracellular histiocytic deposition
of free and esterified cholesterol in skin lesions and rectal mucosa.
Moderate liver enlargement was noted and the patient's tonsils were within
normal limits. The patients's mother died at age 72 following a gall
bladder operation and also had yellow skin discoloration. The patient had
no siblings, and of 6 cousins 2 males had HDL cholesterol values of 29
and 35 respectively. The proband in this kindred had elevated plasma
triglyceride and VLDL cholesterol values, a normal LDL cholesterol level,
and an HDL cholesterol level which was 6% of normal. Plasma apoA-I and
apoA-II levels were 1% and 19% of normal, respectively (63). All other
apolipoproteins were present in approximately normal amounts. The defect

in this condition remains to be defined. The disease differs from Tangier disease, to be subsequently discussed.

Tangier Disease

Tangier disease was originally described by Fredrickson and colleagues in two siblings from Tangier Island in the Chesapeake Bay area of the United States (64,65). These children had enlarged orange tonsils, mild hepatosplenomegaly,and lymphadenopathy. Following the removal of their tonsils, pathology revealed macrophages containing increased amounts of cholesterol ester. Similar pathology has been noted in the bone marrow, skin, rectal mucosa, liver, spleen, lymph nodes, omentum and conjunctiva of Tangier patients. The two original probands subsequently developed mild transient peripheral neuropathy in adolescence. Also mild diffuse corneal opacification was noted, but no xanthomas have been observed. Twenty three other patients from 20 kindreds with clinical and laboratory features consistent with homzygous Tangier disease have been described (66,67). Premature coronary artery disease has been detected in subjects under the age of 60, but not in subjects under age 40 years (67). In addition several homozygotes were noted to have cerebrovascular disease. Biochemical abnormalities noted in Tangier homzygotes include mild hypertriglyceridemia and increased VLDL cholesterol values, decreased LDL cholesterol values, and marked HDL deficiency. Plasma apoA-I and apoA-II concentrations are approximately 1% and 9% of normal, while other apolipoproteins are present in either normal or slighty reduced amounts (68,69). Heterozygotes with this disorder have plasma HDL cholesterol, apoA-I, and apoA-II concentrations that are approximatley 50% of normal. Other apolipoproteins in these subjects are within the normal range.

Despite intensive study, the precise molecular defect in Tangier disease remains to be elucidated. It has been shown that a significant fraction of plasma apoA-I in these subjects is found in the 1.21 g/ml infranate, while a significant fraction of apoA-II is found in the 1.063 g/ml supernatent, suggesting that apoA-I and apoA-II in these patients are not found in the same particle (68,69). Moreover it has been reported that Tangier homozygotes have an increased proportion of pro-apoA-I in their plasma (70). Recent data indicate however that the sequence of the six amino acid pro-sequence in Tangier apoA-I is identical to normal, and that the analysis of cDNA for apoA-I from these subjects indicates no detectable abnormality. It has been reported that these subjects have hypercatabolism of HDL constituents especially apoA-I following either HDL infusion or injection of tracer amounts of labeled HDL (72-74). They also have enhanced catabolism of apoA-II. Abnormalities of lipoprotein lipase, hepatic lipase, or lecithin:cholesterol acyltransferase activity have not been found. Recently it has been suggested that these patients have a defect in the uptake and processing of HDL by macrophages. Further work in this area is required. Tangier disease will be covered in greater detail in the chapter by Dr. Brewer.

Familial Apolipoprotein A-I and C-III Deficiency States

Two variants of this disorder have been described. Probands for the kindred affected with variant 1 as described by Norum and colleagues were two sisters from Detroit, Michigan, (USA), who presented with striking premature coronary artery disease, mild diffuse corneal opacification and yellow-orange indurated plaques on the trunk, neck, eye lids, chest, arms and back (75). Biopsy of skin xanthoma revealed perivascular histiocytic lipid deposition. Tonsils in these patients were normal. One proband also had tuberous xanthomas on the lateral aspect of the left foot. Biochemical abnormalities in the two homozygous probands indicated marked HDL deficiency, with decreased VLDL cholesterol and plasma triglyceride

levels, triglyceride-rich LDL, undetectable plasma apoC-III levels, and
trace amounts of apoA-I. All other apolipoproteins were found to be
present, and apoA-II was 50% of normal in homozygotes. There was a
normal ratio of free to total cholesterol in plasma, but there was an
increased percentage of free cholesterol in the HDL fraction in the HDL
fraction. Following an infusion of normal plasma into each proband the
decay of HDL protein yeilded a plasma half life that was approximately
2-3 days, in contrast to the striking hypercatabolism of HDL constituents
observed in Tangier disease (75). Analysis of the adjacent apoA-I and
apoC-III genes, which are on the long arm of chromosome 11, indicated a
major DNA rearrangement affecting both genes, and inactivating the
synthesis of both proteins (76). The defect in this condition, therefore
appears to be due to an inability to synthesize both apoA-I and apoC-III.
Heterozygotes with this abnormality had HDL cholesterol, apoA-I, and
apoC-III levels in plasma that were approximately 50% of normal (77).
Familial apolipoprotein A-I and C-III, variant I, will be covered in
greater detail in the chapter by Dr. Norum.

The second variant of this disorder was originally described as plasma
apolipoprotein A-I absence by Schaefer and colleagues (42,78,79). The
proband in this kindred was a 45 year white female from north-west Alabama
who developed angina pectoris at age 42, and also had mild corneal
opacification. She has no xanthomas and her tonsils were within normal
limits. She died shortly after coronary artery bypass surgery. Autopsy
revealed significant atherosclerosis in the descending aorta, carotid,
pulmonary, and coronary arteries, normal tonsils, and lack of lipid
depositon in reticulo-endothelial cells in the liver, spleen and bone
marrow. Diffuse extracellular granular lipid depostion was observed in
the corneal stroma.

Of 38 kindred members tested, 17 were heterozygotes, none of whom had
premature coronary artery disease or corneal opacification before age 40.
Two heterozygotes developed clinical evidence of coronary artery disease
before the age of 60 and one died at age 56 of a myocardial infarction.
The homozygous proband for this disorder had marked HDL deficiency and
undetectable plasma levels of apoA-I and apoC-III by radioimmunoassay.
ApoA-II values were 11% of normal, and LDL was apoB rich. In addition
vitamin E and linoleic acid were also noted. Other apoproteins were
found to be present in normal amounts. Heterozygotes had HDL cholesterol.
apoA-I, and apoC-III levels that were 54%, 58%, and 83% of normal,
respectively.

When DNA was isolated from the white cells of the two offspring of
the proband and subjected to restriction enzyme digestion and Southern
blotting analysis utilizing a specific apoA-I gene probe, no apoA-I gene
abnormality was detected (79). The defect in this disorder also appears
to be due to an inability to synthesize apoA-I and apoC-III, however the
precise gene abnormality remains to be defined. Variant II is similar
biochemically to variant I, however it differs in that the variant II
proband lacked xanthomas and had a more striking deficiency of apoA-II
than variant I probands. Moreover these two kindreds do not share the
same gene abnormality.

Conclusion

A variety of HDL deficiency states have been reported. These disorders
are quite rare, in contrast to familial hypoalphalipoproteinemia. Recent
data suggest that a specific gene abnormality is associated with this
latter disorder, and that this disease may be the most common genetic
lipoprotein disorder observed in patients with premature CAD. Why certain
kindreds with HDL deficiency develop premature coronary artery disease

TABLE 2

LABORATORY VALUES IN GENETIC HDL DEFICIENCY STATES*

DISORDER	HDL CHOLESTEROL	APOA-I	APOA-II	APOC-III	LDL CHOLESTEROL	TRI- GLYCERIDE
Familial Hypoalpha- lipoprotein- emia	26	65	–	–	115	113
Familial LCAT deficiency	5	51	–	–	225	400
Fish eye Disease						
Homozygotes	7	38	5.3	–	199	424
Heterozygotes	33	–	–	–	186	132
HDL deficiency with Planar Xanthomas	3	2	7		134	290
Tangier Disease						
Homozygotes	2	1.3	3	6.5	42	200
Heterozygotes	27	56	14	19.5	112	129
Familial Apolipoprotein A-I and C-III deficiency						
Variant I						
Homozygotes	6	0.006	19	ND	120	62
Heterozygotes	33	71	32		97	52
Variant II						
Homozygotes	1	ND	3	ND	106	62
Heterozygotes	27	79	28		136	131

*Mean values reported in mg/dl as reported in literature, and in the case of apoA-II adjusted to account for elevated normal range of one laboratory, ND is not detectable; –, data not reported.

TABLE 3

APOLIPOPROTEIN A-I VARIANTS

NAME	SUBSTITUTION	RESIDUE	CHARGE DIFFERENCE
A-I$_{Milano}$	Arg → Cys	173	–1
A-I$_{Munster\ 3A}$	Asp → Asn	103	+1
A-I$_{Munster\ 3B}$	Pro → Arg	4	+1
A-I$_{Munster\ 3C}$	Pro → His	3	+1
A-I$_{Giessen}$	Pro → Arg	143	+1
A-I$_{Munster}$	Lys → 0	107	–1
A-I$_{Marburg}$			

and others do not remains to be elucidated. Factors that clearly may help to answer these questions are the presence of other risk factors, whether the HDL defect is due to decreased production of HDL constituents or enhanced metabolism, and whether there may be cellular defects as well. Patients that have hypercatabolism of HDL constituents such as Tangier homozygotes may not be at the same increased risk as patients with decreased production of HDL constituents, such as homozygotes for familial apolipoprotein A-I and C-III deficiency. Moreover patients with homozygous Tangier disease have significant reduction in LDL cholesterol values, in contrast to the normal levels observed in the familial apoA-I and apoC-III deficiency states. The moderate HDL deficiency that is observed in subjects with familial hypoalphalipoproteinemia may only be of clinical significance in populations that have moderately elevated LDL levels due to fairly high cholesterol diets, and where the prevelance of smoking is significant. Future research will undoubtedly uncover new genetic mutations associated with HDL deficiency, and define the precise defects in the disorders that have been described. Additional information is required as to the precise relationship between HDL deficiency and atherosclerosis, and research on the therapy of these disorders, especially familial hypoalphalipoproteinemia, is required.

REFERENCES

1. R. J. Havel, H. A. Eder, J. H. Bragdon, The distribution and chemical composition of ultracentrifugally separated lipoproteins in human serum, J Clin Invest 34:1345-1353 (1955).

2. D. W. Anderson, A. V. Nichols, S. S. Pan, F. T. Lindgren, High density lipoprotein distribution:resolution and determination of three major components in a normal population sample, Atherosclerosis 29:161-179 (1978).

3. A. J. Scanu, J. Toth, C. Edelstein, E. Stiller, Fractionation of human serum high density lipoprotein in urea solutions:evidence of polypeptide heterogeneity, Biochemistry 8:3309-3316 (1969).

4. B. Shore, V. Shore, Isolation and characterization of polypeptides of human serum lipoproteins, Biochemistry 8:4510-4516 (1969).

5. A. Heiberg, K. Berg, On the relationship between Lp(a) lipoprotein, "sinking pre-beta-lipoprotein" and inherited hyper-beta-lipoproteinemia, clin Genet 5:144-150 (1974).

6. G. Kostner, P. Alaupovic, Studies of the composition and structure of plasma lipoproteins. Separation and quantitation of the lipoprotein families occurring in high density lipoproteins of human plasma, Biochemistry 11:3429-3433 (1972).

7. A. Gustafson, P. Alaupovic, R. H. Furman, Studies of the composition and structure of serum lipoproteins. Separation and characterization of phospholipid-protein residues obtained by partial delipidation of very low density lipoproteins of human serum, Biochemistry 5:632-640 (1966).

8. W. V. Brown, R. I. Levy, D. S. Fredrickson, Further separation of the apoproteins of the human plasma very low density lipoproteins, Biochem Biophys Acta 280:573-575 (1970).

9. W. J. McConathy, P. Alaupovic, Studies on the isolation and partial characterization of apolipoprotein and lipoprotein D in human plasma, Biochemistry 15:515-552 (1976).

10. F. Shelburne, S. Quarfordt, A new apoprotein of human plasma very low density lipoproteins, J Biol Chem 249:1428-1432 (1974).

11. S. O. Olofsson, W. J. McConathy, P. Alaupovic, Isolation and partial characterization of a new acidic apolipoprotein (apolipoprotein F) from high density lipoproteins of human plasma, Biochemistry 17:1032-1036 (1978).

12. M. Ayrault-Jarrier, J. Alix, J. Polonovski, Une nouvelle proteine des lipoproteines du serum hamain:isolement et caracterisation partielle d'une apolipoproteine, G Biochimie 60:65-71 (1978).

13. C. J. Fielding, V.G. shore, P.E. Fielding, A protein co-factor of lecithin:cholesterol acyltransferase, Biochem Biophy Res Commun 46:1493-1498 (1972).

14. A. K. Soutar, G. W. Garner, G. N. Baker, et al, Effect of the human plasma apolipoproteins and phosphatidyl-choline acyl donor on the activity of lecithin:cholesterol acyltransferase, Biochemistry 14:3057-3064 (1975).

15. C. E. Jahn, J. O. Osborne, E. J. Schaefer, H. B. Brewer Jr., ApoA-II specific activation of hepatic lipase enzymatic activity:identification of a major HDL apoprotein as the activating plasma component in vitro, Eur J Biochem 131:25-29 (1983).

16. J. C. LaRosa, R. I. Levy, P. N. Herbert, S. E. Lux, D. S. Fredrickson, A specific apoprotein activator for lipoprotein lipase, Biochem Biophy Res Comm 41:57-62 (1970).

17. F. Shelburne, J. Hanks, W. Myers, S. Quarfordt, Effects of apoproteins on hepatic uptake of triglyceride emulsions in the rat, J Clin Invest 65:652-658 (1980).

18. M. S. Brown, S. E. Dana, J. L. Goldstein, Regulation of 3-hydroxy-3-methylglutaryl coenzyme A reductase activity in cultured human fibroblasts:Comaprison of cells from a normal subject and a patient with homozygous familial hypercholesterolemia, J Biol Chem 249:789-796 (1974).

19. B. C. Sherrill, T. L. Innerarity, R. W. Mahley, Rapid hepatic clearance of the canine lipoproteins containing only the E apoprotein by a high affinity receptor, J Biol Chem 255:1804-1807 (1980).

20. R. Biesbroeck, J. F. Oram, J. J. Albers, E. L. Bierman, Specific high affinity binding of high density lipoproteins in cultured human skin fibroblasts and arterial smooth muscle cells, J Clin Invest 71:525-539 (1983).

21. A. L Wu, H. G. Windmueller, Relative contributions by liver and intestine to individual apolipoproteins in the rat, J Biol Chem 254:7316-7322 (1979).

22. D. W. Anderson, E. J. Schaefer, T. J. Bronzert, F. T. Lindgren, T. Forte, T. E. Stanze, G. D. Nibalck, L. A. Zech, H. B. Brewer Jr., Transport of apolipoprotein A-I and A-II by human thoracid duct lymph, J Clin Invest 67:857-866 (1981).

23. J. Marsh, Apoproteins of the lipoproteins in a nonrecirculating perfusate of rat liver, J Lipid Res 17:85-90 (1976).

24. T. E. Felker, M. Fairnaru, R. L. Hamilton, R. J. Havel, Secretion of the arginine-rich and A-I apolipoproteins by the isolated perfused rat liver, J Lipid Res 18:465-473 (1977).

25. D. W. Bilheimer, S. Eisenberg, R. I. Levy, The metabolism of very low density lipoprotein proteins.1.Preliminary in vitro and vivo observations, Biochim Biophy Acta 260:212-221 (1972).

26. S. Eisenberg, D. W. Bilheimer, R. I. Levy, F. T. Lingren, On the metabolic conversion of human plasma very low density lipoprotein to low density lipoprotein, Biochem Biophys Acta 326:361-377 (1973).

27. M. C. Clangeaud, S. Eisenberg, T. Olivecrona, Very low desnity lipoprotein dissociation of apolipoprotein C during lipoprotein lipase induced lipolysis, Biochim Biophys Acta 48:23-35 (1977).

28. R. M. Glickman, P. H. R. Green, The intestine as a source of apolipoprotein A-I, Pro Natl Acad Sci USA 74:2569-2573 (1977).

29. E. J. Schaefer, L. J. Jenkins, H. B. Brewer Jr., Human chylomicron apolipoprotein metabolism, Biochem Biophs Res Commun 80:405-412 (1978).

30. T. G. Redgrave, D. M. Small, Quantitation of the transfer of surface phospholipid of chylomicrons to the high density lipoprotein fraction during the catabolism of chylomicrons in the rat, J Clin Invest 64:162-171 (1979).

31. E. J. Schaefer, M. G. Wetzel, G. Bengtsson, R. B. Scow, H. B. Brewer Jr., T. Olivecrona, Transfer of human lymph chylomicron constituents to other lipoprotein density fractions during in vitro lipolysis, J Lipid Res 23:1259-1273 (1982).

32. E. J. Schaefer, L. A. Zech, L. L. Jenkins, R. A. Aamodt, J. J. Bronzert, E. A. Rubalcaba, F. T. Lindgren, H. B. Brewer Jr., Human apolipoprotein A-I and A-II metabolism, J Lipid Res 23:850-862 (1982).

33. C. K. Glass, R. C. Pittman, G. A. Keller, D. Steinberg, Tissue sites of degradation of apoprotein A-I in the rat, J Biol Chem 258:7161-7167 (1983).

34. L. A. Zech, E. J. Schaefer, L. L. Jenkins, E. A. Rubalcaba, T. J. Bronzert, R. L. Aamodt, H. B. Brewer Jr., Metabolism of human apolipoproteins A-I and A-II, compartmental models, J Lipid Res 34:60-71 (1983).

35. D. P. Barr, E. M. Russ, H. A. Eder, Protein lipid relationships in human plasma.II. In atherosclerosis and related conditions, Am J Med 11:480-492 (1951).

36. G. J. Miller, N. F. Miller, Plasma-high density lipoprotein concentration and development of ischemic heart disease, Lancet 1:16-20 (1975).

37. G. G. Rhoads, C. L. Gulbrandsen, A. Kagan, Serum lipoproteins and coronary heart disease in a population of Hawaii Japaneses men, N Engl J Med 294:293-298 (1976).

38. N. E. Miller, O. H. Forde, D. S. Thelle, The Tromso Heart Study:high density lipoprotein and coronary heart disease:a prospective case control study, Lancet 55:767-772 (1977).

39. W. Castelli, J. T. Doyle, T. Gordon, HDL cholesterol and other lipids in coronary heart disease, The Cooperative Lipoprotein Phenotyping Study, Circulation 55:767-772 (1977).

40. R. M. Krauss, Regulation of high density lipoprotein levels in lipid disorder, R. Havel, ed., Medical Clin North American 66(2):403-430 (1982).

41. E. J. Schaefer, D. A. Foster, L. A. Zech, H. B. Brewer Jr., R. I. Levy, The effect of estrogen administration of plasma lipoprotein metabolism in premenopausal females, J Clin Endocrinal Metab 57:262-267 (1983).

42. E. J. Schaefer, Clinical, biochemical and genetic features in familial disorders of high density lipoprotein deficiency, Arteriosclerosis 4:303 (1984).

43. C. Vergani, A. Bettale, Familial hypoalphalipoproteinemia, Clin Chim Acta 114:45-52 (1981).

44. C. J. Glueck, S. R. Daniels, S. Bates, C. Benton, T. Tracy, J. L. H. C. Third, Pediatric victims of unexplained stroke and their families: familial lipid and lipoprotein abnormalities, Pediatrics 69:308-316 (1982).

45. S. R. Daniels, S. Bates, R. R. Lukin, C. Benton, J. L. H. C. Third, C. J. Glueck, Cerebrobascular arteriopathy (arteriosclerosis) and ischemic childhood stroke, Stroke 13:360-365 (1982).

46. J. H. L. C. Third, J. Montag, M. Flynn, J. Freidel, P. Laskarzewski, C. J. Glueck, Primary and familial hypoalphalipoproteinemia, Metabolism 33:136-146 (1984).

47. N. Fidge, P. Nestel, T. Ishikawa, M. Reardon, T. Billington, Turnover of apoproteins A-I and A-II of human high density lipoprotein and the relationahip to other lipoproteins in normal and hyperlipidemic individuals, Metabolism 29:643-653 (1980).

48. J. M. Ordovas, E. J. Schaefer, D. Salem, R. Ward, C. J. Glueck, C. Vergani, P. W. F. Wilson, S. K. Karathanasis, Apolipoprotein A-I gene polymorphism is the 3' flanking region associated with familial hypoalphalipoproteinemia and premature coronary artery disease, N Engl J Med (in press).

49. K. R. Norum E. Gjone, Familial plasma lecithin:cholesterol acyltransferase deficiency. Biochemical study of a new inborn error of metabolism, Scand J Clin Lab Invest 20:231-236 (1967).

50. E. Gjone, K. R. Norum, Familial serum cholesterol ester deficiency: clinical study of a patient with a new syndrome, Acta Med Scand 183:107-115 (1968).

51. J. A. Glomset, K. R. Norum, E. Gjone, Familial lecithin:cholesterol acyltransferase deficiency, In: J. B. Stanburg, J. B. Wyngaarden, D. S. Fredrickson, J. L. Goldstein, M. S. Brown, eds. The Metabolic Basis of Inherited Disease, 5th ed., New York:McGraw-Hill 643-654 (1983).

52. L. S. Carlson, B. Philipson, Fishe-eye disease:a new familial condition associated with massive corneal opacitites and dyslipoproteinemia, Lancet 8149:921-925 (1979).

53. L. A. Carlson, Fish-eye disease: a new familial condition with massive corneal opacities and dyslipoproteinemia, Eur J Clin Invest 12:41-52 (1982).

54. G. Grancheschini, C. R. Sirtori, A. Capurso, K. H. Weisgraber, R. W. Mahley, A-I$_{Milano}$ apoprotein. Decreased high density lipoprotein cholesterol levels with significant lipoprotein modifications and without clinical atherosclerosis in an Italian family, J Clin Invest 66:892-990 (1980).

55. K. H. Weisgraber, T. P. Bersot, R. W. Mahley, G. Francheschini, C. R. Sirtori, A-I$_{Milano}$ apoprotein, isolation and characterization of a cysteine-containing variant of the A-I apoprotein from human high density lipoprotein, J Clin Invest 66:901-909 (1980).

56. G. Utermann, G. Feussner, G. Franceschini, J. Haas, A. Steinmetz, Genetic variants of group A apolipoproteins. Lipid methods for screening and characterization without ultracentrifugation, J Biol Chem 257:501-507 (1982).

57. S. C. Rall Jr., K. H. Weisgraber, R. W. Mahley, Y. Ogawa, C. J. Fielding, G. Utermann, J. Haas, A. Steinmetz, H. J. Menzel, G. Assmann, Abnormal lecithin:cholesterol acyltransferase activation by a human apolipoprotein A-I variant in which a single lysine is deleted, J Biol Chem 259:10063-10070 (1984).

58. H. J. Menzel, R. G. Kladetzky, G. Assmann, One step screening method for the polymorphism of apolipoproteins A-I, A-II, and A-IV, J Lipid Res 23:915-922 (1982).

59. H. T. Mezel, G. Assmann, S. C. Rall Jr., K. H. Weisgraber, R. W. Mahley, Human apoliporptein A-I polymorphism, J Biol Chem 259:3070-3076 (1984).

60. G. Ghiselli, J. A. Summerfield, E. J. Schaefer, C. Sirtori, E. A. Jones, H. B. Brewer Jr., Abnormal catabolism of apoA-I$_{Milano}$ (abstr) Clin Res 30:291A (1982).

61. A. Rees, C. C. Shoulders, J. Stocks, D. J. Galton, F. E. Baralle, DNA polymorphism adjacent to the human apoprotein A-I gene:relation to hypertriglyceridemia, Lancet 1:444-446 (1983).

62. G. R. Lindeskog, A. Gustafson, L. Enerback, Serum lipoprotein deficiency in diffuse normolipemic plan xanthoma, Arch Dermatol 106:529-532 (1972).

63. A. Gustafson, W. McConathy, P. Alaupovic, M. D. Curry, B. Persson, Identification of apoprotein families in a variant of human plasma apolipoprotein A deficiency, Scand J Clin Lab Invest 39:377-388 (1979).

64. D. S. Fredrickson, P. H. Altrocchi, L. C. Avioli, Tangier disease: combined clinical staff conference at the National Institutes of Health, Ann Intern Med 55:1016-1031 (1961).

65. D. A. Frederickson, The inheritance of high density lipoprotein deficiency (Tangier disease) J Clin Invest 43:228-236 (1964).

66. P. N. Herbert, G. Assmann, A. M. Gotto Jr., D. S. Fredrickson, Familial lipoprotein deficiency:abetalipoproteinemia, hypobetalipo-proteinemia, and Tangier disease In: J. B. Wyngaarden, D. S. Fredrickson, J. Goldstein, M. Brown eds. The Metabolic Basis of Inherited Disease 5th ed. New York:McGraw-Hill 589-621 (1982).

67. E. J. Schaefer, L. A. Zech, D. S. Schwartz, H. B. Brewer Jr., Coronary heart disease prevalence and other clinical features in familial high density lipoprotein deficiency (Tangier disease), Ann Inter Med 93:261-266 (1980).

68. L. O. Henderson, P. N. Herbert, D. S. Fredrickson, R. J. Heinen, Abnormal concentration and anomalous distribution of apolipoprotein A-I in Tangier disease, Metabolism 27:165-174 (1978).

69. P. Alaupovic, E. J. Schaefer, W. J. McConathy J. S. Fesmire, H. B. Brewer Jr., Plasma apolipoprotein concentrations in familial apolipoprotein A-I and A-II deficiency (Tangier disease), Metabolism 30:809-816 (1981).

70. V. I. Zannis, A. M. Lees, R. S. Lees, J. L. Breslow, Abnormal apoA-I isoprotein composition in patients with Tangier disease, J Biol Chem 257:4978-4986 (1982).

71. H. B. Brewer Jr., T. Fairwell, L. Kay, M. Meng, R. Ronan, S. Law, J. A. Light, Human plasma proapoA-I:isolation and amino-terminal sequence, Biochem Biophys Res Comm 113:626-632 (1983).

72. E. J. Schaefer, C. B. Blum, R. I. Levy, et al, Metabolism of high density lipoprotein in Tangier disease, N Engl J Med 299:905-910 (1978).

73. E. J. Schaefer, D. W. Anderson, L. A. Zech, et al, Metabolism of high density lipoprotein subfractions in Tangier disease following the infusion of high density lipoproteins, J Lipid Res 22:217-218 (1981).

74. G. Assmann, A. Capurso, E. Smootz, K. Wellner, apoprotein A metabolism in Tangier disease, Atherosclerosis 30:321-332 (1978).

75. R. A. Norum, J. B. Lakier, S. Goldstein, A. Angel, R. B. Goldberg, W. D. Black, D. K. Noffze, P. J. Dolphin, J. Edelglass, D. D. Borograd, P. Alaupovic, Familial deficiency of apolipoproteins A-I and apoC-III and precocious coronary artery disease, N Engl J Med 306:1513-1519 (1983).

76. S. K. Karathanasis, R. A. Norum, V. I. Zannis, J. L. Breslow, An inherited polymorphism in the human apolipoprotin A-I gene locus related to the development of atherosclerosis, Nature 301:718-720 (1983).

77. R. A. Norum, P. J. Dolphin, P. Alaupovic, Familial deficiency of apolipoproteins A-I and C-III and precocious coronary artery disease: family study in Latent Dyslipoproteinemias and Atherosclerosis J. L. DeGennes ed, Raven Press, New York 289-295 (1984).

78. E. J. Schaefer, W. H. Heaton, M. G. Wetzel, H. B. Brewer Jr., Plasma apolipoprotein A-I absence assoicated with a marked reduction of high density lipoproteins and premature coronary artery disease, Arteriosclerosis 2:16-26 (1982).

79. E. J. Schaefer, J. M. Ordovas, S. Law, G. Ghiselli, M. L. Kashyap, L. S. Srivastava. W. H. Heaton, J. J. Albers, W. E. Conners, H. B. Brewer Jr., Familial apolipoprotein A-I and C-III deficiency, variant II, J Lipid Res 26:1089–1101 (1985).

METABOLIC INTERRELATIONSHIPS OF HDL SUBCLASSES

Ronald M. Krauss and Alex V. Nichols

Donner Laboratory, Lawrence Berkeley Laboratory
University of California, Berkeley, CA 94720

HDL Subpopulations

Levels of plasma HDL are determined by a complex array of metabolic processes involving synthesis, transfer, recycling, and tissue uptake of individual apoprotein and lipid components[1]. At equilibrium, these components are organized within a variety of particulate forms, and detailed understanding of HDL metabolism requires consideration of factors responsible for this macromolecular heterogeneity. The first description of the HDL particle spectrum identified three major forms of HDL based on flotation rates ($F^o_{1.20}$) in the analytic ultracentrifuge[2]: HDL_1, of $F^o_{1.20} > 9$ and density overlapping with LDL; HDL_2, of $F^o_{1.20}$ 3.5-9 (d 1.063-1.125 g/ml), and HDL_3, of $F^o_{1.20}$ 0-3.5 (d 1.125-1.200 g/ml). Figure 1 shows the analytic ultracentrifuge pattern of HDL_2 and HDL_3 in a representative normal subject, as well as the HDL profile as revealed by zonal ultracentrifugation[3] and electrophoresis of HDL in native polyacrylamide gradient gels[4]. Each of these procedures has revealed further heterogeneity within the HDL_2 and HDL_3 density subclasses. In the case of gradient gel electrophoresis, heterogeneity is manifest as multiple electrophoretic bands which have been shown to represent distinct HDL subspecies of differing particle size. There are at least two such species within the HDL_2, and three within HDL_3[4]. A summary of the properties of these subspecies is given in Table 1.

Recently, immunoaffinity chromatography combined with gradient gel electrophoresis has demonstrated variation in content of the major HDL

Abbreviations used in this paper: apo = apoprotein; HDL = high density lipoproteins; IDL = intermediate density lipoproteins; L:CAT = lecithin:cholesterol acyltransferase; LDL = low density lipoproteins; VLDL = very low density lipoproteins.

Ultracentrifugation (size and density)

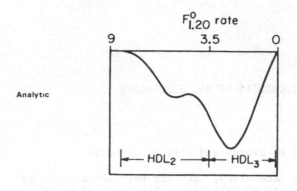

$F^0_{1.20}$ rate

9   3.5   0

Analytic

HDL$_2$ — HDL$_3$

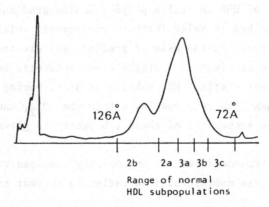

DENSITY (g/ml)

1.10   1.15   1.20

Zonal

ABSORBANCE (280 nm)

0.3

0.2

0.1

0

HDL$_2$ HDL$_{3L}$

HDL$_{3D}$

1   2   3

0   100   200   300   400

EFFLUENT VOLUME (ml)

Gradient Gel Electrophoresis (size)

126Å   72Å

2b   2a 3a 3b 3c

Range of normal
HDL subpopulations

Figure 1.   Analysis of HDL particle distribution.   Top -- analytic ultracentrifugation[2]; middle -- zonal ultracentrifugation[3]; bottom -- gradient gel electrophoresis[4].

Table 1. HDL Subpopulations and Their Properties

HDL Subpopulation*	HDL_{2b}	HDL_{2a}	HDL_{3a}	HDL_{3b}	HDL_{3c}
Particle size interval (nm)	9.7–12.9	8.8–9.7	8.2–8.8	7.8–8.2	7.2–7.8
Density interval (g/ml)	1.063–1.100	1.100–1.125	1.125–1.147	1.147–1.167	1.167–1.200
Flotation rate interval ($F^o_{1.20}$)**	3.5–9	2.5–5.5	1.5–3.5	0.5–2.0	0–1.5

* HDL subpopulations previously have been designated with the subscript (gge) to indicate their identification by gradient gel electrophoresis[4]. This convention is not used in the present report.

** Data obtained from analyses described in Ref. 5.

apoproteins A–I and A–II among the HDL subspecies[6]. Particles containing both apo A–I and apo A–II are found in subspecies across the HDL_2–HDL_3 spectrum, while particles containing apo A–I but not apo A–II have been identified in two distinct size subpopulations, one corresponding to HDL_{2b}, and the other included within HDL_{3a}. Among other major HDL proteins, the majority of apo E and apo D are found in particles containing apo A–I with apo A–II, while the distribution of the C-apoproteins among the various HDL subpopulations has not been described.

Metabolism of HDL Apoproteins and HDL Subpopulations

Since content of apo A–I and apo A–II appears to contribute to HDL particle heterogeneity, studies of the turnover of these proteins may shed light on factors affecting the HDL particle distribution. The major findings may be summarized as follows: residence time of apo A–I is slightly shorter than that of apo A–II in normal subjects[7-9], and the fractional catabolic rate of apo A–I is usually the primary determinant of its plasma concentration[9]. An exception to this is the case of estrogen administration, which results in increased plasma apo A–I levels due to increased synthetic rate[10]. Catabolism of apo A–I is correlated with plasma levels of HDL cholesterol and phospholipid and of HDL_{2b} mass[9].

A number of observations have linked parameters of HDL metabolism to the metabolism of triglyceride-rich lipoproteins. Fractional catabolic rates of apo A–I[9,11] and apo A–II[11] have been correlated inversely with various parameters of VLDL catabolism in normal and hyperlipidemic subjects.

19

Perhaps related to this are the observations that levels of HDL_2 cholesterol are directly correlated[12], and catabolism of HDL apoproteins is inversely correlated[11] with activity of adipose tissue lipoprotein lipase, a rate-limiting factor in plasma triglyceride clearance. Possible dissociation of apo A-I and apo A-II metabolism in hypertriglyceridemic patients is suggested by the report that in 11 patients, the fractional catabolic rate of apo A-II was greater than that of apo A-I, and correlated with plasma triglyceride levels[7]. Conceivably, in such patients, catabolism of particles with both apo A-I and apo A-II is faster than for particles with apo A-I and no apo A-II.

Relation of Triglyceride-rich Lipoprotein Metabolism to HDL Subspecies Patterns

It has been suggested that levels of the larger, more buoyant HDL_2 species in hypertriglyceridemic patients are reduced as a result of triglyceride-cholesteryl ester exchange, subsequent triglyceride hydrolysis and surface lipid depletion, and conversion of the HDL to smaller, more dense particles[1]. While it has been reported that HDL_3-like particles may be generated by in vitro hydrolysis of triglyceride-rich HDL_2 in vivo following a fat meal[13], the relation of these particles to native HDL_2 and HDL_3 subspecies has not been established. We have recently analyzed HDL by gradient gel electrophoresis in four subjects with varying degrees of hypertriglyceridemia (Figure 2). In all subjects, there are two major bands, corresponding to HDL_{3a} and HDL_{3b}; these bands appear to represent the HDL_{3L} and HDL_{3D} species identified by zonal ultracentrifugation (Figure 1). A similar pattern of relative preponderance of HDL species with particle diameters corresponding to HDL_{3a} and HDL_{3b} in hypertriglyceridemic subjects recently has been reported[14]. The data from this study suggest that the relative amount of HDL_{3b} tends to increase as a function of plasma triglyceride level, but this tendency is not uniform, and does not seem to involve significant depletion of HDL_{3a}. Furthermore, as shown in Figure 2, even with the highest triglyceride levels, levels of smaller HDL (HDL_{3c}) do not increase at the expense of the larger species.

Measurements of HDL cholesterol in subjects with a spectrum of triglyceride concentrations (Figure 3) also suggest the operation of factors that may limit the influence of triglyceride-rich lipoproteins on HDL metabolism. While HDL cholesterol and triglyceride concentrations are correlated in subjects with triglyceride levels up to 300-400 mg/dl, above this level HDL cholesterol does not drop below a limiting value of approximately 20 mg/dl. The findings in Figures 2 and 3 suggest that triglyceride metabolism may exert an influence on plasma HDL predominantly within the range of normal to

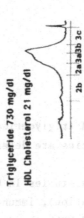
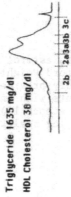
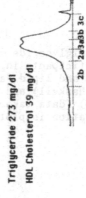

Particle Size Distributions of HDL from Selected Hypertriglyceridemic Subjects (Males, 51-64 yr)

Triglyceride 1077 mg/dl
HDL Cholesterol 19 mg/dl

2b 2a3a3b 3c

Triglyceride 730 mg/dl
HDL Cholesterol 21 mg/dl

2b 2a3a3b 3c

Triglyceride 1635 mg/dl
HDL Cholesterol 38 mg/dl

2b 2a3a3b 3c

Triglyceride 273 mg/dl
HDL Cholesterol 39 mg/dl

2b 2a3a3b 3c

Figure 2. Gradient gel electrophoresis of HDL from four male subjects with hypertriglyceridemia, aged 51-64 yr. HDL was isolated, electrophoresis was performed, and results were analyzed as described in reference 4.

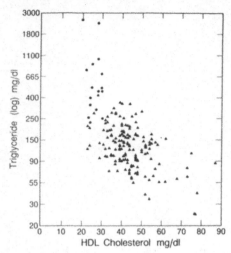

Figure 3. Plasma triglyceride vs. HDL-cholesterol concentrations in 155 male participants in the Stanford Weight Control Project (▲) (samples and lipid values kindly provided by Drs. Peter Wood and William Haskell) and 16 patients with untreated hypertriglyceridemia (●) (data taken from reference 15). Plasma triglyceride concentration is plotted on a log scale.

moderately elevated triglyceride levels and that above these levels, levels of the HDL_3 subspecies are determined by independent processes.

In preliminary studies in hypertriglyceridemic subjects (Cheung and Nichols, in preparation), immunoaffinity chromatography has revealed that the major HDL_{3a} and HDL_{3b} bands consist predominantly of particles with both apo A-I and apo A-II. In normal and hypertriglyceridemic subjects, the apparent peak diameters of these particles show variation across the $HDL_{2a/3a}$ size spectrum. The possibility that such variation reflects changes in lipid content is suggested by the results of recent in vivo and in vitro studies: In normal subjects, changes in composition and density of HDL subfractions after a fat meal suggest that the largest HDL_3 species

22

(HDL_{3a}) may be converted to particles of HDL_{2a} size and density by lipid enrichment, without change in content of apo A-I or apo A-II[16]. Similar conversions of HDL_{3a} to HDL_{2a} have been observed in vitro with incubation of plasma[17,18] or model complexes[19] in the presence of active L:CAT and a cholesterol source.

It is possible that reversal of such a process by lipolysis contributes to the relative depletion of HDL_{2a} and the formation of one or more of the HDL_3 species in the hypertriglyceridemic subjects shown in Figure 2. As noted above, hypertriglyceridemia does not appear to result in further depletion of HDL_{3a} and HDL_{3b} or a significant shift to yet smaller species, and it may be that other pathways are involved in the formation and metabolism of these species in both normal and hypertriglyceridemic subjects.

Interrelated Changes of HDL Subspecies and Other Lipoproteins

Differing metabolic relationships of HDL subspecies have also been suggested by recent findings that changes in levels of HDL_2 and HDL_3 show differing correlations with changes in components of the VLDL-IDL-LDL spectrum[20]. In Table 2 are shown correlations involving HDL changes over one year in 31 normal men who were randomly assigned to the sedentary

Table 2. Spearman Correlations of One-Year Changes in HDL Subfractions with Changes in VLDL, IDL, and LDL Subfractions in 31 Men

Major HDL Subpopulation	$F^o_{1.20}$	3.5–9 2b/2a	2.5–3.5 2a/3a	1.5–2.5 3a	0–1.5 3b/3c
Flotation Rate (S^o_f) Interval	Major Species				
200–400	Large VLDL	−0.06	−0.07	−0.14	0.31
20–100	Small VLDL	−0.25	−0.45*	−0.24	0.31
14–20)	IDL-1	−0.04	−0.17	0.13	0.50**
12–14)	IDL-2	0.11	0.02	0.30	0.45*
10–12		0.35	0.25	0.37*	0.29
9–10)	LDL-I	0.44*	0.36*	0.22	−0.05
8–9)	LDL-II	0.46**	0.41*	0.18	−0.15
7–8		0.40*	0.43*	0.19	−0.16

Significance of correlations: * p < 0.05; ** p < 0.01.

control group in a study of exercise effects on plasma lipids and lipoproteins carried out in collaboration with investigators at Stanford University[21]. The HDL mass distribution between $F^o_{1.20}$ 0-9 as measured by analytic ultracentrifugation is subdivided using the results summarized in Table 1 to provide non-overlapping flotation rate intervals most closely corresponding to the HDL subpopulations identified by gradient gel electrophoresis[5]. In the case of HDL_{2a}, the portion of the flotation interval overlapping with HDL_{3a} ($F^o_{1.20}$ 2.5-3.5) is listed separately from that overlapping with HDL_{2b} ($F^o_{1.20}$ > 3.5).

There is a significant inverse correlation between changes in $HDL_{2a/3a}$ and small VLDL of flotation rate (S^o_f) 20-100, a relationship similar to that seen cross-sectionally at baseline[20]. On the other hand, there are also positive correlations of HDL changes with changes in mass within intervals of the IDL-LDL flotation range that include recently described subspecies of IDL^{22} and LDL^{23}. Changes in $HDL_{3b/3c}$ are significantly positively correlated with changes in IDL (predominantly the larger IDL-1 fraction), while changes in HDL_{3a} are positively correlated with the smaller IDL fraction, which includes the cholesterol-enriched IDL-2 subspecies[22]. Finally, changes in HDL_{2b} and HDL_{2a} are positively correlated with changes in LDL-I and LDL-II, with slightly stronger relationships of LDL-I with HDL_{2b} and LDL-II with HDL_{2a}.

On the basis of in vitro[24] and in vivo studies[25], we have recently suggested the existence of at least two metabolic pathways which involve the production of different forms of LDL from different lipolytic precursors. These pathways may be summarized as follows:

A. Smaller VLDL - IDL-1 (triglyceride-enriched) ---> LDL-II
B.   IDL-2 (cholesterol-enriched)   ---> LDL-I

From the correlations in Table 2, it might be hypothesized that factors affecting levels of HDL_{3b} and HDL_{3c} are linked to metabolism of IDL-1, while the metabolism of HDL_{3a} is linked to that of IDL-2. If the associations of the HDL_2 species with LDL-I and LDL-II are also considered to represent metabolic links, pathways A and B may be amplified as follows:

A. Smaller VLDL - IDL-1 (triglyceride-enriched) ---> LDL-II
                                $HDL_{3b/3c}$ --?--> $HDL_{2a/3a}$

B.   IDL-2 (cholesterol-enriched)   ---> LDL-I
                                HDL_{3a} --?--> $HDL_{2b/2a}$

Since both HDL_{3a} and HDL_{2b} include particles with apo A-I and no apo A-II, it is perhaps reasonable to suggest that these are metabolically related. By analogy, the apo (A-I with A-II) particles in HDL_{3b} and HDL_{3c} might be related to the larger (A-I with A-II) species within the $HDL_{2a/3a}$ spectrum.

Based on this hypothetical model, and the data reviewed above, the HDL subspecies pattern in hypertriglyceridemia could be accounted for by:

1) reduction in concentration of HDL particles in pathway B in conjunction with enhanced apo A-I catabolism; and

2) a relative preponderance of HDL_{3b+3c} and $HDL_{2a/3a}$ particles containing apo (A-I with A-II) in pathway A, and a shift in the $HDL_{2a/3a}$ spectrum to particles of greater density and smaller diameter (HDL_{3a}) as a result of lipid enrichment and subsequent lipolytic processing.

The operation of these hypothetical pathways in hypertriglyceridemia, and the extent to which they might contribute to the HDL particle distribution in the general population, await further study.

ACKNOWLEDGEMENTS

This work was supported by NIH Program Project Grant HL 18574 from the National Heart, Lung, and Blood Institute of the National Institutes of Health. We thank Frank Lindgren and his staff for performing analytic ultracentrifugation measurements, Elaine Gong and Patricia Blanche for carrying out gradient gel electrophoresis analyses, Karen Vranizan for help with statistical procedures, and Mary Lou Olbrich and Linda Abe for preparation of the manuscript.

REFERENCES

1. Eisenberg, S., 1984, High density lipoprotein metabolism, J. Lipid Res., 25:1017-1058.
2. Lindgren, F.T., Jensen, L.C., and Hatch, F.T., 1972, The isolation and quantitative analysis of serum lipoproteins, in: "Blood Lipids and Lipoproteins, Nelson, G.J., ed., John Wiley-Interscience, New York, pp. 181-274.
3. Patsch, W., Schonfeld, G., Gotto, A.M., Jr., and Patsch, J.R., 1980, Characterization of human high density lipoproteins by zonal ultracentrifugation, J. Biol. Chem., 255:3178-3185.
4. Nichols, A.V., Blanche, P.J., and Gong, E.L., 1983, Gradient gel electrophoresis of human plasma high density lipoproteins, in: "CRC Handbook of Electrophoresis", Vol. III, Lewis, L. and Opplt, J. eds., CRC Press, Boca Raton, pp. 29-47.

5. Lindgren, F.T., Nichols, A.V., Wood, P.D., Adamson, G.L., Austin, M.A., Glines, L.A., Martin, V., and Krauss, R.M., 1985, Intercorrelations of subfractions of LDL and HDL by gradient gel electrophoresis (GGE) and analytic ultracentrifugation (AnUC), submitted to A.H.A. 58th Scientific Sessions.

6. Cheung, M.C. and Albers, J.J., 1984, Characterization of lipoprotein particles isolated by immunoaffinity chromatography, J. Biol. Chem., 259:12201-12209.

7. Rao, S.N., Magill, P.J., Miller, N.E., and Lewis, B., 1980, Plasma high density lipoprotein metabolism in subjects with primary hypertriglyceridemia: altered metabolism of apoproteins A-I and A-II, Clin. Sci., 59:359-367.

8. Fidge, N., Nestel, P., Ishikawa, T., Reardon, M., and Billington, T., 1980, Turnover of apoproteins A-I and A-II of human high density lipoprotein and the relationship to other lipoproteins in normal and hyperlipidemic individuals, Metabolism, 29:643-653.

9. Schaefer, E.J., Zech, L.A., Jenkins, L.L., Bronzert, T.J., Rubalcaba, E.A., Lindgren, F.T., Aamodt, R.L., and Brewer, H.B., Jr., 1982, Human apolipoprotein A-I and A-II metabolism, J. Lipid Res., 23: 850-862.

10. Schaefer, E.J., Foster, D.M., Zech, L.A., Lindgren, F.T., Brewer, H.B., Jr., and Levy, R.I., 1983, The effects of estrogen administration on plasma lipoprotein metabolism in premenopausal females, J. Clin. Endocrinol. Metab., 57:262-267.

11. Magill, P., Rao, S.N., Miller, N.E., Nicoll, A., Brunzell, J., St.Hilaire, J., and Lewis, B., 1982, Relationships between the metabolism of high-density and very-low-density lipoproteins in man: studies of apolipoprotein kinetics and adipose tissue lipoprotein lipase activity, Eur. J. Clin. Invest., 12:113-120.

12. Nikkila, E.A., Taskinen, M.R., and Kekki, M., 1978, Relation of plasma high density lipoprotein cholesterol to lipoprotein-lipase activity in adipose tissue and skeletal muscle of man, Atherosclerosis, 29:497-501.

13. Patsch, J.R., Prasad, S., Gotto, A.M., Jr., and Bengtsson-Olivecrona, G., 1984, Postprandial lipemia: a key for the conversion of high density lipoprotein$_2$ into high density lipoprotein$_3$ by hepatic lipase, J. Clin. Invest., 74:2017-2023.

14. Chang, L.B.F., Hopkins, G.J., and Barter, P.J., 1985, Particle size distribution of high density lipoproteins as a function of plasma triglyceride concentration in human subjects, Atherosclerosis, 56:61-70.

15. Eisenberg, S., Gavish, D., Oschry, Y., Fainaru, M., and Deckelbaum, R.J., 1984, Abnormalities in very low, low, and high density lipoproteins in hypertriglyceridemia, J. Clin. Invest., 74:470-482.

16. Tall, A.R., Blum, C.B., Forester, G.P., and Nelson, C.A., 1982, Changes in the distribution and composition of plasma high density lipoprotein after ingestion of fat, J. Biol. Chem., 257:198-207.

17. Nichols, A.V., Gong, E.L., and Blanche, P.J., 1981, Interconversion of high density lipoproteins during incubation of human plasma, Biochem. Biophys. Res. Commun. 100: 391-399.

18. Dieplinger, H., Zechner, R., and Kostner, G.M., 1985, The in vitro formation of HDL$_2$ during the action of L:CAT: the role of triglyceride-rich lipoproteins, J. Lipid Res., 26:273-2812.

19. Nichols, A.V., Blanche, P.J., Gong, E.L., Shore, V.S. and Forte, T.M., 1985, Molecular pathways in the transformation of model discoidal lipoprotein complexes induced by lecithin:cholesterol acyltransferase, Biochim. Biophys. Acta, 834: 285-300.

20. Krauss, R.M., Williams, P.T., Lindgren, F.T., and Wood, P.D., 1985, Coordinate changes in levels of human serum low and high density lipoprotein subclasses, submitted.

21. Wood, P.D., Haskell, W.L., Blair, S.N., Williams, P.T., Krauss, R.M., Lindgren, F.T., Albers, J.J., Ho, P.H., and Farquhar, J.W., 1983, Increased exercise level and plasma lipoprotein concentrations, Metabolism, 32: 31-39.

22. Musliner, T.A., Giotas, C., and Krauss, R.M., 1986, Presence of multiple subpopulations of lipoproteins of intermediate density in plasma from normal and hyperlipidemic subjects, Arteriosclerosis, in press.

23. Krauss, R.M. and Burke, D.J., 1982, Identification of multiple subclasses of plasma low density lipoproteins in normal humans, J. Lipid Res., 23: 97-104.

24. Krauss, R.M., Musliner, T.A., and Giotas, C., 1983, Differing content of low density lipoprotein precursors in subclasses of triglyceride-rich lipoproteins, Arteriosclerosis, 3: 510a.

25. Musliner, T.A., McVicker, K.M., Iosefa, J.F., and Krauss, R.M., 1985, Metabolism of intermediate and very low density lipoprotein subfractions from normal and dysbetalipoproteinemic plasma: in vivo studies in rats, submitted.

SPECIATION OF HDL

John P. Kane

Cardiovascular Research Institute
University of California
San Francisco, CA

There is considerable inferential evidence that the high density lipoproteins (HDL) of mammalian plasma should have greater structural and functional heterogeneity than has been so far established. It is likely that at least transitory progenitors of HDL originating from hepatic and intestinal cells will differ from one another, perhaps forming discrete series of particles. The diverse metabolic processes in which HDL participate in plasma are likely to require still further determinative speciation. Some speciation will no doubt reflect low free energy combination states determined by the interactive energetics and stereochemistry of their constitutive elements. However, kinetic speciation is likely to account for accumulation of some particle species, that is, in concatenated processes, intermediaries with the lowest rate constants of removal tend to accumulate. An accurate assessment of particle populations in plasma must take into account the fact that chemical processes will tend to move toward thermodynamic equilibrium after sampling, thus altering the distribution of species. Approaches to the study of lipoprotein speciation must therefore include the use of techniques which allow for the isolation of transitory or thermodynamically unstable particles and which do not introduce perturbations of structure. In the following we will discuss the plurality of roles for HDL and evidence that these lipoproteins comprise a number of discrete subpopulations.

Within the sphere of lipoprotein metabolism per se several discrete roles are now identified for HDL. It is clearly established that HDL serve to provide C apoproteins to nascent VLDL and chylomicrons[1-3]. These apoproteins return to HDL as lipolysis proceeds. It is recognized, however, that the ratios of the individual C apoproteins to one another are different in VLDL and HDL, suggesting that certain C apoproteins may function in HDL complexes[4].

Apolipoprotein E is transferred to chylomicrons from HDL in the intestinal lymphatics[5,6]. In the case of both VLDL and chylomicrons, exchange of amphipathic lipids parallels the transfer of soluble apoproteins from HDL. Unesterified cholesterol is transferred to the triglyceride-rich lipoproteins, and phospholipids are acquired by HDL, especially during lipolysis when the acquisition of VLDL-derived lipid ultimately changes the particle diameters and densities of HDL as they can be perceived by ultracentrifugation, i.e., formation of increased amounts of HDL_2[2-7,12]. The formation of HDL_2 no doubt reflects in part the direct acqusition of VLDL-derived lipids, but also requires the esterification of cholesterol, mediated by LCAT.

Other important metabolic roles of HDL lie in the centripetal transport of cholesterol. Though this process is incompletely understood, several roles are emerging for what appear to be different subspecies of HDL [13-15]. In the initial phase of centripetal transport, unesterified cholesterol is acquired from tissues by relatively small HDL particles which contain apo A-I but no apo A-II. The process of esterification of cholesterol and transfer of the ester product appears to reside in either one or a limited number of subspecies of HDL [13-15]. It is probable that a portion of the metabolism of triglycerides and certain phospholipid species also takes place in HDL complexes.

The presence of the less abundant apolipoproteins in HDL creates still another, obligatory, basis for speciation. For example, the total content of apo C-II in normal human plasma is about 3.9×10^{-6} M, of which about 40% is complexed to HDL [16]. In contrast, apo A-I in normal males is present at about 4.6×10^{-5} M. Because it can be calculated that total human HDL, on a basis of number averaged particle weights, contain approximately 81×10^5 daltons of protein per particle [17] of which apo A-I comprises approximately two-thirds by mass, only a small fraction of HDL particles can contain apo C-II, even if it were dispersed as a single copy per particle. Similar calculations can be made for other minor constituents, such as apo D, for instance, forcing the conclusion that speciation of HDL must occur on the basis of molecular stoichiometry alone. Such speciation must be viewed in the kinetic or statistical sense, however, to the degree that such minor elements are subject to equilibrium transfer among particles.

The variety of emerging roles for HDL outside of the classical processes of lipid transport also suggest that specialized HDL might exist in support of these other biological objectives. For instance, the role of HDL is recognized in the growth of endothelial cells [18], a function which might afford protection against atherosclerosis. Another role possibly related to the expression of vascular disease is the inhibition of one phase of blood clotting. HDL have been shown to inhibit the activation of factor X by factor VIIa and tissue factor [19,20]. A number of roles for HDL which appear to confer protection against parasitic or infectious disease are now recognized. HDL are apparently capable of lysing certain trypanosomes [21-23] and probably account for the poor ability of those species to infect humans. Extremely potent inhibition of the infectivity of mouse xenotropic C viruses has been attributed to HDL and chylomicrons [24,25]. This activity appears to require the presence of a minor apolipoprotein which is capable of transferring between those lipoproteins. HDL bind bacterial lipopolysaccharide endotoxins [26] and the serum amyloid associated proteins [27], as well as playing a role in immunoregulation and in complement fixation [28]. Thus, though none of these non-transport roles for HDL is yet known to demand the existence of structurally discrete particles, it is a strong possibility that dedicated species do exist, and that their levels in plasma may be regulated independently of other HDL species.

Though the methods previously available have not permitted the isolation and characterization of discrete HDL species, data from a number of analytical techniques present strong inference that a number of species will be found in circulating blood. The emerging awareness that the native structures of HDL particles are severely perturbed by ultracentrifugation provides the explanation for the failure of efforts to identify them heretofore. Since nearly all previous efforts at isolation of HDL have commenced with preparative ultracentrifugation, speciation can be assumed to have been obscured prior to application of secondary techniques, leading to the belief that HDL are highly polydisperse. The evidence that ultracentrifugation causes degenerative polydispersity of HDL particles is twofold. First, several investigations have shown the dissociation of appreciable amounts of at least two apolipoproteins, apo A-I and apo E [29-32], during ultracentrifugation.

Second, patterns of multiple discrete HDL bands, seen upon direct electrophoresis of whole serum, are reduced to two broad featureless zones if the
HDL are subjected to ultracentifugation before electrophoresis[33]. The separation of apolipoproteins from lipoprotein particles during ultracentrifugation obviously depends upon a dissociative event, followed by separation of
the free apolipoproteins by sedimentation. Self-association of apolipoproteins, in the case of apo A-I, may facilitate this process. The dissociation
of apo A-I from HDL is increased at low ionic strengths, indicating that the
high ionic strength of commonly employed ultracentrifugal media such as KBr,
is not the primary cause of disruption of the complexes[33]. Likewise, interaction with tube walls in angle head rotors is not responsible either, as
equal amounts of dissociation occur in swinging bucket rotors[33]. The high
pressures exerted in the fluid column are the most likely cause; hence, no
ultracentrifugal technique can be expected to permit the isolation of HDL
from which native species can subsequently be isolated. Nonetheless, analytical ultracentrifugation has provided inference that subspecies exist
even under such dissociative conditions[34] as does rate zonal ultracentrifugation[35,36]. Perhaps some of the strongest evidence for speciation of HDL
comes from electrophoretic techniques. HDL, stained in whole plasma with a
lipophilic dye, shows distinct speciation when subjected to electrophoresis
in gels[37-39], though this method did not permit isolation of the particles.
Likewise, separation into distinct zones is apparent upon isotachophoresis[40]
and upon gradient gel electrophoresis[41]. On gel permeation chromatography,
HDL fractions of different composition are observed[42] and apo E containing
HDL can be separated by affinity chromatography on concanavalin A-sepharose[43]
or on heparin sepharose[44,45]. Though each of the methods discussed here has
presented evidence of molecular heterogeneity, suggesting discrete speciation
of HDL, none provides the basis for the separation of all such species. A
practical strategy for such an objective would begin with the sequestration
of all HDL from plasma by a minimally perturbing method to be followed by
secondary separation techniques for isolation of individual species.

The technique of selected affinity immunosorption[46], which we have
recently developed, allows the quantitative sequestration and recovery of
apolipoprotein A-I-containing lipoproteins under minimally perturbing conditions. The principle of this technique is the preselection from a polyclonal antiserum of a pauciclonal immunoglobulin fraction, in which all the
immunoglobulin species are dissociated from their ligands by a relatively
non-denaturing eluant such as dilute bicarbonate, dilute acetic acid, or
perhaps peptides derived from the ligand itself. Such a pauciclonal antibody
preparation is not of low-affinity, but operates under gentle conditions of
elution because of the uniformity and specificity of response to the eluant.
Using this technique, we have isolated apo A-I containing lipoprotein particles quantitatively from human serum and plasma. Such particles contain
more protein than HDL obtained by ultracentrifugation. Further, some particles up to 300 Å diameter and enriched in triglycerides are isolated in
this fashion. Such particles have not been identified in ultracentrifuged
lipoprotein fractions, suggesting that they are not stable in the ultracentrifuge. The presence of discrete speciation of HDL has now been confirmed
by others, using the selected affinity immunosorption method[47]. We have
found up to eight bands when apo A-I-containing lipoproteins obtained by
the selected affinity method are subjected to electrophoresis at neutral pH
or to isoelectric focussing in polyacrylamide gels. That the bands which
we observe probably represent native species resulting from normal metabolic
processes is suggested by our observation that distinctive departures from
normal band patterns are seen in individuals with major derangements of
lipoprotein metabolism, such as lipoprotein lipase deficiency and abetalipoproteinemia. If the immunosorbed lipoproteins are subjected to ultracentrifugation, however, they then show degenerate polydispersity, with distribution into two broad, diffuse bands by either electrophoresis or isoelectric
focussing. These bands appear to correspond to HDL_2 and HDL_3, suggesting

that these fractions are partly dissociated HDL or low free energy recombinants of HDL constituents which are formed from native HDL species in the ultracentrifuge.

The first of the species we have isolated from immunosorbed apo A-I-containing particles and characterized in detail is distinguished by prebeta mobility in agarose gel electrophoresis. We have isolated it by starch block electrophoresis of immunosorbate lipoproteins at pH 8.4[48]. This separation yields two distributions, of alpha, and prebeta mobility. The "prebeta HDL" contain from 5 to 20% of the total apo A-I of plasma. The instability of the prebeta HDL to ultracentrifugation and their lack of affinity for lipophilic stains apparently account for the fact that a lipoprotein species of this abundance and discriminating electrophoretic mobility could have escaped clearcut recognition heretofore. The protein-rich prebeta HDL have a particle weight of approximately 80 kilodaltons. They contain two molecules of apo A-I per particle with phospholipid and free and unesterified cholesterol, but triglycerides are undetectable.

At 37° in plasma the prebeta HDL are transformed into lipoproteins of alpha mobility within two hours. Complete inhibition of this transformation by 1.4 mM 5,5'-dithiobis-(2-nitrobenzoic acid) and by 20 mM menthol, both inhibitors of LCAT, suggests that prebeta HDL are normally a substrate for that enzyme[49]. An important metabolic discrimination between the prebeta HDL and the particles of alpha mobility may be reflected by the obseration that the affinity of hepatic membrane binding sites for the prebeta particles is two orders of magnitude lower than for the remainder of HDL[50]. A relationship of prebeta HDL to the catabolism of triglyceride-rich lipoproteins is suggested by the recent observation that the prebeta fraction is essentially undetectable in subjects with abetalipoproteinemia[51]. Taken together, these preliminary findings suggest that the prebeta HDL may be formed during the catabolism of chylomicrons, and possibly VLDL. Their conversion by LCAT to some component of alpha electrophoretic mobility suggests a major transformation, perhaps involving fusion with larger HDL particles. The prebeta HDL may also serve as couriers of cholesterol from peripheral tissues to the particle or particles on which esterification takes place.

Progress to date suggests that selected affinity immunosorption may provide the means for isolation of the native molecular species of HDL and perhaps other lipoproteins as well. Elucidation of the structures of the naturally occurring lipoproteins offers the promise that modelling of the metabolic processes involving lipoproteins in plasma can ultimately be carried out at the molecular level.

ACKNOWLEDGMENTS

The research described in this chapter was supported by grants from the National Institutes of Health, HL 14237 (Arteriosclerosis SCOR), and the Wine Institute.

REFERENCES

1. R. L. Hamilton, Synthesis and secretion of plasma lipoproteins, in "Phamacological Control of Lipid Metabolism," W. L. Holmes, R. Paoletti, and D. Kritchevsky, eds., Plenum Press, New York (1972).
2. R. J. Havel, J. P. Kane, and J. L. Kashyap, Interchange of apolipoproteins between chylomicrons and high density lipoproteins during alimentary lipemia in man, J. Clin. Invest. 52:32 (1973).
3. S. Eisenberg, and R. I. Levy, Lipoprotein metabolism, Adv. Lipid Res. 13:1 (1975).

32

4. E. Polz, L. Kotite, R. J. Havel, J. P. Kane, and T. Sata, Human apo-lipoprotein C-I: concentration in blood serum and lipoproteins, Biochem. Med. 24:229 (1980).

5. K. Imaizumi, M. Fainaru, and R. J. Havel, Composition of proteins of mesenteric lymph chylomicrons in the rat and alterations produced upon exposure of chylomicrons to blood serum and serum lipoproteins, J. Lipid Res. 19:712 (1978).

6. K. Imaizumi, R.J. Havel, M. Fainaru, and J.-L. Vigne, Origin and trans-port of the A-I and arginine-rich apolipoprotein mesenteric lymph of rats, J. Lipid Res. 19:1038 (1978).

7. J. R. Patsch, A. M. Gotto, T. Olivecrona, and S. Eisenberg, Formation of high density lipoprotein -like particle during lipolysis of very low density lipoproteins in vitro, Proc. Natl. Acad. Sci. USA 75: 4519 (1978).

8. M. R. Taskinen, M. L. Kashyap, L. S. Srivastava, M. Ashraf, J. D. Johnson, G. Perisutti, D. Brady, C. J. Glueck, and R. L. Jackson, In vitro catabolism of human plasma very low density lipoproteins: effects of VLDL concentration on the interconversions of high den-sity liporotein subfractions, Atherosclerosis 41:381 (1982).

9. A. R. Tall, P. H. R. Green, R. M. Glickman, and J. W. Riley, Metabolic fate of chylomicron phospholipids and apoproteins in the rat, J. Clin. Invest. 64:977 (1979).

10. T. G. Redgrave, and D. M. Small, Quantitation of the transfer of sur-face phospholipid of chylomicrons to the high density lipoprotein fraction during the catabolism of chylomicrons in the rat, J. Clin. Invest. 64:162 (1979).

11. A. R. Tall, C. B. Blum, G. P. Forester, and C. A. Nelson, Changes in the distribution and composition of plasma high density lipoproteins after ingestion of fat, J. Biol. Chem. 257:198 (1982).

12. M. Mattock, A. Salter, T. Omer, and H. Keen, Changes in high density lipoprotein fractions during alimentary lipemia, Experientia 37:945 (1981).

13. C. J. Fielding, and P. E. Fielding, Cholesterol transport between cells and body fluids, Med. Clin. North Am. 66:363 (1982).

14. T. Chajek, and C. J. Fielding, Isolation and characterization of a human serum cholesteryl ester transfer protein, Proc. Natl. Acad. Sci. USA 75:3445 (1978).

15. C. J. Fielding, and P. E. Fielding, Metabolism of cholesterol and lipo-proteins, in "Biochemistry of Lipids and Membranes," D. E. Vance and J. E. Vance, eds., Benjamin/Cummings Publishing Co., Menlo Park, CA (1985).

16. G. Schonfeld, P. K. George, J. Miller, P. Reilly, and J. Witztum, Apo-lipoprotein C-II and C-III levels in hypolipoproteinemia, Metabolism 28:1001 (1979).

17. J. P. Kane, Plasma lipoproteins: structure and metabolism, in "Lipid Metabolism in Mammals, Plenum Press, New York, (1977).

18. J. P. Tauber, J. Cheng, S. Massoglice, and D. Gospodarowicz, High den-sity lipoproteins and the growth of vascular endothelial cells in serum-free medium, In Vitro 17:519 (1981).

19. S. D. Carson, Plasma high density lipoproteins inhibit the activation of coagulation factor X by factor VIIa and tissue factor, FEBS Lett 132:37 (1981).

20. T. W. Bareowcliffe, C. A. Eggleton, and J. Stocks, Studies of anti Xa activity in human plasma V: the role of lipoproteins, Thromb. Res. 27:185 (1982).

21. W. E. Ormerod, and S. Venkatesan, Similarities of lipid metabolism in mammalian and protozoan cells: An evolutionary hypothesis for the prevalence of atheroma, Microbiol. Rev. 46:296 (1982).

22. M. R. Rifkin, Interaction of high density lipoprotein with Trypanosoma Brucei: effect of membrane stabilizers, J. Cell. Biochem. 23:57 (1983).

23. J. Dhondt, N. Van Meirvenne, L. Moeus, and M. Kondo, Ca2+ is essential cofactor for trypanocidal activity of normal human serum, Nature 282:613 (1979).

24. J. P. Kane, D. A. Hardman, J. C. Dimpfl, and J. A. Levy, Apolipoprotein is responsible for neutralization of xenotropic type C virus by mouse serum, Proc. Natl. Acad. Sci. USA 76:5957 (1979).

25. J. A. Levy, J. Dimpfl, D. Hardman, and J. P. Kane, Transfer of mouse anti-xenotropic virus neutralizing factor to human lipoproteins, J. Virology 42:365 (1982).

26. R. J. Ulevitch, A. R. Johnston, and D. B. Weinstein, New function for high density lipoproteins—isolation and characterization of a bacterial pepopolysaccharide high density lipoprotein complex formed in rabbit plasma, J. Clin. Invest. 67:827 (1981).

27. C. L. Malmendier, and J. P. Ameryckx, Apoprotein S, a family of human serum lipoprotein polypeptides, Atherosclerosis 42:161 (1982).

28. T. F. Lint, C. L. Behrends, and H. Gewurz, Serum lipoproteins and C-567-INH activity, J. Immunol. 119:883 (1977).

29. M. Fainaru, M. C. Glangeaud, and S. Eisenberg, Radioimmunoassay of human high density lipoprotein apoprotein A-I, Biochim. Biophys. Acta 386:432 (1975).

30. M. Fainaru, R. J. Havel, and T. E. Felker, Radioimmunoassay of apolipoprotein A-I of rat serum, Biochim. Biophys. Acta 446:56 (1976).

31. M. D. Curry, P. Alaupovic, and C. A. Suenram, Determination of apolipoprotein A-I and A-II polypeptides by separate electroimmunoassays, Clin. Chem. 22:315 (1976).

32. R. W. Mahley, and K. S. Holcombe, Alterations of the plasma lipoproteins and apoproteins following cholesterol feeding in the rat, J. Lipid Res. 18:314 (1977).

33. S. T. Kunitake, and J. P. Kane, Factors affecting the integrity of high density lipoproteins in the ultracentrifuge, J. Lipid Res. 23:936 (1982).

34. D. W. Anderson, A. V. Nichols, T. M. Foote, and F. T. Lindgren, Particle distribution of human serum high density lipoproteins, Biochim. Biophys. Acta 493:55 (1977).

35. G. M. Kostner, J. R. Patsch, S. Sailer, H. Braunsteiner, and A. Holasek. Polypeptide distribution of the main lipoprotein density classes separated by ratezonal ultracentrifugation, Eur. J. Biochem. 45:611 (1974).

36. B. H. Chung, J. P. Segrest, J. T. Cone, J. Pfau, J. C. Geer, and L. A. Duncan, High resolution plasma lipoprotein cholesterol profiles by a rapid, high volume semiautomated method, J. Lipid Res. 22:1003 (1981).

37. G. Utermann, Disc-electrophoresis patterns of human serum high density lipoproteins, Clin. Chim. Acta 36:521 (1972).

38. J. Janecki, and A. Fijalkowska, A simple method of quantitative estimation of the subfractions of human small molecular diameter lipoproteins, J. Clin. Chem. Clin. Biochem. 17:789 (1979).

39. S. L. Mackenzie, G. S. Sundaram, and H. S. Sodhi, Heterogeneity of human serum high density lipoprotein (HDL3), Clin. Chim. Acta 43:223 (1973).

40. P. J. Blanche, E. L. Gong, T. M. Forte, and A. V. Nichols, Characterization of human high density lipoproteins by gradient gel electrophoresis, Biochim. Biophys. Acta 665:408 (1981).

41. G. Bittolo Bon, G. Cazzolato, and P. Avogaro, Preparative isotachophoresis of human plasma high density lipoproteins HDL2 and HDL3, J. Lipid Res. 22:998 (1981).

42. L. L. Rudel, J. A. Lee, M. D. Morris, and J. M. Felts, Characterization of plasma lipoproteins separated by agarose-column chromatography. Biochem. J. 139:89 (1974).

43. T. Mitamura, Separation of cholesterol-induced high density lipoprotein (HDLc) by concanavalin A-sepharose affinity chromatography, J. Biochem. 91:25 (1982).

44. Y. L. Marcel, C. Vezina, D. Emond, and G. Suzue, Heterogeneity of human high density lipoprotein: presence of lipoproteins with and without apo E and their roles as substrates for lecithin:cholesterol acyltransferase reaction, Proc. Natl. Acad. Sci. USA 77:2969 (1980).

45. K. H. Weisgraber, and R. W. Mahley, Subfractionation of human high density lipoproteins by heparin-sepharose affinity chromatography. J. Lipid Res. 21:316 (1980).

46. J. P. McVicar, S. T. Kunitake, R. L. Hamilton, and J. P. Kane, Characteristics of human lipoproteins isolated by selected-affinity immunosorption of apolipoprotein A-I, Proc. Natl. Acad. Sci. USA 81:1356 (1984).

47. M. C. Cheung, and J. J. Albers, Characterization of lipoprotein particles isolated by immunoaffinity chromatography, J. Biol. Chem. 259:12201 (1984).

48. S. T. Kunitake, K. J. La Sala, and J. P. Kane, Apoprotein A-I-containing lipoproteins with pre-beta electrophoretic mobility, J. Lipid Res. 25:549 (1985).

49. S. T. Kunitake, B. Ishida, C. J. Fielding, and J. P. Kane, Apoprotein (apo) A-I-containing lipoproteins with prebeta electrophoretic mobility: conversion to lipoproteins with alpha mobility, Circulation 72:III:92 (1985).

50. C. M. Mendel, S. T. Kunitake, and J. P. Kane, Discrimination between subclasses of human high density lipoproteins by the HDL binding sites of bovine liver, Biochim. Biophys. Acta 875:59 (1986).

51. M. J. Malloy, K. J. La Sala, S. T. Kunitake, and J. P. Kane, unpublished results

MODIFICATIONS AND DEGRADATION OF HIGH DENSITY LIPOPROTEINS

A. Angel and B. Fong

Department of Medicine and Institute of Medical Science
University of Toronto and Division of Endocrinology and
Metabolism, Toronto General Hospital, Toronto, Ontario
Canada M5S 1A8.

INTRODUCTION

Plasma high density lipoproteins (HDL) attract wide interest because of their putative antiatherogenic role in man (1). It has been repeatedly demonstrated that the plasma concentration of HDL cholesterol correlates inversely with coronary disease risk and this is true for both men and women (2). This protective effect of HDL is independent of total plasma cholesterol levels and is thus viewed as a risk predictor independent of all other lipoprotein classes. Reduced plasma HDL cholesterol levels are associated with a large number of clinical states and the list continues to grow. In Table 1, both genetic and acquired diseases where HDL cholesterol are frequently reduced with increased risk for coronary heart disease is shown. Tempting as it is to attribute causality to the HDL particle per se, the precise mechanism or mechanisms responsible for the protection associated with this lipoprotein has not been determined. It is evident that until all the relevant functional properties of HDL as it circulates in the plasma and extravascular compartments are well understood, the mechanism underlying its antiatherogenic properties shall remain a mystery. Furthermore, because HDL are formed and degraded by complex mechanisms involving turnover of constituent elements, an understanding of these processes is also necessary as background to a discussion of degradation of this species.

HDL METABOLISM

A good deal has been learned over the past decade about HDL metabolism and function in vivo (Table 2), particularly events within the plasma compartment and several recent studies concerning the physiological function of HDL with respect to cholesterol flux have appeared. A major review was published recently covering the subject (3). In contrast, very little is known about HDL metabolism in the extravascular space and almost nothing is known about the degradation processes in humans. A simplified view of the dynamic character of HDL turnover and metabolism based on a large number of studies (3,4,5) is shown schematically in Figure 1.

Briefly, high density lipoprotein particles are synthesized de novo as precursors in the liver and small intestinal epithelial cells. These particles are secreted in discoidal form consisting of bimolecular leaflets composed of polar phospholipids and unesterified cholesterol along with apoproteins (Apo A) and the enzyme lecithin-cholesterol acyl transferase (LCAT) embeded within. Upon entry into the systemic circulation, these particles rich in free cholesterol undergo a progressive physical transformation with the accumulation of neutral lipids and assume a more spherical shape. The core, now composed of cholesterol ester and triglyceride, is surrounded by phospholipids, free cholesterol and a number of apoproteins particularly Apo AI, Apo AII, Apo C´s, Apo D as well as some Apo E and LCAT. This particle designated HDL$_3$ plays an important role as a cholesterol efflux acceptor at cell surfaces and is involved in facilitating the clearance of glyceride-rich plasma particles (chylomicrons and VLDL) during their catabolism at endothelial surfaces by the action of lipoprotein lipase. A significant

TABLE 1

Clinical States Associated with Reduced HDL Cholesterol

GENETIC	Familial Hypoalphalipoproteinemia
	Tangier Disease
	Familial LCAT Deficiency
	Apo AI-Apo CIII Deficiency
	Apo AI Milano
	Etc.
ACQUIRED	Obesity
	Hyperprebetalipoproteinemia (↑ VLDL)
	Uncontrolled Diabetes Mellitus
	Smoking
	Renal Diseases – Nephrotic Syndrome
	Chronic Dialysis
	Transplant Patients
	GI Diseases - Malabsorption Syndromes
	Regional Enteritis
	Premature Menopause
	Androgen Therapy

amount of redundant coat material including phospholipids, cholesterol, Apo Cs and Apo E from chylomicrons and VLDL transfers to HDL$_3$ which expands spherically (6). This enlarged particle circulates as HDL$_2$ which becomes enriched with core cholesterol ester and undergoes metabolic transformations primarily in the liver by an exchange reaction accomplished through the action of hepatic lipase (7). Cholesterol ester transfer from HDL$_2$ to VLDL also occurs in plasma in exchange for triglyceride (8). Transfer of LDL cholesterol ester to HDL is another important mechanism of HDL cholesterol-ester accumulation (9,10). While it is generally assumed that HDL$_3$ is involved as an efflux cholesterol acceptor from a variety of cells in the interstitial compartment (11), the precise HDL species involved in cholesterol efflux in various interstitial spaces is not known. All may be involved as interstitial fluids contain all species of lipoprotein known to be present in plasma (12).

APOPROTEIN EXCHANGES AND LIPID TRANSFER

The various lipid and apoprotein exchanges between HDL species and other lipoproteins indicate the dynamic character of these particle and the difficulty in defining their catabolic fate (13). Furthermore there is considerable evidence that the turnover and tissue distribution of HDL cholester-ester differs substantially from the major apoprotein Apo AI (14). Thus, any notion about the catabolic fate of the intact HDL particle presents conceptual difficulties as complete metabolism of the intact holoparticle may not represent the major clearance process. Indeed HDL interactions with various cells differ depending on the tissue´s metabolic role in cholesterol homeostasis (15). This contrasts with LDL and its interaction with the classical Apo B receptor which is specific, well defined and involves adsorptive endocytosis of the intact particle (16). Thus HDL does not appear to have a single ubiquitous catabolic process and this point is schematically represented in Figure 1 where peripheral degradation is indicated in terms of matabolic fate of particle constituents. It is possible that removal of the intact particle as such is a minor catabolic mechanism whereas the independent clearance of free Apo AI (and minor apoproteins) and cholesterol ester core may be more significant (14,17).

TABLE 2

Metabolic Functions of HDL

PLASMA	Cholesterol esterification (LCAT)
	Facilitates CHYLO and VLDL glyceride clearance
	Apoprotein shuttle (Apo C & Apo E)
INTERSTITIAL & CELLULAR	Cholesterol delivery (STEROIDOGENESIS)
	Efflux acceptor of cholestrol
	Modulates LDL binding & metabolism

HDL CLEARANCE AND TISSUE UPTAKE OF CONSTITUENTS

Initial studies of HDL catabolism centered on kinetic clearance of radioiodinated HDL and largely reflected the catabolic clearance of Apo I (18,19). While these studies provided insight into rates of irreversible clearance from plasma and uptake of labeled apoproteins by various tissues, the results could not be interpreted as an accurate quantitative measure of cellular internalization because of the confounding effects of liberated free iodide, liberated iodopeptides, and tissue accumulation of untransformed apoproteins. Furthermore, uptake of radioiodine cannot be

equated with cellular uptake of the intact particle because the
apoproteins readily dissociate from core lipids which can be taken up by
cells independent of apoproteins (14).

A major advance in the analysis of HDL metabolism in vivo was achieved
with the introduction of cumulative markers identifying cellular
distribution of the major apoprotein (Apo AI) (labeled with
^{125}I-tyramine cellobiose) and core lipid (^{14}C-cholesteryl-ether)
(14). Using HDL doubly labeled with these cumulative markers in rats, the
liver appeared to be a major site of terminal catabolism of
cholesterol-ether followed closely by the adrenal and gonads. Almost all
other tissues took up small amounts of cholesterol-ether. However,
steroidogenic tissues displayed the highest cholesterol specific
activity. The tissue distribution of Apo AI, on a proportional basis, was
significantly different than that of cholesterol-ether. While liver took

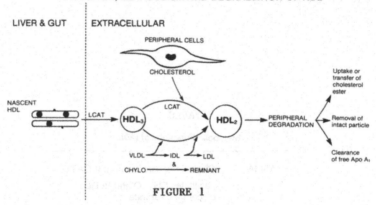

FORMATION, METABOLISM AND DEGRADATION OF HDL

FIGURE 1

up a significant amount of Apo AI, kidney uptake of Apo AI very much
exceeded that of cholesterol-ether. The kidney accumulated very little
cholesterol in contrast with steroidogenic tissues which selectively
removed the lipid core (15). Thus a significant proportion of HDL is
cleared not as an intact particle but by selective clearance of
constituents. This differential catabolism of HDL constituents varies
tissue by tissue and the plasma clearance characteristics thus reflect the
sum of all the tissues combined. For this reason it is important to
determine the role of each tissue independently. This has been approached
in three ways. Using cells in propagated culture, significant binding and
metabolism of HDL has been demonstrated in cultured hepatocytes, adrenal
cortical cells, ovarian cells, testes, endothelial cells and skin
fibroblasts (3,20,21,22). Alternatively tissue cells have been freshly
isolated from surgical biopsies and examined in short term culture ex
vivo. With the latter approach, liver cells, small intestinal cells, and
human adipocytes have been studied in some detail (20,23,24). In situ
organ perfusion has also been explored using autoradiographic methods
(15.)

HDL DEGRADATION

High density lipoproteins undergo a variety of modifications in plasma and in the extravascular space. It is convenient to examine metabolic processing in these compartments separately. For the present discussion of catabolic transformations we will restrict our considerations to proteolytic and covalent chemical modifications rather than changes in size and composition secondary to transfer or uptake of lipid and apoproteins. In Table 3 the various modifications of HDL are listed in relation to fluid compartments and cell surfaces. This compartmental approach helps focus on the principle that the metabolic fate of HDL is probably determined by a variety of reactions in plasma, in the interstitial compartment and at various cell surfaces.

TABLE 3

Modifications and Degradation of Lipoproteins

1. PLASMA
   1. Post translational modifications
      Pro Apo AI → Apo AI
   2. Proteolysis of mature apolipoproteins

Apo AI	11,000 Kd (plasmin)
Apo AII	7,000 + 4,000 Kd (PMN enzyme esterase)
Apo E (HTG-VLDL)	12,000 & 22,000 Kd (Thrombin)

   3. Chemical modification – In Vivo Glucosylation

2. INTERSTITIUM
   1. Increased mobility of lymph lipoproteins on agarose
   2. Inflammatory fluid

3. CELLULAR MODIFICATIONS
   1. Adsorptive endocytosis
   2. Retro endocytosis
   3. Surface proteases
   4. Lipid exchange
   5. Peroxidation

PROTEOLYTIC MODIFICATIONS OF HDL

Lipoproteins undergo proteolytic modifications in vivo and several studies have reported experimental modifications of lipoproteins using proteolytic enzymes. In vivo proteolytic action on HDL apoproteins occur intracellularly as post-translational modifications in the conversion of preproapo AI to proapo AI prior to assembly of nascent HDL particles (25). Apo AI is secreted into plasma in precursor form and this proapo AI is converted to mature Apo AI within the plasma compartment by a protease intrinsic to HDL. Preproapo AII is similarly synthesized and converted to proapo AII intracellularly. While it is thought that proapo AII is secreted into the plasma where it undergoes cleavage, proapo AII has not been identified in the plasma compartment (25). Similar prepeptides have been identified as signal peptide sequences of the Apo C´s and apo E, and will undoubtedly be found for apo B (26).

Mature apoproteins are sensitive to proteolytic enzymes and have been studied using a number of plasma proteases. In vitro, Apo AI can be broken to release smaller peptides by the action of plasmin (27). Additionally HDL displays elastase activity and promotes the release of this proteolytic activity from polymorphonuclear cells (26). The significance of proteolytic enzymes released from polymorphonuclear leukocytes in the presence of HDL has not been established but may be relevant in inflammatory reactions with extensive extravasation of plasma and blood leukocytes. Apo AII is particularly susceptible to elastase degradation (26). Apo B in LDL is also susceptible to degradation by elastase released by human polymorphonuclear cells (28).

Apo E is sensitive to thrombin action with release of two fragments with molecular weights of 12,000 and 22,000 daltons. The 12,000 dalton fragment is similar to that released from plasma VLDL of patients with hypertriglyceridemia and these particles display proteolytic activity (29). Thus hypertriglyceridemic VLDL as well as HDL contain proteolytic activities directed at intrinsic apoproteins.

The significance of the intrinsic proteolytic enzymes with respect to HDL function remains to be established. It is tempting to suggest that protease activities alter lipoprotein binding characteristics and lipid transport function of these particles. Alternatively, they may program the apoprotein for its irreversible catabolism following progressive and sequential modifications in structural integrity. Recent studies in our laboratory have shown that proteolytic modification of adipocyte membrane by pronase enhance both LDL and HDL binding (unpublished). Apo B of LDL is also subject to proteolytic degradation in plasma of many persons. Upon standing, Apo B100 is converted to complementary fragments B74 and B26. This degradation can be prevented with proteolytic inhibitors. Furthermore treatment of Apo B with kallikrein will also produce B74 and B26 peptides (30). The biological significance of this particular transformation is not evident at present as LDL so altered by kallikrein treatment binds normally to the high affinity LDL receptor.

In considering the structural remodelling of lipoproteins, the role of phospholipases deserves consideration. Phospholipase A_2 has been shown to degrade LDL phospholipids and could alter apoprotein orientation on the lipoprotein surface and thus alter lipoprotein binding to its receptor. The physiologic significance of this process remains to be established.

CHEMICAL MODIFICATIONS OF HDL

Chemical modification of lipoproteins alter their binding characteristics and these modifications may be biological or experimental. In plasma, glucosylation of LDL and HDL occurs in chronic hyperglycemic states and this can affect and modify high affinity receptor clearance of LDL (31). Glucosylation of HDL has not resulted in altered binding to cultured fibroblasts but HDL binding is markedly reduced by modification of its tyrosine residues using tetranitramethane (TNM) (32). However cross linking of apoproteins due to TNM treatment could have been responsible for the altered HDL binding. Reductive methylation, acetylation, and cyclohexanedione modification of HDL are without effect on binding to cultured skin fibroblasts (32).

42

INTERSTITIAL MODIFICATIONS OF LIPOPROTEINS

Modifications of lipoproteins can occur in the extravascular space and studies of interstitial fluids from various tissue compartments indicate changes in lipoprotein composition as well as physical characteristics. This is quantitatively significant because 30% to 40% of LDL and HDL are distributed in extra vascular compartments. The mobility of lipoproteins isolated from interstitial compartments by lymphatic cannulation reveal increased mobility on agarose electrophoresis (8). The metabolic significance of this property has not been established but similar changes have been noted in lipoproteins oxidatively modified by incubation with cultured endothelial cells (33).

Significant alterations in lipoproteins occur in inflammatory fluids (34). Inflammatory fluids differ appreciably from interstitial lymph as the former is an exudate containing extravasated plasma proteins, lipoproteins and polymorphonuclear leukocytes and other inflammatory cells. Thus lipoproteins in inflammatory fluids are exposed to the degradative action of many proteolytic enzymes which can alter their metabolic fate.

CELL SURFACE MODIFICATION OF LIPOPROTEINS

Under physiological conditions extensive modifications probably occur at or close to cell surfaces where lipoproteins interact directly as they deliver or accept cholesterol for return transport to the liver. Lipoproteins can bind, enter the cell and exit again by retroendocytosis (35) It is likely that subtle changes in apoprotein structure and organization occur as a consequence of these events. Cells also display surface proteases which in theory can alter exposed lipoproteins (24, 36). This is particulrly evident with LDL and adipocytes which contain surface proteolytic activity that can degrade LDL to a significant extent (24, and Fig.2).

HDL METABOLISM IN HUMAN ADIPOSE CELLS

Adipose tissue is a dynamic cholesterol storage pool and a significant site of lipoprotein metabolism. The diagram in Figure 2 schematically indicates lipoprotein interactions with human fat cells and demonstrates how cholesterol enters and leaves adipose tissue. Lipoproteins undergo significant modification as a consequence of these interactions. Under steady state conditions adipose tissue contains 1.0-2.0 mg cholesterol per gm triglyceride with a turnover half-time in vivo of sixty days or so. Virtually all the cholesterol in human fat is present in unesterified form within the central oil droplet and none of it is derived from in situ synthesis. A metabolic block in cholesterol synthesis at the level of squalene exists in this tissue. Thus almost all the cholesterol in adipose tissue is delivered in the form of lipoproteins that interact with specific recognition sites on the cell surface. Both low density lipoproteins (LDL) and high density lipoproteins (HDL_2 and HDL_3) are present in interstitial fluid and easily accessible to adipocyte surfaces in vivo.

The human fat cell freshly isolated and studied ex vivo readily binds human LDL and HDL particles with specificity, high affinity and high capacity (37,38). A portion of LDL bound to human adipocytes undergoes internalization and intracellular metabolism. Human fat cells also degrade LDL to a significant extent and unlike most cultured cell systems a significant amount of apo B degradation occurs at or near the cell surface presumably by surface proteases (24,and Fig.2). The binding characteristics

of LDL differ in several respects from those of the classical apo B
receptor of Brown and Goldstein. In the fat cell, LDL binding is calcium
independent, protease insensitive, insensitive to high salt concentrations
and completely inhibited by HDL (37). LDL binding is dose dependent
within the concentration range found in the peripheral lymph in contrast
to the concentrations of LDL in cardiac and lung interstitial fluid which
are 40-50% that of plasma and would completely saturate cell surface
exposed in these compartments. Modification of LDL by methylation of
lysine residues only partially inhibits binding. Thus human fat cell can
bind and metabolize both normal and modified LDL (37).

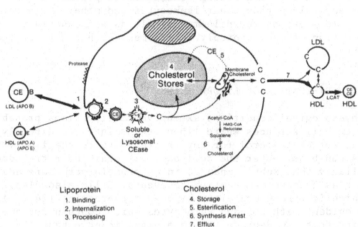

FIGURE 2

High density lipoproteins (HDL_2 and HDL_3) also bind reversibly to
human fat cells in a dose dependent and saturable manner with high
specificity and capacity (38). HDL binding is also independent of calcium
and protease treatment but high salt concentrations partially inhibit both
HDL_2 and HDL_3. Unlike LDL, the apoproteins of HDL are not degraded
significantly by human fat cells. High density lipoprotein binding is
partially (50-80%) inhibited by LDL in contrast to LDL binding which is
completely inhibited by HDL. This partial cross inhibition of LDL and HDL
implies different binding characteristics or possibly different binding
sites. While two distinct recognition sites could explain the findings, a
single domain with multiple potential configurations capable of
recognizing a variety of lipoproteins is currently a favored model as
shown (Figure 2).

While recognition of apo E is a significant factor in lipoprotein
recognition by the B-E receptor in cultured fibroblasts its presence is
not a requirement for lipoprotein binding to human adipocytes. In recent
studies using a variety of mutant HDL lipoproteins deficient in apo E,
binding to human fat cells and purified membranes occurred at rates and
affinities indistinguishable from normal HDL_2 and HDL_3 (Figure 3).
These findings favor the view that apo AI and/or apo AII are the primary
recognition ligands. Support for this suggestion is evident in studies

44

using mutant HDL obtained from a patient with familial apo AI-CIII deficiency. Binding of this mutant HDL (Table 4) is much reduced compared to normal HDL_2. Similarly, HDL_3 treated to displace apo AI with apo II also results in marked reduction in HDL_3 (39). This indicates a significant role for apo AI in HDL recognition and binding.

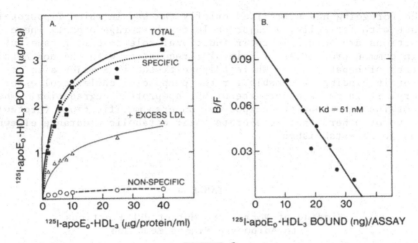

FIGURE 3

^{125}I-apoE deficient (apo E_0) HDL_3 binding to human adipocyte plasma membranes. Plasma membranes were purified from adipocytes that were isolated from properitoneal fat obtained from a 44 years old male. (A) Purified plasma membranes (10 µg protein) were incubated with varying concentrations of ^{125}I-apoE_0-HDL_3 in the absence (●) or presence of 100 fold excess unlabeled apo E_0 HDL_3 (O) or in the presence of 100 fold protein excess unlabeled LDL (Δ). Specific binding (■) was calculated by subtracting non-specific (O) from total binding (●) and was saturable. Excess unlabeled LDL partially inhibited ^{125}I-apo E_0-HDL_3 specific binding. Each point representes the average of duplicate assays. (B) Scatchard plot of ^{125}I-apoE_0-HDL_3 specific binding. B/F represents the amount of specifically bound ^{125}-apoE_0-HDL_3 (nanogram protein) divided by the amount of unbound (nanogram protein) in the assay. Abscissa shows the amount of ^{125}I-apoE_0-HDL_3 specifically bound (nanogram protein) to the membranes per assay. The data were best fitted to a straight line by computer analysis. The intercept of the line at the x axis represents maximum binding. The Kd equals -1/slope of the plot and measures the affinity of apoE_0-HDL_3 binding. The molecular weight of 1.75×10^5 (55% protein) was used in the calculation.

While both LDL and HDL_3 are able to bind to fat cells and presumably deliver cholesterol ester lipids to the fat cell, the efflux acceptor processes remains uncharacterized. Evidence has been deduced that both LDL and HDL particles can participate in the process, however the mechanism of cholesterol efflux in adipose tissue is poorly defined (40).

Net mobilization of cholesterol from adipose tissue is a slow process. In studies of obese patients maintained on severe weight reducing diets adipose tissue cholesterol content is maintained despite significant mobilization of triglycerides (41). Indeed the cholesterol/triglyceride ratio increase in adipose tissue during prolonged weight reduction (42). This could be interpreted several ways including failure to mobilize cholesterol stores or mobilization of cholesterol but continuing uptake of lipoprotein cholesterol and therefore no apparent net depletion.

The foregoing has summarized briefly the way in which lipoproteins interact with fat cells, a major cholesterol storage organ. These interactions are unique in their functional role in this tissue but show certain common properties of molecular regulation with the non B and non E receptors of hepatocytes. While adipose tissue is a major site of lipoprotein binding presumably for the purpose of cholesterol ester delivery, it is not a major site of HDL apoprotein degradation. Whether lipoproteins once bound to adipocytes undergo modifications (proteolytic, oxidative or other) that accelerate their catabolic clearance elsewhere remains to be established.

TABLE 4

Binding of Normal and Mutant HDL Fractions
to Adipocyte Plasma Membrane

^{125}I-HDL	HDL Binding (μg P/mg Membrane P)	
	Total	Non-specific
Normal HDL_2	2.25 ± 0.09	0.11 ± 0.03
$(AI\text{-}CIII)_o$ HDL	0.39 ± 0.10	0.05 ± 0.001
$E_o\text{-}HDL_2$	3.02 ± 0.36	0.15 ± 0.02
$E_o\text{-}HDL_3$	1.99 ± 0.16	0.13 ± 0.02
Normal HDL_3	1.10 ± 0.02	0.06 ± 0.01

HDL fractions were isolated from a normal donor, from a subject with apo AI-CIII deficiency (courtesy of Dr. R. Norum) and a subject with apo E deficiency (courtesy of Dr. R. Gregg). Human adipocyte plasma membranes obtained from properitoneal fat were incubated for 1 hour at 4° C with 40 μg P of each ^{125}I-labeled HDL. Following incubation, the membrane was pelleted by centrifugation, washed once and counted. Non specific binding was determined in a parallel reaction by addition of 100 fold excess unlabeled ligand. Each value is the average of duplicates range. Note the marked reduction in binding of $(AI\text{-}CIII)_o$ HDL compared to others.

SUMMARY

It is evident that lipoprotein modifications, degradation and clearance from plasma and interstitial compartments involves both cellular and extracellular processing. Cellular uptake of the intact particle as a whole and/or selective removal of constituent apoproteins and lipids by various parechymal cells goes on continuously. Regulation of these processes undoubtedly varies tissue to tissue and much remains to be clarified in human tissues in vivo. The metabolic effects of chemical, proteolytic, and lipolytic modification of lipoproteins secondary to transient cellular encounters (e.g. during transit through endothelial barriers, or reversible binding to cells) on apolipoprotein clearance remains to be defined. It is likely that multiple post-secretory modifications occur and together represent subtle regulatory events that modulate lipid shuttle functions and cellular targetting properties of HDL particles.

ACKNOWLEDGEMENTS

This work is supported by the Medical Research Council of Canada and the Heart and Stroke Foundation of Ontario. We are grateful to Dr. R. Norum and Dr. R. Gregg for providing the plasma from the AI-CIII deficient patient and from their apo E deficient patient respectively. We thank Drs. B. Goldman, R. Baird, L. Mickleborough, H. Sculley, R. Weisel and the staff of the Division of Cardiac Surgery, Toronto General Hospital, for providing adipose tissue biopsies. We also like to thank Ms. Laura Sheu for excellent technical assistance and Mrs. Tina Lagopoulos for her help in the preparation of this manuscript.

REFERENCES

1. G.J. Miller and N.F. Miller, Plasma high density lipoproteins concentration and development of ischemic heart disease. Lancet. 1:16-20 (1975).
2. Gordon, T., Castelli, W.P., Hjortland, M.C., Kamel, W.B., and Dawber, T.R., High density lipoprotein as a protective factor against coronary heart disease: The Framingham Study, Am. J. Med. 62:707-730 (1977).
3. S. Eisenberg, High density lipoprotein metabolism, J. Lipid Res. 25:1017-1058 (1984).
4. R.L. Hamilton, M.C. Williams, C.J. Fielding, and R.J. Hajel, Discoidal bilayer structure of nascent high density lipoproteins from perfused rat liver, J. Clin. Invest. 58:667-680 (1976).
5. P.H.R. Green, A.R. Tall, and R.M. Glickman, Rat intestine secretes discoidal high density lipoprotein, J. Clin. Invest. 61:528-534 (1978).
6. T.G. Redgrave, and D.M. Small. Quantitation of the transfer of surface phospholipid of chylomicrons to the high density lipoprotein fraction during the catabolism of chylomicrons in the rat, J. Clin. Invest. 64:162-171 (1979).
7. K. Shirai, R.L. Barnhart, R.L. Jackson, Hydrolysis of human plasma high density lipoprotein 2, phospholipids and triglycerides by hepatic lipase. Biochem. Biophys. Res. Commun. 100:591-599 (1981).
8. P.J. Nestel, M. Reardon, and T. Billington, In vivo transfer of cholesteryl esters from high density lipoproteins to very low density Lipoproteins in Man, Biochim. Biophys. Acta. 573:403-407.
9. A. Sniderman, B. Teng, C. Vezina, and Y.L. Marcel, Cholesterol ester exchange between human plasma high and low density lipoproteins mediated by a plasma protein factor, Atherosclerosis. 31:327-333 (1978).

10. C.C. Schwartz, M. Beiman, Z.R. Ulahcevic, and L. Swell, Multicompartmental analysis of cholesterol metabolism in man. <u>J. Clin. Invest</u>. 70:863-876 (1982).

11. M.C. Phillips, L.R. McLean, G.W. Stoudt and G.H. Rothblat, Mechanism of cholesterol efflux from cells, <u>Atherosclerosis</u>. 36:409 (1980).

12. P. Julien, B. Fong, and A. Angel, Cardiac and peripheral lymph lipoproteins in dogs fed cholesterol and saturated fat, <u>Arteriosclerosis</u>. 4:435 (1984).

13. R.C. Pittman and D. Steinberg, Sites and mechanisms of uptake and degradation of high density and low density lipoproteins, <u>J. Lipid Res</u>. 25:1577-1585 (1984).

14. C. Glass, R.C. Pittman, M. Civen, and D. Steinberg, Uptake of high density lipoprotein associated Apoprotein A-I and cholesterol esters by 16 tissues of the rat in vivo and by adrenal cells and hepatocytes in vitro, <u>J. Biol. Chem</u>. 260:744-750 (1985).

15. E. Reaven, Y.I. Chen, M. Spicher, and S. Azhar, Morphological evidence that high density lipoproteins are not internalized by steroid-producing cells during in situ organ perfusion, <u>J. Clin. Invest</u>. 1384-1397 (1984).

16. J.L. Goldstein, and M.S. Brown, The low-density lipoprotein pathway and its relation to atherosclerosis. <u>Ann. Rev. Biochem</u>. 46:897-930 (1977).

17. G. Ponsin, J.T. Sparrow, A.M. Gotto Jr., and H.J. Pownall, In vivo interaction of synthetic acetylated apopeptides with high density lipoproteins in rat, <u>J. Clin. Invest</u>. 77:559-567 (1986).

18. P.S. Roheim, D. Rachmilewitz, O. Stein, and Y. Stein, Metabolism of iodinated high density lipoproteins in the rat. I. Half-life in the circulation and uptake by organs. <u>Biochim. Biophys. Acta</u> 248:315-329 (1971).

19. Rachmilewitz, D., O. Stein, P.S. Roheim, and Y. Stein, Metabolism of iodinated high density lipoproteins in the rat. II. Autoradiographic localization in the liver, <u>Biochim. Biophys. Acta</u> 270:414-425 (1972).

20. G.A. Drevon, T. Berg, and K.R. Norum, Uptake and degradation of cholesterol ester labelled rat plasma lipoproteins in purified rat hepatocytes and non parenchymal liver cells, <u>Biochem Biophys Acta</u>, 487:122-136 (1977).

21. P.S. Bachorik, F.A. Franklin, D.G. Virgil, and P.O. Kwiterovich Jr., High-affinity uptake and degradation of apolipoprotein E free high density lipoprotein and low density lipoprotein in cultured porcine hepatocytes, <u>Biochemistry</u>. 21:5675-5684 (1982).

22. J.T. Gwynne, and B. Hess, The role of high density lipoproteins in rat adrenal cholesterol metabolism and steroidogenesis, <u>J. Biol. Chem</u>. 255:10875-10833 (1980).

23. N. Suzuki, N. Fidge, P. Nestel, and J. Yin, Interaction of serum lipoproteins with the intestine. Evidence for specific high density lipoproteins binding sites on isolated rat intestinal mucusal cells, <u>J. Lipid Res</u>. 24:253-264 (1983).

24. A. Angel, M.A. D´Costa, R. Yuen, Low density lipoprotein binding, internalization, and degradation in human adipose cells, <u>Can. J. Biochem</u>. 57:578-587 (1979).

25. W. Stoffel, Transport and processing of apolipoproteins of high density lipoproteins, <u>J. of Lipid Res</u>. 25:1586-1592 (1984).

26. A. Scanu, R.E. Byrne, and C. Edelstein, Proteolytic events affecting plasma apolipoproteins at the co-and post-translational levels and after maturation, <u>J. Lipid Res</u>. 25:1593-1602 (1984).

27. H.R. Lijnen, and D. Collen, Degradation of the apoprotein A-1 polypeptide chain of human high density lipoprotein by human plasmin, <u>Thromb. Res</u>. 24:151-156, (1981).

28. D. Polacek, R.E. Byrne, G.M. Fless, and A.M. Scanu, In vitro proteolysis of human plasma low density lipoproteins by an elastase released from human blood polymorphonuclear cells, J. Biol. Chem. 261:2057 (1986).

29. W.A. Bradley, E.B. Gillian, A.M. Gotto Jr., and S.H. Gianturco, Apolipoprotein E degradation in human very low density lipoproteins by plasma protease(s): Chemical and biological consequences, Biochem. Biophys. Res. Commun. 109:1360-1367 (1982).

30. M. Yamamoto, S. Ranganathn, and B.A. Kottke, Structure and Function of Human Low Density Lipoproteins, J. Biol. Chem. 8509-8513 (1985).

31. U.P. Steinbricher, J.L. Witztum, Y.A. Kesaniemi, and R. Elam, Comparison of glucosylated low density lipoprotein with methylated or cyclohexanedione treated low density lipoprotein in the measurement of receptor independent low density lipoprotein catabolism, J. Clin. Invest. 71:960-964 (1983).

32. E.A. Brinton, J.F. Oram, C.H. Chen, J.J. Albers, and E.L. Bierman, Binding of high density lipoprotein to cultured fibroblasts after chemical alteration of apoprotein amino acid residues, J. Biol. Chem. 261:495-503 (1986).

33. U.P. Steinbrecher, S. Parthasarathy, D.S. Leake, J.L. Witztum, and D. Steinberg, Modification of low density lipoprotein by endothelial cells involves lipid peroxidation and degradation of low density lipoprotein phospholipids, Proc. Nat'l Acad. Sci U.S.A. 81:3883-3887 (1984).

34. T.L. Raymond, and S.A. Reynolds, Lipoprotein of the extravascular space: Alterations in low density lipoproteins of interstitial inflammatory fluid, J. Lipid Res. 24:113 (1983).

35. T.H. Avlinskas, D.R. van der Westhuyzen, E.L. Bierman, W. Gevers and G.A. Coetzee. Retro-endocytosis of low density lipoproteins by cultured bovine aortic smooth muscle cells, Biochim. Biophys. Acta 664:255-265 (1981).

36. K. Tanaka, T. Nakamura, and A. Ichihara, A unique trypsin-like protease associated with plasma membranes of rat liver, J. Biol. Chem. 261, 2610-2615, (1986).

37. B.S. Fong, P.O. Rodrigues, and A. Angel, Characterization of low density lipoprotein binding of human adipocytes and adipocyte membranes, J. Biol. Chem. 259:10168-10174 (1984).

38. B. Fong, P.O. Rodrigues, A.M. Salter, B.P. Yip, J.P. Despres, R.E. Gregg, and A. Angel, J. Clin. Invest.75:1804-1812 (1985).

39. A. Salter, B. Fong, J. Jimenez, and A. Angel, Involvement of apolipoprotein A-1 in binding of high density lipoprotein to adipocyte plasma membranes. Arteriosclerosis 4:539A, (1984).

40. A. Angel, S. Thanabalasingham, D. Reichl, J.J. Pflug, G.R. Thompson, N.B. Myant, Effects of Starvation and plasma exchange on lecithin:cholesterol acyl transferase activity and cholesterol efflux and cholesterol fed pigs, Res. Exp. Med. 184:231-242 (1984).

41. E.J. Schaefer, R. Wood, M.Kibata, L. Bjornson, and P.H. Schreibman, Mobilization of triglyceride but not cholesterol or tocopherol from human adipocytes during weight reduction, Am. J. Clin. Nutrition 37:749-754 (1983).

42. A. Angel, and G.A. Bray, Synthesis of fatty acids and cholesterol by liver, adipose tissue, and intestinal mucosa from obese and control patients, Eur. J. Clin. Invest. 355-362 (1979).

EFFECT OF DIABETES MELLITUS AND END-STAGE RENAL DISEASE

ON HDL METABOLISM

M.H. Tan

Division of Endocrinology & Metabolism
Department of Medicine, Dalhousie University
Halifax, Nova Scotia, Canada

ABSTRACT

Different types of diabetes mellitus have different effects on high density lipoprotein (HDL) metabolism. Impaired glucose tolerance may be associated with no change or a slight decrease in HDL cholesterol. Type I diabetes may have normal or elevated HDL cholesterol levels. This HDL elevation may be due to an increase in HDL_2 or HDL_3. Apo A-I/Apo A-II ratio is also higher in these diabetics. Type II diabetics may have normal or low HDL cholesterol levels as well as normal or decreased Apo A-I levels. In gestational diabetics, the mean HDL cholesterol is lower than controls. Dietary therapy resulting in > 10% weight loss in obese diabetics leads to an increase in their HDL-cholesterol levels, although the effect on the latter is controversial. Intensive insulin therapy (for 2 - 3 weeks) increases serum apo A-I and HDL-cholesterol levels.

End-stage renal disease also affects HDL metabolism. In general, patients with this disorder have a decrease of cholesterol and an increase in triglyceride in their HDL. There is an increase in apo E and a decrease in apo CII in their HDL. Apo A-I levels are unaffected whereas apo A-II levels are decreased. Renal transplant patients may have low, normal or high HDL cholesterol and normal or high apo-I levels. In non-diabetic, normotriglyceridemic patients peritoneal dialysis increases their HDL-cholesterol. In non-diabetic hypertriglyceridemic and diabetic patients, peritoneal dialysis causes no change in their HDL-cholesterol. Hemodialysis can increase HDL-cholesterol levels in these patients.

In contrast to uremic humans, uremic rats have elevated HDL-cholesterol and this accounts for most of the increase in their serum cholesterol. The dialysed and undialysed plasma of these uremic rats have inhibitory action on hepatic lipase activity.

Diabetes mellitus and uremia due to end-stage renal disease are common metabolic disorders which can affect high density lipoprotein (HDL) metabolism. In this paper the many changes in HDL metabolism associated with these syndromes will be presented.

CLASSIFICATION OF DIABETES MELLITUS

Diabetes mellitus is a metabolic disorder characterized, on the one hand, by derangements in carbohydrate, lipid and protein metabolism, and, on the other hand, by structural abnormalities in small (microangiopathy) and big (macroangiopathy) blood vessels as well as in other tissues. These two aspects of diabetes mellitus share in common hyperglycemia. Recently, a classification of diabetes was proposed (Table 1) (1). The changes in HDL metabolism associated with the various types of diabetes mellitus will now be described.

TABLE 1

CLASSIFICATION OF DIABETES MELLITUS AND
OTHER CATEGORIES OF GLUCOSE INTOLERANCE

A.   DIABETES MELLITUS

  1.   Type I (Insulin dependent diabetes)
  2.   Type II (Non-insulin dependent diabetes)
  3.   Other types (secondary diabetes)

B.   IMPAIRED GLUCOSE TOLERANCE

C.   GESTATIONAL DIABETES MELLITUS

D.   PREVIOUS ABNORMALITY OF GLUCOSE TOLERANCE

E.   POTENTIAL ABNORMALITY OF GLUCOSE TOLERANCE

TYPE I DIABETES (TABLE 2)

At the time of initial diagnosis, these insulin deficient diabetics have low HDL cholesterol levels (2). Patients with diabetic ketoacidosis characterized by insulin deficiency, also have decreased HDL cholesterol levels (3). After initiation of insulin therapy these patients often have peripheral hyperinsulinemia due to exogenous insulin. They, then, have either normal (4, 5) or high (6, 7) HDL cholesterol levels. However, Type I diabetics with end-stage renal disease can have decreased HDL cholesterol levels (8).

The increase in HDL has been reported to be due to an increase in HDL_2 (HDL_{2a}) subfraction (7) or HDL_3 subfraction (9), the latter report showing a decrease in the HDL_2/HDL_3 ratio in both diabetic males and females. The changes in serum apo A-I and apo A-II in these patients are variable. An increase (5, 6) and a decrease (10, 11) in serum apo A-I have been reported. No change (6, 10) or decreased (5, 11) apo A-II levels have also been reported.

TYPE II DIABETES (TABLE 2)

In contrast to Type I diabetics, Type II diabetics either have normal (4, 12) or decreased (13, 14) HDL cholesterol. Newly diagnosed Type II diabetics have lower HDL cholesterol than controls matched for sex, age, obesity, alcohol and cigarette smoking (14). The presence of obesity and hypertriglyceridemia, common in Type II diabetics, tend to lower HDL cholesterol. When matched for obesity and hypertriglyceridemia Type II diabetics and controls have similar HDL cholesterol levels (15).

The decrease in HDL is present in the HDL_3 subfraction (16). Again an increase (13), a decrease (11), or no change (14) in serum apo A-I levels may occur in these diabetics. Similarly, no change (13) or a decrease (11, 13) in serum apo A-II levels have also been reported.

TABLE 2

HDL SUBFRACTION AND APOPROTEIN CHANGES IN DIABETES MELLITUS

A.   IMPAIRED GLUCOSE TOLERANCE

     1.   No change (17).
     2.   Decreased (16).

B.   TYPE I DIABETES MELLITUS

     1. HDL_2 (HDL_{2a}) increased (7).
     2. HDL_3 increased in males (9).
     3.   Apo A-I increased in males (5).
         Apo A-II decreased in females (5).
     4.   Apo A-I increased (6).
     5.   Apo A-I decreased (10).
     6.   Apo A-I and Apo A-II decreased (11).

C.   TYPE II DIABETES MELLITUS

     1. HDL_3 decreased (16).
     2.   Apo A-II decreased in males (13).
         Apo A-I increased in females (13).
     3.   No change in Apo A-I and AII (10).
     4.   Decrease in Apo A-I and Apo A-II (11).

IMPAIRED GLUCOSE TOLERANCE AND GESTATIONAL DIABETES (TABLE 2)

Italians with impaired glucose tolerance have similar levels of HDL cholesterol as their controls (17). In contrast, Pima Indians with impaired glucose tolerance have decreased HDL cholesterol (16).

Gestational diabetics and Type II pregnant diabetics have lower HDL cholesterol than controls and Type I pregnant diabetics (18).

EFFECT OF TREATMENT (TABLE 3)

Weight loss in newly diagnosed Type II diabetics increases their HDL cholesterol levels (19). A high fibre, high carbohydrate diet decreases HDL cholesterol when compared with a low fibre, low carbohydrate diet eaten by diabetics (20).

Treatment with glipizide leads to an increase in serum HDL: total cholesterol in Type II diabetics (21). Treatment with metformin or glibenclamide leads to an increase in serum apo A-I but not HDL cholesterol in female Type II diabetics (22).

Patients with diabetic ketoacidosis increase their HDL cholesterol 24 hours after initiation of insulin and fluid therapy (3). Their mean serum apo A-I in plasma and HDL was normal before therapy and they decreased during the first 12 hours of therapy before returning to normal by 24 hours. Intensive insulin therapy for 2 - 3 weeks increases both serum apo A-I and HDL cholesterol in male, but only apo A-I in female Type I diabetics (23). Treatment with continuous subcutaneous insulin infusion for 5 - 14 months results in an increase in HDL cholesterol (24).

THE DIABETES - HDL CONNECTION

HDL connection in serum is affected by many factors, including the rate of secretion of nascent HDL, lipoprotein lipase (LPL) activity in certain extrahepatic tissues, hepatic lipase activity, diet and exercise. There is a positive correlation between HDL_2 and LPL activity (25). Insulin deficiency and excess may affect HDL concentration via its effect on the activities of LPL and hepatic lipase. In patients with diabetic ketoacidosis, a decrease in LPL activities result in impaired clearance of triglyceride-rich lipoproteins which is associated with a decrease in the production of HDL_2. In insulin treated Type I diabetics, peripheral hyperinsulinemia often occurs and can lead to increased LPL activities, enhanced clearance of triglyceride-rich lipoproteins and inceased production of HDL. Type II obese diabetics may have decreased LPL activity, leading to decreased HDL cholesterol.

TABLE 3

EFFECT OF THERAPY ON HDL LEVELS IN DIABETICS

A. DIETARY THERAPY

    1. Obese Type II diabetics increase their HDL when they lose > 10% body weight (19).
    2. A high CHO/high fibre diet decreases HDL (20).

B. ORAL HYPOGLYCEMIC AGENTS

    1. Patients on oral agents have lower HDL-C (21).
    2. Glipizide increases HDL-C (21).
    3. Metformin increases APO A-I in obese females (20).
    4. Glibenclamide increases APO-I in NON-OBESE FEMALES (22).

C. INSULIN THERAPY

    1. DKA (24 hrs): Slight increase (3).
    2. Intensive therapy (2 - 3 weeks) (23).
        Males: APO A-I and HDL-C
        Females: No Change.
    3. Pump therapy (5 - 14 months: HDL-C (24).

EFFECT OF END-STAGE RENAL DISEASE - HUMANS

The increased prevalence of hyperlipidemia in uremic patients with end-stage renal disease on dialysis is well documented (26, 27). Whereas the predominant lipid disorder in hemodialysis patients is hypertriglyceridemia, that associated with renal transplant patients is hypercholesterolemia. Changes in HDL in these patients will now be presented.

In general, HDL cholesterol is decreased and HDL triglyceride is increased in hemodialysis patients when compared with controls (28, 30). However, a decrease in HDL triglyceride has also been reported (31). A decrease in HDL_2 and HDL_3 may account for the decrease in the HDL cholesterol (32). Serum apo A-I level is normal in dialysis patients (29, 33), but total apo A proteins are decreased (26). There is an increase in apo C-III/apo C-II ratio due to a decrease in apo C-II and an increase in apo C-III in HDL (32, 34). An increase in apo E in HDL in these patients has also been reported (32, 33).

The treatment for these patients consists of dialysis (peritoneal dialysis or hemodialysis) or renal transplantation. In non-diabetic, normotriglyceridemic patients peritoneal dialysis increased HDL cholesterol (35). In contrast, in non-diabetic hypertriglyceridemic patients and diabetics, peritoneal dialysis caused no change in the HDL cholesterol (35). Hemodialysis also caused an increase in the HDL cholesterol (36). Renal transplant patients have higher serum HDL cholesterol but no change in their apo A-I levels during the first five years after the transplant (37).

EFFECT OF END-STAGE RENAL DISEASE - RATS

Unlike uremic humans, uremic rats (1 2/3 hephrectomized) have elevated HDL cholesterol, triglycerides and phospholipids (Tables 4, 5), accounting for the major increase in these serum lipids in these rats. The accumulation of HDL may be due to the decreased renal mass as the kidneys are believed to be major organs involved in the catabolism of HDL in rats (38, 39). The accumulation of HDL may also be due to decreased hepatic lipase activity. But, hepatic lipase activities are similar in these uremic and normal rats (unpublished results). Instead, uremic plasma causes a greater inhibition of hepatic lipase activity than normal rat plasma (Figures 1 and 2). Whether this inhibition is due to the high HDL levels remains to be established.

TABLE 4

SERUM LIPID PROFILE OF UREMICAND CONTROL RATS (mean ± SEM)

	CONTROL (n=14)	UREMIC (n=13)	P VALUE
Tg (mg/dL)	20.9 ± 6.8	29.8 ± 12.6	<0.05
TC	41.8 ± 7.5	80.9 ± 18.7	<0.001
FC	9.3 ± 2.4	24.1 ± 7.6	<0.001
EC	32.5 ± 5.5	67.9 ± 19.3	<0.001
PL	51.4 ± 8.0	98.3 ± 26.5	<0.001
C/PL	0.18 ± 0.04	0.25 ± 0.04	<0.001

TABLE 5

LIPID PROFILE OF SERUM HDL OF UREMICAND CONTROL RATS (mean ± SEM)

	CONTROL (n=15)	UREMIC (n=13)	P VALUE
Tg (mg/dL)	2.26 ± 0.86	5.78 ± 1.89	<0.001
TC	20.96 ± 7.03	48.05 ± 13.05	<0.001
FC	2.06 ± 0.89	7.01 ± 2.29	<0.001
EC	18.90 ± 6.49	41.04 ± 11.46	<0.001
PL	17.80 ± 7.25	51.36 ± 14.67	<0.001
C/PL	0.11 ± 0.05	0.14 ± 0.03	--

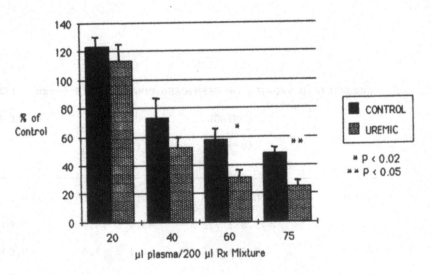

FIGURE 1

Inhibition of hepatic lipase activity caused by uremic and normal
undialysed plasma (mean   SEM).

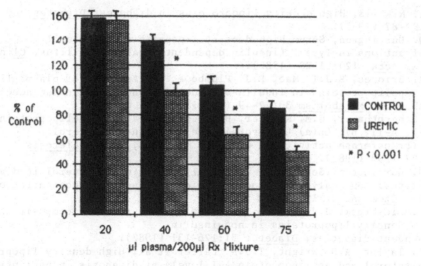

FIGURE 2

Inhibition of hepatic lipase activity caused by uremic and
normal dialysed plasma (mean   SEM).

REFERENCES

1. National Diabetes Data Group: Classification and diagnosis of diabetes
    mellitus and other categories of glucose tolerance, Diabetes
    28:1039-1057 (1979).
2. K. Asayama, S. Amemiya, K. Kato, Serum lipids and postheparin plasma
    lipase activity in Japanese children with ketosis-prone diabetes
    mellitus, Tohoku J. Exp. Med. 141(Suppl):627-630, (1983).
3. S.W. Weidman, J.B. Ragland et al, Effects of insulin on plasma
    lipoproteins in diabetic ketoacidosis: evidence for a change in high
    density lipoprotein composition during treatment, J. Lip. Res.
    23:171-182 (1982).
4. G.D. Calvert, J.J. Graham, T. Mannik et al, Effects of therapy on
    plasma high density lipoprotein cholesterol concentration in diabetes
    mellitus, Lancet. 2:66-68 (1978).
5. R.H. Eckel, J.J. Albers, M.C. Cheung et al, High density lipoprotein
    composition in insulin-dependent diabetes mellitus, Diabetes 30:132-138
    (1981).
6. U. Ewald, S. Gustafson, T. Tuvemo et al, Increased high density
    lipoproteins in diabetic children, Eur. J. Pediatr. 142:154-156 (1984).
7. M.B. Mattock, A.M. Salter, J.H. Fuller et al, High density lipoprotein
    subfractions in insulin-dependent diabetic and normal subjects,
    Atherosclerosis 45:67-79 (1982).

8. E.A. Nikkila, High density lipoproteins in diabetes, _Diabetes_ 30(Suppl 2):82-87 (1981).

9. P.N. Durrington, Serum high density lipoprotein cholesterol subfractions in Type I (insulin dependent) diabetes mellitus, _Clin. Chem. Acta._ 120:21-28 (1982).

10. E.R. Briones, S.J.T. Mao, P.J. Plumbo et al, Analysis of plasma lipids and apolipoproteins in insulin-dependent and non-insulin dependent diabetics, _Metabolism_ 33:42-49 (1984).

11. G. Schernthaner, G.M. Kostner, H. Dieplinger et al, Apolipoproteins (A-I, A-II, B), Lp(a) Lipoprotein and Lecithin-cholesterol acyltransferase activity in diabetes mellitus, _Atherosclerosis_ 49:277-293 (1983).

12. P.N. Durrington, Serum high density lipoprotein cholesterol in diabetes mellitus: an analysis of factors which influence its concentration, _Clin. Chem. Acta._ 104:11-23 (1980).

13. R.C. Biesbrock, J.J. Albers, P.W. Wahl et al, Abnormal composition of high density lipoproteins in non-insulin dependent diabetics, _Diabetes_ 31:126-131 (1982).

14. K.G. Taylor, A.D. Wright, T.J.N. Carter et al, High density lipoprotein cholesterol and apolipoprotein A-I levels at diagnosis in patients with non-insulin dependent diabetes, _Diabetologia_ 20:535-539 (1981).

15. M.R. Taskinen, E.A. Nikkila, T. Kuusi et al, Lipoprotein lipase activity and serum lipoproteins in untreated Type II (insulin independent) diabetes associated with obesity, _Diabetologia_ 22:46-50 (1982).

16. B.V. Howard, W.C. Knowler, B. Vasquez et al, Plasma and lipoprotein cholesterol and triglyceride in the Pima Indian population. Comparison of diabetics and non-diabetics, _Atherosclerosis_ 4:462-471 (1984).

17. B. Capaldo, L. Tutino, L. Patti et al, Lipoprotein composition in individuals with impaired glucose tolerance, _Diabetes Care_ 6:575-578 (1983).

18. D.R. Hollingsworth, S.M. Grundy, Pregnancy-associated hypertriglyceridemia in normal and diabetic women, _Diabetes_ 31:1092-1097 (1982).

19. L. Kenney, K. Walshe, D.R. Hadden et al, The effect of intensive dietary therapy on serum high density lipoprotein cholesterol in patients with Type II (non-insulin-dependent) diabetes mellitus. A prospective study, _Diabetologia_ 23:24-27 (1982).

20. G. Riccardi, A. Rivellese, D. Pacioni et al, Separate influence of dietary carbohydrate and fibre on the metabolic control in diabetes, _Diabetologia._ 26:116-121 (1984).

21. M.S. Greenfield, L. Doberne, M. Rosenthal et al, Lipid metabolism in non-insulin-dependent diabetes mellitus, _Arch.Int. Med._ 142:1498-1500 (1982).

22. K.G. Taylor, W.G. Joh, K.A. Matthews et al, A prospective study on the effect of 12 months treatment on serum lipids and apolipoproteins A-I and B in Type II (non-insulin-dependent) diabetes, _Diabetologia_ 23:507-510 (1982).

23. M.F. Lopes-virella, H.J. Wohltmann, R.K. Mayfield et al, Effect of metabolic control on lipid, lipoprotein, and apoprotein levels in 55 insulin-dependent diabetic patients. A longitudinal study, _Diabetes_ 32:20-25 (1983).

24. J.M. Folks, T.M. O´Dorisio, S. Cataland, Improvement of high-density lipoprotein-cholesterol levels. Ambulatory Type I diabetics treated with the subcutaneous insulin pump, _JAMA_ 247:37-39 (1982).

25. E.A. Nikkila, M.R. Taskinen, M. Kekki, Relation of plasma high-density lipoprotein cholesterol to lipoprotein-lipase activity in adipose tissue and skeletal muscle of man, _Atherosclerosis_ 29:497-501 (1978).

26. T. Driieke, B. Lacour, J.B. Roullet et al, Recent advances in factors that alter lipid metabolism in chronic renal failure, _Kid. Int._ 24(Suppl 16):S134-S138 (1983).

27. M.H. Tan, Hyperlipidemia associated with end-stage renal disease, <u>N. S.</u> <u>Med. Bull.</u> 62:79-81 (1983).

28. J.D. Bagdade, J.J. Albers, Plasma high-density lipoprotein concentration in chronic hemodialysis and renal transplant patients, <u>N.</u> <u>Engl. J. Med.</u> 296:1436-1439 (1977).

29. J.D. Brunzell, J.J. Albers, L.B. Haas et al, Prevalence of serum lipid abnormalities in chronic hemodialysis, <u>Metabolism</u> 26:903-910 (1977).

30. H.E. Norbeck, Serum lipoprotein in chronic renal failure, <u>Acta Med.</u> <u>Scand.</u> 649:(Suppl):1-49 (1981).

31. A. Pasternack, T. Leino, T. Solakivi-Jaakkola et al, Effect of furosemide on the lipid abnormalities in chronic renal failure, <u>Acta</u> <u>Med. Scand.</u> 214:153-157 (1983).

32. J. Rapoport, M. Aviram, C. Chaimovitz et al, Defective high-density lipoprotein composition in patients on chronic hemodialysis, <u>N. Engl.</u> <u>J. Med.</u> 299:1326-1329 (1978).

33. P.J. Nestel, N.H. Fidge, M.H. Tan, Increased lipoproteion-remnant formation in chronic renal failure, <u>N.Eng. J. Med.</u> 307:329-333 (1982).

34. I. Staprans, J.M. Felts, B. Zacherle, Apoprotein composition of plasma lipoproteins in uremic patients on hemodialysis, <u>Clin. Chem. Acta</u> 93:135-143 (1979).

35. W.C. Breckenridge, D.A.K. Roncari, R. Khanna et al, The influence of continuous ambulatory peritoneal dialysis on plasma lipoproteins, <u>Atherosclerosis</u> 45:249-258 (1982).

36. J.K. Huttunen, A. Pasternack, T. Vanttinen et al, Lipoprotein metabolism in patients with chronic uremia, <u>Acta Med. Scand.</u> 204:211-218 (1978).

37. N. Kobayashi, M. Okubo, F. Marumo et al, De novo development of hypercholesterolemia and elevated high density lipoprotein cholesterol: apoprotein A-I ratio in patients with chronic renal failure following kidney transplantation, <u>Nephron.</u> 35:237-240 (1983).

38. F.M. van't Hooft, G.M. Dallinga-Thie, A. van Tol, Leupeptin as a tool for the detection of the sites of catabolism of rat high-density lipoprotein apolipoproteins A-I and E, <u>Biochem. Biophys. Acta</u> 388:75-84 (1985).

39. C.K. Glass, R.C. Pittman, G.A. Keller et al, Tissue sites of degradation of apoprotein A-I in the rat, <u>J. Biol. Chem.</u> 258:7161-7167 (1983).

HDL BINDING TO HUMAN ADIPOCYTE PLASMA MEMBRANES: REGIONAL VARIATION IN OMENTAL AND SUBCUTANEOUS DEPOTS

Bessie Fong, Andrew Salter, Jose Jimenez,
Jean-Pierre Despres and Aubie Angel

Department of Medicine and Institute of Medical Science
University of Toronto and Division of Endocrinology and
Metabolism, Toronto General Hospital, Toronto, Ontario
Canada M5S 1A8

Human adipose tissue contains one of the largest pools of exchangeable cholesterol in the body, accounting for about 25 percent of total body cholesterol in normal man (1,2). This adipose cholesterol pool is much expanded in obesity and in the massively obese state, well over half of total body cholesterol can be found in the fat tissue (1,2). Obesity is also commonly associated with a number of lipoprotein abnormalities. Increased synthesis and/or turnover of LDL (3) and VLDL (4,5) have been reported. In addition, HDL-cholesterol in plasma is significantly reduced (6-13) and has been attributed primarily to a decrease in the HDL_2 subfraction (7).

We have previously demonstrated that human adipocytes and adipocyte plasma membranes possess specific binding sites for LDL (14) and HDL (15). The characteristics of LDL binding to fat cell membranes differ markedly from the classical receptor of cultured fibroblasts (14). The binding of HDL has also been shown to be independent of apo B or apo E (15). We have suggested that due to its large size, adipose tissue in humans is an important site for lipoprotein metabolism and is proportionally more important in the obese individuals with a much increased adipose mass. A question could be raised as to whether the lipoprotein abnormalities observed in obesity (3-13) may be related to the interaction of lipoproteins with fat cells.

In the present report we demonstrate regional variation in adipocyte-lipoprotein binding in that plasma membranes from abdominal subcutaneous fat consistently bind HDL_2 and HDL_3 to a greater extent than plasma membranes from omental fat.

MATERIALS AND METHODS

Adipose tissue biopsies were obtained from non-diabetic massively obese patients ($BMI>40$ kg/m^2) undergoing gastroplasty. Omental fat tissue was removed from the major omentum and subcutaneous fat tissue from the abdominal wall at the site of incision. Adipocytes were isolated by collagenase digestion in accordance to the method of Rodbell (16). Mean fat cell size was measured with a microscope equipped with a graduated ocular (17).

Plasma membranes were prepared from the freshly isolated omental and subcutaneous fat cells in accordance to a modified procedure of McKeel and Jarett (20) by ultracentrifugation in sucrose density gradients (14). Human lipoproteins were prepared by sequential ultracentrifugation (21) and radioiodinated by the monochloride method (22). The binding of lipoproteins to adipocyte plasma membranes was standardized and carried out in 75 µl of buffer containing 100 mM NaCl, 0.5 mM $CaCl_2$, 50 mM Tris-HCl, pH 7.5, 2 mg/ml of BSA, 10 µg purified plasma membrane protein and varying amounts of ^{125}I-labeled lipoprotein as previously described (1). The membranes were re-isolated by centrifugation after 1 hr of incubation at 4^oC and the radioactivity associated with the membrane pellet was determined.

RESULTS AND DISCUSSION

Table 1 shows that abdominal subcutaneous fat cells are significantly larger than those from the omental depot. Similar observations have been reported for both moderately obese (BMI>27 kg/m^2) and lean subjects by Gurr et al (18) and for massively obese subjects by Livingston et al (19).

Table 1. Fat cell size of massively obese subjects

Subject	BMI (kg/m^2)	Fat cell weight (µg lipid/cell)	
		Omental	Abdominal Subcut.
V.M.	40.0	0.75	0.91
W.S	47.0	0.76	0.96
C.F	45.2	0.73	0.84
M.M.	49.6	0.68	1.23
Mean SEM	45.5 ± 2.03	0.73 ± 0.02	0.99 ± 0.09*

*significantly greater than omental, $p<0.05$ by paired t test

A 50 µl aliquot of isolated fat cells was resuspended in 1 ml of 0.4% trypan blue-0.9% saline solution. The diameter of at least 500 cells was measured under a microscope equipped with a graduated ocular. Fat cell volume was calculated from each individual fat cell diameter by using the equation of a sphere. The density of triolein (0.915) was used in the calculation of fat cell weight from the volume of the adipocyte.

Dose response curves of ^{125}I-HDL$_3$ binding to adipocyte plasma membranes isolated from omental and abdominal subcutaneous fat depots of the same individual are shown in figure 1. Non-specific binding, measured as binding in presence of 100 fold excess unlabeled ligand, accounted for less than 5% of total binding. The results in figure 1 show that binding of ^{125}I-HDL$_3$ to subcutaneous adipocyte membranes was higher than that to omental membranes at all concentrations of ^{125}I-HDL$_3$ used. Scatchard analysis of specific binding (difference between total and non-specific binding) data yielded a 1.75 fold higher Bmax (maximum binding capacity) for subcutaneous membranes compared to omental. Similar results were obtained using ^{125}I-HDL$_2$ as the binding ligand.

Table 2 compares the specific binding of ^{125}I-HDL₂ and
^{125}I-HDL₃ of omental and abdominal subcutaneous fat cell membranes
from a number of obese individuals. These results show that HDL₂ and
HDL₃ specific binding to subcutaneous membranes is on average about
1.5 fold higher than that to omental in all subjects studied. Thus
along with a difference in fat cell size (Table 1), omental and
subcutaneous adipose tissues also differ in respect to their interaction
with high-density lipoproteins. Regional variation in lipoprotein
binding is not a unique difference as other metabolic activities such as
glucose metabolism and lipolysis have previously been reported for human
adipose tissue. The sensitivity of the antilopolytic effect of insulin,

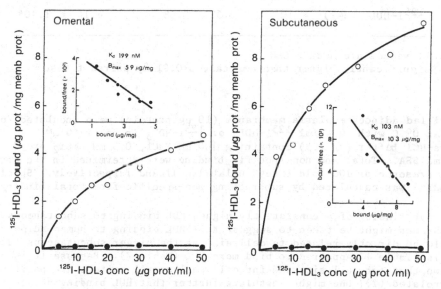

Figure 1. ^{125}I-HDL₃ binding to omental and subcutaneous adipocyte
plasma membranes from a massively obese subject. Adipocytes were
isolated by collagenase digestion and plasma membranes were prepared.
Membranes (10 μg protein) were incubated with varying concentrations of
^{125}I-HDL₃ in the absence (total binding; ○——○) or presence
(non-specific binding; ●——●) of 100 fold excess unlabeled HDL₃.
Specific binding was calculated by subtracting non-specific from total.
Each point represents the average of duplicate assays. Inset shows the
Scatchard plot of the specific binding data and was best-fitted to a one
component binding system by computer analysis. The intercept of the line
at the x axis represents maximum binding (Bmax). The Kd equals −1/slope
of the plot and measures the affinity of the HDL₃ binding. To convert
from micrograms of HDL₃ protein to moles, the molecular weight of 1.75
× 10⁵ (55% protein) was used in the calculation. The results show that
Bmax was significantly higher for subcutaneous membranes compared to
omental.

for example, is higher in abdominal subcutaneous than in omental fat tissue (23) while the basal cAMP concentration and rate of glycerol release is greater in epigastric subcutaneous than in omental fat cells of identical size (24). Metabolic differences are observed even among the various subcutaneous fat depots such as a higher lipoprotein lipase (LPL) activity in the femoral than in the abdominal region in females (25) except during lactation when a marked decrease in LPL activity occurs in femoral fat but remains unchanged in the abdominal region (26).

Table 2. Comparison of HDL_2 and HDL_3 specific binding between omental and subcutaneous fat cell plasma membranes of massively obese subjects

	μg HDL prot. bound/mg memb. prot.	
	Omental	Abdominal Subcut.
^{125}I-HDL_2 (n=6)	3.87 ± 0.38	6.37 ± 0.65*
^{125}I-HDL_3 (n=7)	2.51 ± 0.68	4.14 ± 1.10*

all values are mean ± SEM
*significantly higher than omental, p<0.01 by paired t test

Purified adipocyte plasma membranes (10 μg protein) were incubated for 1 hr at 0° C with 40 μg/ml ^{125}I-HDL_2 or ^{125}I-HDL_3 in 75 μl 50 mM Tris-HCl buffer (pH 7.5) containing 100 mM NaCl, 0.5 mM $CaCl_2$ and 2 mg/ml BSA. Total and non-specific binding were determined in the absence and presence of 100 fold excess unlabeled ligand respectively. Specific binding was calculated by subtracting non-specific from total binding.

Our finding of a consistently higher HDL binding to subcutaneous membranes might be taken to suggest that HDL binding to human adipocytes could be directly related to cell size as the bigger subcutaneous fat cells (Table 1) appeared to bind more HDL (Table 2). Because it is also known that fat cell size and fat cell cholesterol content are positively correlated (2), one might postulate further that HDL binding and cellular cholesterol content are related. HDL binding has been shown to be up-regulated in cultured human skin fibroblasts (27) upon cholesterol loading of cells and is thought to play a role in cholesterol efflux. It might therefore be inferred that in adipocytes, the high HDL binding also reflects up-regulation of HDL binding sites in response to increased cellular cholesterol content and functions to facilitate cholesterol removal from the cells.

The high HDL binding capacity of human adipose tissue may have yet another important implication. In obesity, the increase in body fat in subcutaneous depots is relatively greater than in other sites (28). The higher HDL binding to abdominal subcutaneous as compared to omental fat cell membranes (Table 2) could imply enhanced metabolism of HDL at this site which could contribute disproportionately to the low plasma HDL cholesterol frequently associated with obesity (6-13). Such an idea is consistent with the recent finding that serum HDL-cholesterol shows a higher negative correlation with subscapular and abdominal skinfold thickness than other fat depots (29). Alternatively, one might argue that HDL serves to deliver cholesterol to the adipocytes for storage. This hypothesis is supported by studies in which plasma HDL cholesterol levels increase with weight reduction (10, 30-34).

From a clinical perspective, it is worth pointing out the possibility that the direct correlation between waist/hip ratio and increased risk for coronary heart disease (34-36) may in fact be linked to regional difference in HDL metabolism in the various fat depots.

REFERENCES

1. Angel, A. and G.A. Bray. 1979. Eur. J. Clin. Invest. 9:355-362
2. Schreibman, P.H. and R.B. Dell. 1975. J. Clin. Invest. 55:986-993
3. Kesaniemi, Y.A. and S.M. Grundy. 1983. Arteriosclerosis 3:170-177
4. Nestel, P.J. and H. M. Whyte. 1968. Metabolism 17:1122-1128
5. Nestel, P. and B. Goldrick. 1976. Clin. Endocrinol. Metabol. 55:313-335
6. Carlson, L.A. and M. Ericsson. 1975. Atherosclerosis 21:417-433
7. Albrink, M.J., R.M. Krauss and F.T. Lindgren. 1980. Lipids 15:668-676
8. Phillips, N.R., R.J. Havel and J.P. Kane. 1981. Arteriosclerosis 1:13-24
9. Rossner, S. and D. Halberg. 1978. Acta Med. Scand. 204:103-110
10. Contaldo, F., P. Strazzullo and A. Postiglione. 1980. Atherosclerosis 37:163-167
11. Matter, S., A. Weltman and B.A. Stamford. 1980. J. AM. Diet Assoc. 77:149-152
12. Berchtold, P., M. Berger and V. Jorgens. 1981. Int. J. Obesity 5:1-10
13. Wechsler, J.G., V. Hutt and H. Wenzel. 1981. Int. J. Obesity 5:325-331
14. Fong, B.S., P.O. Rodrigues and A. Angel. 1984. J. Biol. Chem. 259:10168-10174
15. Fong, B.S., P.O. Rodrigues, A.M. Salter, B.P. Yip, J.P. Despres, R.E. Gregg and A. Angel. 1985. J. Clin. Invest. 75:1804-1812
16. Rodbell, M. 1964. J. Biol. Chem. 239:375-380
17. Despres, J.P., C. Bouchard, A. Tremblay, R. Savard and M. Marcotte. 1985. Med. Sci. Sports and Exercise 17:113-118
18. Gurr, M.I., R.T. Jung, M.P. Robinson and W.P.T. James. 1982. Int. J. Obesity 6:419-436
19. Livingston, J.N., K.M. Lerea, J. Bolinder, L.Kager, L. Backman and P. Arner. 1984. Diabetologia 27:447-453
20. McKeel, D.W. and L. Jarett. 1970. J. Cell Biol. 44:417-432
21. Havel, R.J., H. Eder and H.F. Bragden. 1955. J. Clin. Invest. 34:1345-1353
22. Shepherd, J., D.K. Bedford, H.G. Morgan and E. Scott. 1976. Clin. Chim. Acta 66:97-109
23. Bolinder, J., L. Kager, J. Ostman and P. Arner. 1983. Diabetes 32:117-123
24. Ostman, J., P. Arner, P. Engfeldt and L. Kager. 1979. Metabolism 28:1198-1205
25. Lithell, H. and J. Boberg. 1978. Int. J. Obesity 2:47-52
26. Rebuffe-Scrive, M., L. Enk, N. Crona, P. Lonnroth, L. Abrahamsson, V. Smith and P. Bjorntorp. 1985. J. Clin. Invest. 75:1973-1976
27. Oram, J.F., E.A. Brinton and E.L. Bierman. 1983. J. Clin. Invest. 72:1611-1621
28. Lohman, T.G. 1981. Hum. Biol. 53:181-225

29. Despres, J.P., C. Allard, A. Tremblay, J. Talbot and C. Bouchard. 1985. Metabolism 34:967-973
30. Wolf, R.N. and S.M. Grundy. 1983. Arteriosclerosis 3:160-169
31. Gonen, B., J.D. Halverson and G. Schonfeld. 1983. Metabolism 32:492-496
32. Zimmerman, J., N.A. Kaufman, M. Fainaru, S. Eisenberg, Y. Oschry, Y. Friedlander and Y. Steion. 1984. Arteriosclerosis 2:15-123
33. Sorbis, R., B.G. Petersson and P. Nilsson-Ehle. 1981. Eur. J. Clin. Invest. 11:491-498
34. Friedman, C.I., J.M. Falko, S.T. Patel, M.H. Kim, H.A.I. Newman and H. Barrows. 1982. J. Clin. Endocrinol. Metabl. 55:258-262
35. Kissebah, A.H., N. Vydelingum, R. Murray, D.J. Evans, A.J. Hartz, R.K. Kalkhoff and P.W. Adams. 1982. J. Clin. Endocrinol. Metab. 54:254-260
36. Larsson, B., K. Svardsudd, L. Welin, L. Wilhelmsen, P. Bjorntorp and G. Tibblin. 1984. Brit. Med. J. 288-1401-1404

ACKNOWLEDGEMENTS

This work is supported by the Medical Research Council of Canada and the Heart and Stroke Foundation of Ontario. We are grateful to Dr. Rotstein for providing adipose tissue biopsies and for his interest in this work. We also like to thank Ms. Laura Sheu for excellent technical assistance and Mrs. Tina Lagopoulos for her help in the preparation of this manuscript.

CLINICAL, NUTRITIONAL AND BIOCHEMICAL

CONSEQUENCES OF APOLIPOPROTEIN B DEFICIENCY

Herbert J. Kayden and Maret G. Traber

Department of Medicine
New York University School of Medicine
New York, New York 10016

Abetalipoproteinemia (ABL) was originally described 35 years ago (1) and to date approximately 100 patients have been recognized to have this inborn metabolic abnormality. We have continued to follow some 25 patients with the homozygous form of either ABL or hypobetalipoproteinemia. Our efforts are directed toward the clinical and biochemical evaluation of patients with particular stress on delineating changes in the neurological system, along with ocular changes, both retinopathy and opthalmoplegia. We have carefully documented the degree of neurological abnormality in these patients and have measured their adipose tissue levels of tocopherol (vitamin E). We have also continued to characterize the apolipoprotein and lipoprotein abnormalities in these patients. This report will stress a number of facets of the disease of ABL; these include published work of our own and others, and also reflect the work in progress in a number of areas, the most important of which is the recognition that the neurological abnormalities seen in ABL are a result of vitamin E deficiency and that these neurological abnormalities can be prevented by the oral administration of vitamin E from infancy and early childhood.

GENETICS OF ABL

The primary defect in ABL is the inability of these patients to synthesize apolipoprotein B (apo B), which results in the complete absence of plasma lipoproteins containing apo B—chylomicrons, VLDL and LDL (very low and low density lipoproteins, respectively) (2,3,4). The patterns of inheritance show that the parents of patients with ABL have normal levels of apo B, while the parents of patients with the homozygous form of hypobeta-lipoproteinemia (who are clinically indistinguishable from ABL patients) have reduced levels of apo B (5,6). These observations reinforce the requirement for research on the apo B gene. Presently, there are several groups trying to identify the defect in ABL that results in the failure of homozygotes to synthesize apo B, and to investigate the heterozygote carriers for ABL.

CLINICAL DIAGNOSIS

The clinical diagnosis in these patients is well known and patients are now recognized in infancy because their problems in malabsorption and diet

Table 1. Tocopherol Content of Plasma and Adipose
Tissue of Two Siblings with ABL

Subject	Age	Date	Plasma µg E/ml	Adipose Tissue ng E/mg TG
LL	4 months	12/82	0.20	17.3
	8	4/83	0.32	12.0
	10	6/83		6.6
	13	9/83		37.5
	18	2/84	0.75	21.3
	22	6/84	0.88	38.0
	29	1/85	0.64	66.0
DL	birth	6/84	0.20 cord blood	
	7 months	1/85	0.56	32.0
Father		12/82	4.6	
		7/84	10.7	
Mother		12/82	4.4	
		7/84	13.7	
			6.3 breast milk	

E = tocopherol, TG = triglyceride

intolerance bring them to the attention of pediatric gastroenterologists.
We have added 5 patients in recent years based on requests by pediatricians
for assistance in the diagnosis of ABL, and for guidance in clinical management, especially in terms of vitamin E therapy. Table 1 shows data on two
siblings with ABL. The first child in this family was diagnosed as having
ABL at age 4 months and has received oral vitamin E since that time; both
her plasma and her adipose tissue tocopherol levels have been increased by
this therapy. Now, at almost age 3, she has no detectable neurological
abnormalities, although her size is still below normal at the 55th percentile for weight and the 5th percentile for height. The patient's brother
was diagnosed as having ABL by analysis of the cord blood at birth. The
tocopherol content of the cord blood was 0.20 µg/ml, striking lower than the
tocopherol content of cord blood from a normal baby and as reported (7). DL
has been given vitamin E since birth; he has no detectable neurological
abnormalities. His weight at 7 months is at the 25th percentile and his
height is between the 10th and 25th percentile. Both parents have normal
plasma levels of tocopherol; the table also shows that the mother's breast
milk had a normal level of tocopherol.

Many of the characteristics described in the Bassen-Kornzweig syndrome
of ABL are not directly due to the defect in the synthesis of apo B and the
resulting lack in transport of plasma lipids in VLDL, and LDL, but rather
are due to the inability of these patients to synthesize chylomicrons,
leading to the malabsorption of lipids and fat soluble vitamins. Specifically, the sensitivity of erythrocytes to hemolysis, the retinopathy and the
peripheral neuropathy are a result of vitamin E deficiency (8)(which is
discussed further below). There have been occasional reports of patients
with ABL having vitamin K insufficiency, resulting in prolonged prothrombin
times (9,10), or having vitamin A deficiency with resultant ocular lesions
(11,12,13). The existence of specific transport proteins for vitamins A
(retinol binding protein), vitamin D and vitamin K probably aids in the
absorption of these vitamins, perhaps preventing the universal appearance of
these vitamin deficiencies in patients with ABL. However, vitamin E
apparently does not have a specific transport protein, but is carried within

the plasma lipoproteins (14,15), and vitamin E absorption is dependent on chylomicron synthesis (16); thus, the ABL patient is particularly at risk for malabsorption of this vitamin.

We have also evaluated the fatty acid pattern of the plasma and erythrocytes from patients with abetalipoproteinemia, as well as analyzing the fatty acids in adipose tissue. The analysis of fatty acids of plasma lipids and erythrocytes was measured in samples obtained prior to any treatment in one 27 year old woman (who weighed only 40 kg), in whom the diagnosis of ABL had not previously been made, and who had followed a most rigid diet restricted in fat. This data provided evidence that ABL patients do absorb some exogenous fatty acids and demonstrated that although the level of the essential fatty acid linoleic acid was low, the arachidonic acid levels were within the normal range (17). Furthermore, there were no abnormalities in platelet function, nor other evidence of essential fatty acid deficiency.

NEUROLOGICAL ABNORMALITIES, VITAMIN E DEFICIENCY AND TISSUE LEVELS OF TOCOPHEROL

Experimental animals (rats, dogs and monkeys) fed vitamin E deficient diets develop neurological, muscular, and ocular abnormalities (18, 19). Specifically, in the peripheral nervous system, the large fiber, myelinated nerves degenerate with loss of axons and myelin, suggesting an axonal neuropathy. In the spinal cord there is axonal degeneration in the posterior columns. With severe vitamin E deficiency myopathy occurs with resultant muscular atrophy. Retinopathy also occurs resulting from the disruption of the outer segments of the photoreceptors with accumulation of lipofuscin granules, while the pigment epithelium remains relatively intact.

We have been systematically evaluating the neurological lesions in patients with ABL in collaboration with Dr. Mitchell Brin. The abnormalities seen in patients with ABL, which appear to be a result of vitamin E deficiency (20), include: large fiber axonopathy, ascending axonopathy in the spinal cord with medullary involvement of the sensory relay nuclei, retinopathy due to pigmentary degeneration (21), myopathy and autofluorescent lipopigment accumulation in muscles. Neuropatholic examination by Brin et al. of the first two siblings with ABL described by Bassen and Kornzweig demonstrated striking similarities in the neurologic lesions seen in these patients to those observed in animals with experimental vitamin E deficiency, including loss of axons in the peripheral nerves, posterior columns, and spinocerebellar tracts; muscle fiber atrophy and accumulation of lipopigment in large central nervous system neurons and skeltal muscle; and pigmentary degeneration of the retina (20,22).

We have been measuring the level of adipose tissue tocopherol for the past 8 years in order to use these values as an indicator of the adequacy of vitamin E intake in patients with ABL. Measuring tissue levels is necessary as these patients, even when supplemented with gram quantities of vitamin E/day never achieve plasma levels of tocopherol that are more than 10% of normal plasma levels (23,24). Furthermore, we wished to evaluate whether there is a specific abnormality, other than the failure in absorption of dietary vitamin E due to the inability to form chylomicrons, and whether large oral doses could be used to raise the tocopherol content of the tissues. Table 2 demonstrates that oral administration of tocopherol to ABL patients can raise their adipose tissue levels of tocopherol to apparently normal levels, whether measured on the basis of tocopherol/triglyceride ratios or on the basis of tocopherol/cholesterol ratios. Table 2 also shows the extremely low levels of tocopherol measured in some patients. As shown in Table 3 correlation of the tissue level of tocopherol with the degree of

69

Table 2. Tocopherol Levels in Adipose Tissue from ABL Patients

Patient	Age	ng E/mg TG	ng E/μg chol	μg chol/mg TG
LLL	1	42	16	2.4
LLL	3	232	44	15
LL	2	66	30	2.2
MS	>20	117	2	6.9
HoB	>20	21	4	5.7
KuB	>20	30	7	4.5
AL	>20	45	13	3.5
RH	>20	242	24	10.2
KeB	>20	149	34	4.4
AF	>20	142	81	1.8
AFS*	18	510	80	5.8
"normal baby" fat		462	49	9
		441	36	6

E = tocopherol, chol = cholesterol, TG = triglyceride
* = normotriglyceridemic abetalipoproteinemic

neurological abnormality has not been entirely successful; however, it should be noted that those patients with the most severe neurologic disease do have the lowest levels of adipose tissue tocopherol. Neurological abnormalities are generally observed in adult patients who have adipose tissue tocopherol levels of <50 ng tocopherol/mg triglyceride or <20 ng tocopherol/μg cholesterol. In children under 2, however, levels of adipose tissue tocopherol of 50 ng/mg triglyceride are often observed, although no neurological abnormality has been detected (20).

Table 3. Neurological Abnormalities in Patients with ABL: Correlation with Vitamin E Therapy

Patient	Vitamin E Therapy age initiated	duration	Neurological Abnormality Areflexia	Loss of Vibratory Sensation	Retinopathy	Adipose Tissue Level ngE/mgTG
CR	35 yrs	0 yrs	+	+	+	<0.1
BL	28	2	+	+	+	24.8
RI	23	8*	+	+	+	54.2
GP	11	14*	+	+	+	<0.1
RP	8	14*	+	+	+	<0.1
AMV	8	18	+	+	–	378.0
KM	2:11	0:8	+	–	–	22.0
LLL	2:0	1:4	–	nt	–	558.0
CW	0:7	0:1	–	nt	–	30.1

ngE/mgTG= ng tocopherol/mg triglyceride, *poor compliance, nt=not tested. The children's ages are given in years:months.

It appears that the age at which therapy is initiated is related to the degree of neurological damage. As shown in Table 3, both the neurological abnormalities and the ocular lesions seen with vitamin E deficiency are present in adults in whom therapy with vitamin E was not begun until age 8 or older. Hypo- or areflexia was noted in all patients, except those who started vitamin E therapy at age 2 or younger. The loss of vibratory sensation was noted in the patients started on vitamin E therapy at age 8 or older. Finally, the ocular abnormalities have only been documented in the patients who have initiated vitamin E therapy at age 8 or older. This data suggests that the effect of vitamin E deficiency can be detected first in the long tracts of the peripheral nervous system as areflexia. As further damage occurs there is loss of vibration sensation and proprioception. With progressive deterioration of the peripheral nervous system, ataxia is present in both the limbs and trunk, along with wasting of the musculature. The changes in the eye seem to be due to a more prolonged or severe deficiency of vitamin E, as these changes are detected following the appearance of changes in the peripheral nervous system. Similar neurological abnormalities have been documented by Dr. Ron Sokol in children with severe cholestatic liver disease. (Personal communication) In these patients vitamin E absorption is limited or non-existent due to the near or complete absence of bile acids in the small intestine. In 25% of children under two areflexia, but no other neurological abnormality has been noted. By age 4 areflexia, ataxia and peripheral neuropathy are observed. Unlike the patients with ABL, children under 4 also have opthalmoplegia, possibly due to the severity of vitamin E deficiency.

VITAMIN E DEFICIENCY AND ABL

It should be emphasized that the test of the thesis that vitamin E deficiency exists in these patients has rested on the following documentation: 1) That oral therapy with vitamin E normalizes the peroxide hemolysis test despite continued low plasma tocopherol levels and relatively low erythrocyte tocopherol levels (8). 2) That measurement of adipose tissue levels have shown increases in relation to dose that can bring adipose tissue levels into the normal range (23,24), although it has not been possible to sharply correlate the adipose tissue levels and the extent of neurological disease (20). 3) The most critical test is in the clinical symptomatology. We emphasize that no patient we have examined, who has received adequate vitamin E therapy before the age of 8, has shown any, but minor neurological abnormality. Furthermore, the literature is replete with individual examples of ABL patients, who have been given vitamin E therapy with the result that the progressive deterioration of the peripheral nervous system has been arrested, with amelioration of damage in some cases (8, 10, 12, 20, 24, 25, 26, 27). 4) It has been possible to observe patients with ABL from birth, having documented that the B apo is missing in the cord blood, and that in such babies given vitamin E, the neurological development thus far is normal.

What should be noted in addition, is that as one reads the clinical description, although a number of the patients are normal and bright a great percentage of patients fall into the category of low mental development (IQ test) (28,29) and the lower range of growth and development in their final status. These observations underscore the problems of faulty absorption of nutrients and faulty plasma transfer of nutrients, which are essential for optimal growth and development. We include in this category essential fatty acids, other vitamins, particularly the fat soluble vitamins. The management of the nutrition of the infants with ABL in the early neonatal period is not routine and solicitous attention to the diet including nutrients other than vitamin E is required. As a passing commentary, the use of medium chain triglycerides (MCT) in these patients appears to be contra-indicated as there are several reports of increased hepatic fibrosis as a concomitant of MCT therapy (10,30).

TRANSPORT OF TOCOPHEROL AND ITS TRANSFER FROM PLASMA TO TISSUES

We have been studying mechanisms by which tocopherol is transferred from the plasma lipoproteins to the tissues. Thus far we have demonstrated that two mechanisms which serve to deliver lipids to tissues also function to deliver tocopherol to cells in an in vitro system. The two mechanisms

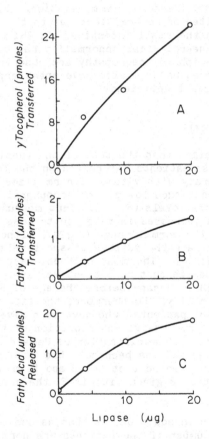

Figure 1. Increasing amounts of lipoprotein lipase result in increasing amounts of tocopherol and fatty acid transferred to cells. Duplicate incubations containing increasing amounts of bovine milk lipase (as shown), Intralipid (120 nmoles tocopherol, 270 μmoles fatty acid, labeled with radioactive triglyceride), 4% BSA (160 mg) and fibroblasts were carried out. Following washing, the fibroblasts were analyzed for cellular tocopherol (panel A), cellular fatty acid radioactivity (panel B) and the medium analyzed for triglyceride hydrolysis (expressed as fatty acids released from the radioactive triglyceride) (panel C).

are: lipoprotein lipase, which hydrolyzes the triglyceride contained in chylomicrons or VLDL to free fatty acids (31), and the LDL receptor, which functions to deliver plasma LDL-cholesterol to cells (32). We will briefly review the experiments, which have been published recently (33,34).

To demonstrate the effect of lipoprotein lipase activity on tocopherol transfer, fibroblasts were incubated with substrate containing triglyceride and tocopherol (in the presence of albumin as a fatty acid acceptor); then lipoprotein lipase (isolated from bovine milk) was added, the cells were incubated for 1 hour at 37°C, and the tocopherol content of the cells was measured. Experiments were carried out using substrate which contained either or both chylomicrons (from lipoprotein lipase deficient patients), containing alpha tocopherol, and Intralipid, containing gamma tocopherol. Both alpha and gamma tocopherol were transferred similarly by lipoprotein lipase. Figure 1 shows the results of a representative experiment in which Intralipid (labeled with radioactive triglyceride) served as the sole source of both triglycerides and gamma tocopherol. The amount of gamma tocopherol transferred to cells is shown in panel 1A; panel 1B shows the fatty acids transferred to cells and panel 1C shows the hydrolysis of triglyceride, expressed as fatty acid released, as determined from the radioactivity present in the cells and from the radioactivity in the free fatty acids extracted from the medium, respectively. The tocopherol content of cells, fatty acids transferred and released were all observed to increase with the addition of increasing amounts of lipoprotein lipase added to the incubation system, demonstrating that the transfer of tocopherol to cells occurred even in the absence of apolipoproteins.

Lipoprotein lipase is known to bind to cell membranes and heparin inhibits this binding (35). To test whether the transfer of gamma tocopherol to cells required the binding of the lipase to the cell membrane, heparin was added (prior to the lipase) to the incubation medium, the cells incubated for 1 hour with the lipase and the Intralipid substrate, then washed and the amount of cellular tocopherol was measured. Both normal fibroblasts, and LDL receptor negative fibroblasts from a subject with the homozygous form of Familial Hypercholesterolemia were used; the results were similar in both types of fibroblasts. The data shown in Figure 2 is from the LDL receptor negative fibroblasts. In the absence of heparin, the amount of tocopherol transferred was the same as that shown for normal fibroblasts incubated with 8 µg lipase. The addition of as little as 0.02 units of heparin to the incubation medium was sufficient to abrogate the transfer of tocopherol. At this concentration of heparin, there was a measureable amount of fatty acid transferred, but this amount was less than 40% of that transferred to those cells incubated without heparin (Figure 2B). As shown in Figure 2C, the rate of triglyceride hydrolysis was constant and was not affected by the presence of heparin in the incubation mixture at any of the concentrations of heparin used. This data demonstrates that the binding of lipase to the cell membrane is required for the transfer of tocopherol, and that the presence of the products of hydrolysis, or the changes in conformation of the remnants of the lipid droplet are not sufficient for the transfer of tocopherol.

In addition to studying the effect of lipoprotein lipase on the transfer of tocopherol, we have also studied the role of the LDL receptor in delivering tocopherol to cells (34). As LDL contains about 50% of the plasma tocopherol with the remaining tocopherol in VLDL and HDL (14,15), it seemed likely that uptake of LDL by the LDL receptor would increase cellular tocopherol. However, Kayden and Bjornson had previously demonstrated that the tocopherol present in LDL and HDL could exchange with the tocopherol in erythrocytes much like cholesterol exchanges (36). To test whether the presence of LDL receptors was necessary for the transfer of tocopherol from LDL to cells, a comparison was made, following incubation with LDL, of the

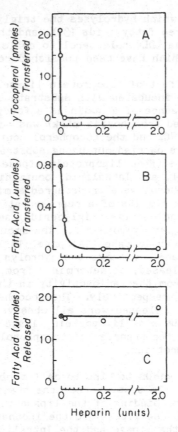

Figure 2. Addition of heparin to the incubation system abrogates
the transfer of tocopherol to cells. Prior to the addition of
lipase (8 µg), heparin, in the indicated amount, was added to the
incubation system (4ml), which contained Intralipid (160 nmoles
gamma tocopherol, 330 µmoles fatty acid), 4% BSA and LDL receptor-
negative fibroblasts. Following a 1 hour incubation at room
temperature, analysis of the cellular tocopherol (panel A) and
fatty acids, as radioactivity (panel B) and the medium for the
amount of triglyceride hydrolysis (panel C) was carried out.
⊖=two separate points too close to draw separately.

tocopherol contents of cells which do or do not have LDL receptors. Fibroblasts, either from a normal subject or from a patient with the homozygous
form of Familial Hypercholesterolemia were incubated overnight with 10% LPDS
(lipoprotein deficient serum, d>1.21 g/ml). Incubation with LPDS results in
the efflux of cellular cholesterol, which in turn stimulates the synthesis
of the rate limiting enzyme in cholesterol synthesis, 3-hydroxy 3-methyl
glutaryl coenzyme A, and the synthesis of LDL receptors (32). As fibroblasts from homozygotes for familial hypercholesterolemia do not have the
capability of synthesizing LDL receptors, incubation with LPDS does not
result in LDL receptor activity in these cells (32). The cells were then
incubated for 24 hours with increasing amounts of LDL (from 10 - 250 µg

protein/ml); the cellular tocopherol contents are shown in Figure 3. The LDL receptor-negative fibroblasts, incubated with 250 µg LDL protein/ml only contained twice as much tocopherol as fibroblasts incubated without LDL. In contrast the normal fibroblasts, incubated with 10 µg/ml doubled their tocopherol content, and exhibited a five fold increase in tocopherol content when incubated with 250 µg LDL/ml. A double reciprocal plot of the LDL concentration vs the cellular tocopherol concentration (minus the level of tocopherol found in the cells not incubated with LDL) resulted in a straight

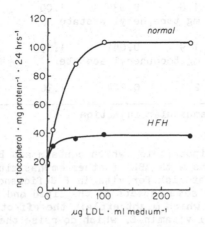

Figure 3. Uptake of LDL via the LDL receptor mechanism results in delivery of tocopherol to fibroblasts. Fibroblasts were incubated overnight with medium containing 10% LPDS, then incubated for 24 hours with medium containing 10% LPDS and an increasing concentration of LDL, as indicated. (The tocopherol content of the LDL was 2.1 µg/mg protein.) The cells were washed, harvested and the tocopherol and protein contents analyzed. Shown is the tocopherol content of fibroblasts from a normal subject (o) and of fibroblasts lacking LDL receptors from a person with the homozygous form of Familial Hypercholesterolemia (HFH, ●).

line (r=0.93) with the calculated maximum velocity (V_{max}) of 113 ng tocopherol/mg protein and the LDL concentration for half maximal velocity (K_m) was 40.8 µg protein/ml. These data demonstrate that uptake of LDL via the LDL receptor mechanism results in a marked uptake of tocopherol by cells and that this increase in the cellular tocopherol content is in excess of that due to exchange or from cells imbibing the medium.

ABSORPTION, TRANSPORT AND DELIVERY OF VITAMIN E IN ABL PATIENTS

Both of the mechanisms we have proposed for delivery of tocopherol to

Table 4. Distribution of Tocopherol in Serum
Lipoproteins with Densities > and <1.063
g/ml Isolated from a Patient with ABL
Following Intramuscular Injections with
Vitamin E

Time	Serum	Tocopherol Distribution µg E/ml		%Recovery
hrs	µg E/ml	<1.063	>1.063	
0	1.5	0.008	1.02	68
im inj 25 mg tocopheryl acetate				
1	1.4	0.015	0.99	72
20	1.8	0.034	1.00	57
im inj 50 mg tocopheryl acetate				
24	1.9	0.041	1.17	65
44 im inj 75 mg tocopheryl acetate				
68	2.1	0.044	1.36	64

im inj = intramuscular injection

tissues in vivo utilize lipoproteins which contain apo B, thus are unlikely
to be effective in patients with ABL, further emphasizing the degree to
which these patients are at risk for vitamin E deficiency.   In the next
sections we will consider some aspects of the lipid and lipoprotein metabolism of these patients, which might explain the effectiveness of large
oral doses of supplemental vitamin E, which do raise the adipose tissue
levels in these patients to those seen in normal individuals (23,24).

To further amplify on the course of events with which vitamin E is
absorbed, in collaboration with Dr. Robert Glickman, we have studied the
changes in plasma levels of tocopherol after oral administration of vitamin
E, looking for its appearance either in circulating HDL or in a new particle
that might contain apo AIV, reflective of intestinal synthesis. This was
unsuccessful as neither new HDL, nor increased amounts of apo AIV was
detected. However, following the intramuscular administration of vitamin E
in a patient with ABL, we have been able to show the appearance of an
increment in the plasma tocopherol both in the density <1.063 region, which
has no apo B, but more importantly in the the density >1.063 fraction,
presumably in the HDL_2 fraction. (Table 4)

From our observations, the usual level of vitamin E in the plasma of
ABL patients, who have been supplemented with 100 mg vitamin E/kg/day, is
about 10% of the level in the LDL fraction of normal subjects. This is an
adequate level of plasma tocopherol to achieve and maintain adequate tissue
concentrations of tocopherol.

LIPOPROTEIN LIPASE ACTIVITY IN ABL

The lipoprotein lipase activity in these patients is usually not
functional, unless the patients are treated with parenteral (IV) lipids
(37). Although lipoprotein lipase activity is ordinarily diminished one can
demonstrate release of enzyme into the plasma after intravenous injection of
heparin (3). This lipoprotein lipase enriched plasma, when tested in an in
vitro system with adequate substrate, contains levels of lipase approxi-

mately at the lower level of normal (3). This is also true of the LCAT
activity (lecithin cholesterol: acyl transferase) in these patients, which
is normal when tested in serum from true LCAT deficient patients (3,38,39).

LDL RECEPTOR ACTIVITY IN ABL

In the early studies in fibroblasts in tissue culture on the regulation
of intracellular cholesterol and the role of LDL in regulating the activity
of 3-hydroxy 3-methylglutaryl CoA reductase on cholesterol synthesis and in
the activity of the ACAT (acylCoA: acyl transferase) enzyme for intracellular esterification of cholesterol, there were some observations that
indicated that ABL serum was indistinguishable from lipoprotein deficient
serum in its effect on LDL receptor activity. At the same time other
reports noted that the lymphoid cell lines from ABL patients showed normal
LDL receptor activities (40), that total body cholesterol synthesis in these
patients was not elevated (41,42,43) and clinical evaluation of the patients
did not reveal gross alteration in any metabolic parameter, specifically
related to cellular cholesterol deficiency. In an extensive study of plasma
(obtained by plasmaphoresis) from 5 ABL patients, Scanu et al. defined a
series of abnormalities of the HDL fraction, both qualitatively, quantitatively , and morphologically--and presented the first evidence of a
previously unrecognized peptide--subsequently identified as the arginine
rich apolipoprotein, currently referred to as apo E (2).

Following those observations, the role of apo E-containing lipoproteins
in ABL patients has been studied extensively. It is apparent, especially
from the detailed studies of Blum et al. that the apo E contained in the
HDL_2 carries out all of the functions of an apo B-LDL in regards to the
delivery of cholesterol to cells (44). Their calculations on the basis of
binding potencies of the apo E rich-HDL and on the concentration and composition of the HDL, would suggest that ABL plasma has the equivalent delivery
for cholesterol of between 50-150 mg/dl of LDL cholesterol. Uptake of HDL_2
into the tissues has been shown by Illingworth to be via the "LDL" or B/E
receptor mechanism (45). Our studies and those of others have documented
the function of the LDL receptor in accepting lipoproteins containing apo E
from patients with ABL.

LIPOPROTEIN AND APOLIPOPROTEINS

There have been a number of studies of the lipoproteins and apolipoproteins of patients with ABL and hypobetalipoproteinemia. Patients who are
heterozygous for hypobetalipoproteinemia appear to have a normal complement
of apolipoproteins and formed lipoproteins, but the amount of circulating
apo B in the VLDL and of LDL is reduced (5,6). Studies carried out by
Sigunderson, Nicoll and Lewis in two patients with hypobetalipoproteinemia
documented a reduced rate of synthesis of VLDL, but a normal fractional
catabolic rate, including a normal conversion rate of apo B in VLDL to LDL
(46). A separate analysis for the synthesis of apo B in LDL again showed a
decreased synthetic rate, but the fractional catabolic rate was normal.

There have been many analyses of the serum lipoproteins in ABL, all
documenting the absence of apo B with the consequent absence of chylomicrons, characteristic VLDL and LDL. Particles have been isolated from ABL
plasma in the density range of VLDL (d<1.006 g/ml) and LDL (d<1.063 g/ml);
these particles lack apo B and have similar apolipoproteins to those found
in the HDL_2 and HDL_3 density range (2,3,4). These "VLDL" and "LDL"
particles have also been isolated by molecular sieving fractionation procedures and would appear to represent aggregates or multimers of particles
isolated in the HDL_2 fraction (47).

One of the reasons to study the lipoproteins and apolipoproteins of ABL patients is that the absence of apo B allows one to interpret the inter-relationships of the lipoproteins in the normal individual. We and others (3) have examined the plasma of a number of patients with ABL for the presence of Lp(a) and have not found it. This adds support to the conclusion that Lp(a) is a concomittant of the secretion of apo B.

In a recent study, Gibson et al. studied the apolipoproteins of the lipoproteins from ABL patients as examples of lipoproteins which contain apo E and occur in the absence of the metabolism of apo B-containing lipoproteins, such as chylomicrons, VLDL or LDL (48). It has been apparent that there is a vigorous exchange and metabolic interrelation between trigly-ceride-rich lipoproteins undergoing hydrolysis and HDL particles. In the ABL subjects, however, there is evidence for an independent origin of apo E-containing HDL, which is not a consequence of the apo B metabolic cascade of triglyceride-rich lipoproteins. That the major fraction of HDL involved in these interactions is the HDL_2 fraction is not surprising as a majority of the apo E was contained in this fraction.

CONCLUSIONS

More than 35 years have passed since the first case of ABL was described. The characteristics of this disease can divided into those features of the clinical syndromes which are a consequence of malabsorption secondary to the primary genetic abnormality and those abnormalities which reflect the primary genetic defect. We have several patients currently in their 20's who have received their vitamin E since the end of their first decade, who aside from very minor neurological abnormalities, are entirely normal. Thus, patients with ABL in whom the problems of the malabsorption syndrome have been overcome by proper nutritional supplementation have protection against the neurological abnormalities, retinopathy, and sensitivity of erythrocytes to hemolysis. One continues to observe the abnormalities of the chylomicrons, VLDL and LDL, which result in low serum lipid values including low plasma tocopherol levels. Their erythrocytes continue to demonstrate abnormal shape (49) and phospholipid composition (50), but yet function normally. Our observations suggest that secretions of adrenal hormones and sex hormones are entirely normal and that patients may conceive and deliver normal children. However, Illingworth et al. have reported low progesterone levels in one homozygous hypobetalipoproteinemic patient (51), but she has also had a normal and uneventful pregnancy (Illingworth-- personal communication).

The absence of LDL in ABL patients, and the low levels of LDL in the hypobetalipoproteinemic patients warrant attention to the development of atherosclerosis, if the thesis of elevated LDL cholesterol levels (and apo B levels) as a cause of vascular disease is correct. It will be important to continue following these patients for the extent of atherosclerosis both during life and at post-mortem examination. The two adult patients, who have come to post-mortem, have been women below the age of 40 and these patients have shown little evidence of vascular disease; but it is essential to recognize that the HDL of these patients are considerably different from the HDL found in normal subjects, and these changes in this lipoprotein fraction may play a role in modulating atherosclerotic disease.

ACKNOWLEDGEMENTS

Support was provided by U.S. Public Health Service grant number HL30842.

REFERENCES

1. Bassen, F.A. and A.L. Kornzweig. 1950. Malformation of the erythrocytes in a case of atypical retinitis pigmentosa. Blood 5:381-
   387.
2. Scanu, A.M., L.P. Aggerbeck, A.W. Kruski, C.T. Lim and H.J. Kayden.
   1974. A study of the abnormal lipoproteins in abetalipoproteinemia. J. Clin. Invest. 53:440-453.
3. Kostner, G., A. Holasek, H.G. Bohlmann and H. Thiede. 1974. Investigation of serum lipoproteins and apoproteins in abetalipoproteinaemia. Clin. Sci. and Mol. Med. 46:457-468.
4. Shepherd, J., M. Caslake, E. Farish and A. Fleck. 1978. Chemical and
   kinetic study of the lipoproteins in abetalipoproteinaemic plasma.
   J. Clin. Path. 31:382-387.
5. Cottrill, C., C.J. Glueck, V. Leuba, F. Millett, D. Puppione and W.V.
   Brown. 1974. Familial Homozygous Hypobetalipoproteinemia.
   Metabolism 23:779-791.
6. Mars, H., L.A. Lewis, A.L. Robertson, A. Butkus and G.H. Williams.
   1969. Familial hypobetalipoproteinemia. A genetic disorder of
   lipid metabolism with nervous system involvement. Am. J. Med.
   46:886-900.
7. Haga, P., J. Ek and S. Kran. 1982. Plasma tocopherol levels and
   vitamin E/beta lipoprotein relationships during pregnancy and in
   cord blood. Am. J. Clin. Nutr. 36:1200-1204.
8. Kayden, H.J. and R. Silber. 1965. The role of vitamin E deficiency in
   the abnormal autohemolysis of acanthocytosis. Trans. Assoc. Am.
   Phys. 78:334-341.
9. Caballero, F.M. and G.R. Buchanan. 1980. Abetalipoproteinemia presenting as severe vitamin K deficiency. Pediatrics 65:161-163.
10. Illingworth, D.R., W.E. Connor and R.G. Miller. 1980. Abetalipopro-
    teinemia; report of two cases and review of therapy. Arch.
    Neurol. 37:659-662.
11. Gouras, P., R.E. Carr and R.D. Gunkel. 1971. Retinitis pigmentosa in
    abetalipoproteinemia: Effects of vitamin A. Invest. Ophthal.
    10:784-793.
12. Bishara, S., S. Merin, M. Cooper, E. Aziz, G. Delpre and R.J.
    Deckelbaum. 1971. Combined vitamin A and E therapy prevents
    retinal electrophysiological deterioration in abetalipoprotein-
    aemia. Brit. J. Ophthal. 12:767-770.
13. Bieri, J.G., J.M. Hoeg, E.J. Schaefer, L.A. Zech and B. Brewer. 1984.
    Vitamin A and vitamin E replacement in abetalipoproteinemia. Ann.
    Int. Med. 100:238-239.
14. Bjornson, L.K., H.J. Kayden, E. Miller and A.N. Moshell. 1976. The
    transport of alpha tocopherol and beta carotene in human blood.
    J. Lipid Res. 17:343-351.
15. Behrens, W.A., J.N. Thompson and R. Madere. 1982. Distribution of
    alpha tocopherol in human plasma lipoproteins. Am. J. Clin. Nutr.
    35:691-696.
16. Kayden, H.J. 1969. Vitamin E deficiency in patients with abetalipo-
    proteinemia. Vitamine A, E und K. Klinische und physiologisch-
    chemische probleme--Symposiom September 1967. H.F. von Kress and
    K.U. Blum, editors. p 301-308.
17. Kayden, H.J. 1980. Is essential fatty acid deficiency part of the
    syndrome of abetalipoproteinemia? Nutr. Rev. 38:244-246.
18. Nelson, J.S. 1983. Neuropathological studies of chronic vitamin E
    deficiency in mammals including humans. R. Porter and J. Whelan,
    editors. Ciba Found. Symp. 101:92-99.
19. Brin, M.F. and M.G. Traber. 1985. Vitamin E deficiency and human
    neurological disease. Laboratory Management. In Press.
20. Brin, M.F., T.A. Pedley, R.G. Emerson, R.E. Lovelace, P. Gouras, C.

MacKay, H.J. Kayden, J. Levy and Herman Baker. 1985. Electro-physiological features of abetalipoproteinemia: functional consequences of vitamin E deficiency. Neurology. Submitted.

21. Cogan, D.G., M. Rodrigues, F.C. Chu and E.J. Schaefer. 1984. Ocular abnormalities in abetalipoproteinemia. A clinicopathologic correlation. Ophthal. 91:991-998.

22. Brin, M.F., J.S. Nelson, W.C. Roberts, M.D. Marquardt, P. Suswankosai and C.K. Petito. 1983. Neuropathology of abetalipoproteinemia: A possbile complication of the tocopherol (vitamin E)-deficient state. Neurology 33(suppl 2):142-143.

23. Kayden, H.J., L.J. Hatam and M.G. Traber. 1983. The measurement of nanograms of tocopherol from needle aspiration biopsies of adipose tissue: normal and abetalipoproteinemic subjects. J. Lipid Res. 24:652-656.

24. Kayden, H.J. 1983. Tocopherol content of adipose tissue from vitamin E-deficient humans. R. Porter and J. Whelan, editors. Ciba Found. Symp. 101:70-91.

25. Hegele, R.A. and A. Angel. 1985. Arrest of neuropathy and myopathy in abetalipoproteinemia with high dose vitamin E therapy. Can. Med. Assoc. J. 132:41-44.

26. Malloy, M.J., J.P. Kane, D.A. Hardman, R.L. Hamilton and K.B. Dalal. 1981. Normotriglyceridemic abetalipoproteinemia; absence of the B-100 apolipoprotein. J. Clin. Invest. 67:1441-1450.

27. Muller, D.P.R., J.K. Lloyd and A.C. Bird. 1977. Long-term management of abetalipoproteinaemia; possible role for vitamin E. Arch. Dis. Child. 52:209-214.

28. Lowry, N.J., M.J. Taylor, W. Belknapp and W.J. Logan. 1984. Electro-physiological studies in five cases of abetalipoproteinemia. Can. J. Neurol. Sci. 11:60-63.

29. Sperling, M.A., F. Hengstenberg, E. Yunis, F.M. Kenny and A.L. Drash. 1971. Abetalipoproteinemia: metabolic, endocrine and electron-microscopic investigations. Pediatrics 48:91-102.

30. Partin, J.S., J.C., Partin, W.K. Schubert,et al. 1974. Liver ultra-structure in abetalipoproteinemia: Evolution of micronodular cirrhosis. Gastroenterology 67:107-118.

31. Nelsson-Ehle, P., A.S. Garfinkel and M.C. Schotz. 1980. Lipolytic enzymes and plasma lipoprotein metabolism. Ann. Rev. Biochem. 49:667-693.

32. Goldstein J.L. and M.S. Brown 1976. The LDL pathway in human fibro-blasts: a receptor mediated mechanism for the regulation of cholesterol metabolism. Curr. Top. in Cell Regul. II:147-81.

33. Traber, M.G. , T. Olivecrona and H.J. Kayden. 1985. Bovine milk lipoprotein lipase transfers tocopherol to human fibroblasts during triglyceride hydrolysis in vitro. J. Clin. Invest. 75:In Press.

34. Traber, M.G. and H.J. Kayden. 1984. Vitamin E is delivered to cells via the high affinity receptor for low density lipoprotein. Am. J. Clin. Nutr. 40:747-751.

35. Chajek-Shaul, T., G. Friedman, O. Stein, T. Olivecrona and Y. Stein. 1982. Binding of lipoprotein lipase to the cell surface is essential for the transmembrane transport of chylomicron chole-steryl ester. Biochim. Biophys. Acta 712:200-210.

36. Kayden, H.J. and L.K. Bjornson. 1972. The dynamics of vitamin E transport in the human erythrocyte. Ann. N.Y. Acad. Sci. 203:127-140.

37. Kostner, G.M., H.G. Bohlmann and A. Holasek. 1976. The recycling of apolipoproteins between lipoproteins of different density classes in the serum of patients with apo B- deficiency after intravenous intralipid administration. J. Mol. Med. 1:311-324.

38. Cooper, R.A. and C. L. Gulbrandsen. 1971. The relationship between serum lipoproteins and red cell membranes in abetalipoproteinemia:

Deficiency of lecithin:cholesterol acyltransferase. J. Lab. Clin. Med. 78:323-335.

39. Subbaiah, P.V. 1982. Requirement of Low Density Lipoproteins for the Lysolecithin Acyl Transferase Activity in human plasma: assay of enzyme activity in abetalipoproteinemic patients. Metabolism 31:294-298.

40. Kayden, H.J., L. Hatam and N.G. Beratis. 1976. Regulation of 3-hydroxy-3-methylglutaryl Coenzyme A reductase activity and the esterification of cholesterol in human long term lymphoid cell lines. Biochem. 15:521-528.

41. Kayden, H.J. 1978. Abetalipoproteinemia: abnormalities of serum lipo-proteins. Protides of the Biological Fluids. H. Peeters, Editor. Pergamon Press. p271-276.

42. Myant, N.B., D. Reichl and J.K. Lloyd. 1978. Sterol balance in a patient with abetalipoproteinaemia. Atherosclerosis 29:509-512.

43. Illingworth, D.R., Connor, W.E., Buist, N.R., B.M. Jhaveri, D.S. Lin, and M.P. McMurry. 1979. Sterol Balance in Abetalipoproteinemia: Studies in a patient with homozygous familial hypobetalipopro-teinemia. Metabolism 28:1152-1160.

44. Blum, C.B., Deckelbaum, R.J., Witte, L.D., Tall, A.R. and J. Cornicelli. 1982. Role of apolipoprotein E-containing lipopro-teins in abetalipoproteinemia. J. Clin. Invest. 70:1157-1169.

45. Illingworth, D.R., N.A. Alam, E.E. Sundberg, F.C. Hagemenas and D.L. Layman. 1983. Regulation of low density lipoprotein receptors by plasma lipoproteins from patients with abetalipoproteinemia. Proc. Natl. Acad. Sci. USA 80:3475-3479.

46. Sigurdsson, G., A. Nicoll and B. Lewis. 1977. Turnover of apolipo-protein B in two subjects with familial hypobetalipoproteinemia. Metabolism 26:25-31.

47. Lim, C.T., J. Chumg, H.J. Kayden and A.M. Scanu. 1976. Apoproteins of human serum high density lipoproteins. Isolation and characteri-zation of the peptides of sephadex fraction V from normal subjects and patients with abetalipoproteinemia. Biochim. Biophys. Acta 420:332-341.

48. Gibson, J.C., A. Rubinstein, W.V. Brown, H.N. Ginsberg, H. Greten, R. Norum and H.J. Kayden. 1985. Apolipoprotein E containing lipo-proteins in low density or high density lipoprotein-deficient states. Arteriosclerosis. In Press

49. Kayden, H.J. and M. Bessis. 1970. Morphology of normal erythrocyte and acanthocyte using nomarski optics and the scanning electron micro-scope. Blood 35:427-436.

50. Phillips, G.B. and J.T. Dodge. 1968. Phospholipid and phospholipid fatty acid and aldehyde composition of red cells of patients with abetalipoproteinemia. J. Lab. Clin. Med. 71:629-637.

51. Illingworth, D.R., D.K. Corbin, E.D. Kemp and E.J. Keenan. 1982. Hormone changes during the menstrual cycle in abetalipoprotein-emia: Reduced luteal phase progesterone in a patient with homo-zygous hypobetalipoproteinemia. Proc. Natl. Acad. Sci. USA 79:6685-6689.

FAMILIAL HYPOALPHALIPOPROTEINEMIA

Charles J. Glueck*, Marc A. Melser*, I.B. Borecki§,
Jane L.H.C. Third*, D.C. Rao§, and Peter M. Laskarzewski*

From the Lipid Research Clinic, General Clinical Research
Center, and CLINFO Centers, Departments of Medicine and
Pediatrics, Lipid Research Division, University of Cincinnati
Medical Center, Cincinnati, Ohio*, and from the Divisions of
Biostatistics, Department of Preventive Medicine, Psychiatry,
and Genetics, Washington University School of Medicine
St. Louis, Missouri§

INTRODUCTION

Primary hypoalphalipoproteinemia, characterized by primary depression
of high density lipoprotein cholesterol (HDL-C) levels below the age-sex-
race-specific 10th percentile (in the absence of other abnormalities of
lipoprotein cholesterols), is closely associated with premature
atherosclerotic coronary heart disease, myocardial infarction, and ische-[1-4]
mic cerebrovascular disease. These premature coronary and cerebrovascular events usually occur in the primary hypoalphalipoproteinemic
subjects at young ages, often before age 55, in keeping with the significant inverse association of HDL-C with coronary heart disease (CHD) in
population groups.[5,6] Primary depression of HDL-C and/or apolipoprotein
Al (apo Al) is a valuable marker in identifying individuals at accelerated
risk for premature atherosclerosis.[1-6]

In free living, unselected population groups, variation in HDL-C
levels reflects a substantial genetic component.[7] Three previous
reports[1-4,8] have suggested that the hypoalphalipoproteinemia phenotype
may be determined by a major gene in kindreds identified by a primary
hypoalphalipoproteinemic proband. These conclusions were based on an
arbitrary definition of hypoalphalipoproteinemia (below the 10th percentile of the relevant age-sex-race-specific distribution) which was applied
in these family studies.[1-4,8] Segregation ratios within sibships were
estimated and compared with the ratios expected under a monogenic
hypothesis.[1] There was good agreement between observed and expected
ratios under the hypothesis of an autosomal dominant mode of
inheritance.[1]

In the current report, our specific aim was to review our previously
published data on familial hypoalphalipoproteinemia,[1,2,9,10] and to present new data better characterizing hypoalphalipoproteinemic kindreds,
including the outcome of therapy designed to elevate HDL-C, and to summarize recently completed studies of the heritability of this disorder
using the unified mixed model of segregation analysis.[11]

RESULTS AND DISCUSSION

Sixteen kindreds with primary and familial hypoalphalipoproteinemia

We have previously reported data on 16 kindreds with primary and/or
familial hypoalphalipoproteinemia whose probands were selected arbitrarily
by virtue of HDL-C $\leq$ the age-sex-race-specific 10th percentile as their
sole dyslipoproteinemia, with the additional requirement that they be normotriglyceridemic (triglyceride levels < 90th percentile).[1,2,9-11] The
probands were also required to have primary hypoalphalipoproteinemia, not
secondary to diseases and/or drugs.[1] Since one of the most important
determinants of HDL-C levels is plasma triglyceride,[12,13] we focused on
probands selected by normal triglyceride levels and primary depression of
HDL-C levels as their sole dyslipoproteinemia, excluding hypertriglyceridemic probands whose hypoalphalipoproteinemia might be secondary to
hypertriglyceridemia.[1,10,11] As previously reported,[1] and as summarized
in Table 1, 15 of our 16 probands were men; 12 were referred because of
premature myocardial infarction, angina, or stroke, 2 because of family
history of premature myocardial infarction or stroke, and 2 because of low
HDL-C observed on routine health examinations. After identifying probands, extensive efforts were made to sample all spouses and all first and
second degree relatives. Of 72 living first degree relatives of the probands, 60 (83%) were sampled. Of the 60 living first degree relatives
sampled, 27 (45%) had HDL-C $\leq$ 10th percentile, while 12 (20%) had triglyceride $\geq$ 90th percentile. We recognized in our initial publication (1)
that our identification of "hypoalphalipoproteinemia" by levels < 10th
percentile was entirely arbitrary, and that, in the strictest "genetic"
sense, one might expect to identify Mendelian disorders at even greater
extremes of the distribution, perhaps below the fifth or first percentiles
for HDL-C, mimicking the presence of familial hypercholesterolemia in subjects with LDL-C $\geq$ 95th and/or 99th percentile.[14]

The kindreds were characterized not only by premature atherosclerotic
CHD and cerebrovascular disease in the probands, but also by presence of
premature CHD and/or CVA in many adult first degree relatives.[1]

Simple segregation analyses suggested that the segregation of low
HDL-C in offspring of critical matings was primarily accounted for by isolated low HDL-C, not by hypoalphalipoproteinemia secondary to
hypertriglyceridemia.[1] We concluded that familial hypoalphalipoproteinemia was a heritable disorder with a pattern of transmission not significantly different from that expected by a hypothesis of Mendelian
dominance.

Approaches to elevate HDL-C levels in familial hypoalphalipoproteinemic subjects by diet, exercise, and Gemfibrozil

Subsequent to our initial diagnosis of primary and familial
hypoalphalipoproteinemia in 16 probands,[1] we combined non-pharmacologic
(weight loss towards ideal body weight, increasing aerobic exercise) and
pharmacologic approaches (Gemfibrozil[15,16]) to increase HDL-C levels.
Table 1 displays lipids, lipoprotein cholesterols, and apolipoproteins A1
and A2, in our previously reported 16 probands with primary and familial
hypoalphalipoproteinemia,[1] on resampling after three months of dietary,
exercise, and drug intervention. Baseline, pre-intervention levels of
triglyceride, HDL-C and total cholesterol/HDL-C are presented for comparison (Table 1). Regrettably, we had only incomplete data on baseline
apolipoprotein A1[17] and A2[18] insufficient for comparison to our values
obtained on resampling (Table 1).

TABLE 1: Total cholesterol (TC), triglyceride (TG), high and low density lipoprotein cholesterols (HDL-C, LDL-C), and apolipoproteins A1 and A2 (mg/dl), in 16 probands with primary and familial hypoalphalipoproteinemia, on resampling, after dietary, exercise, and (in some subjects), drug intervention. For reference, initial baseline TG, HDL-C and TC/HDL-C are presented in [ ].

SUBJECT #	SEX	AGE	TC	TG	[TG]	HDL-C	[HDL-C]	LDL-C	A1	A2	HDL-C/A1	HDL-C/A2	TC/HDL-C	[TC/HDL-C]
1*	M	43	250	263	[159]	39	[29]	158	112	36.9	.348	1.06	6.41	[7.17]
5*	M	49	197	78	[115]	52	[32]	19	114	48.1	.456	1.08	3.79	[4.78]
8*	F	32	221	67	[58]	47	[29]	11	111	51.3	.423	.916	4.70	[5.72]
11*	M	62	181	83	[200]	39	[27]	15	114	39.2	.342	.995	4.64	[3.85]
15*	M	48	202	63	[180]	36	[18]	13	94	25.3	.383	.423	5.61	[10.83]
16*	M	61	187	152	[194]	33	[29]	14	102	32.7	.324	.009	5.67	[6.90]
19*	M	49	259	157	[155]	42	[32]	16	123	49.5	.341	.848	6.17	[6.91]
27*	M	44	150	62	[72]	42	[30]	6	112	32.5	.375	.292	3.57	[4.40]
28*	M	48	217	122	[192]	30	[21]	13	83	34.1	.361	.880	7.23	[10.67]
4	M	36	197	264	[226]	22	[21]	12	85	34.2	.259	.643	8.95	[8.81]
14	M	35	269	295	[231]	31	[32]	19	104	41	.298	.756	8.68	[7.53]
17	M	34	136	180	[165]	23	[28]	7	92	36.2	.250	.635	5.91	[5.71]
18	M	42	181	124	[174]	30	[31]	16	96	37.5	.313	.800	6.03	[5.45]
23	M	49	174	106	[89]	34	[28]	19	109	35	.312	.971	5.12	[6.43]
25	M	38	170	182	[199]	28	[30]	16	93	35	.301	.800	6.07	[5.63]
26	M	19	186	127	[104]	36	[28]	15	98	40.7	.367	.885	5.17	[6.21]
X̄		43	199	145	[158]	35	[28]	134	103[1]	38[2]	0.341	0.927	5.86	[6.69]
(SD)		11	37	74	[54]	8	[4]	30	12	7	0.055	0.211	1.49	[2.00]
Median		43	192	126	[170]	35	[29]	126	103	37	0.342	0.900	6.69	[6.32]

[1,2] (X̄ ± SD) values for A1 and A2 in 38 healthy controls were 141 ± 19, 51 ± 9.5 mg/dl at time of resampling

*Subjects receiving Gemfibrozil 1200 mg/day at time of resampling

TABLE 2: Mean (SD) lipids and lipoprotein cholesterols (mg/dl) in 16 familial hypoalphalipoproteinemic probands at baseline and after 3 months on therapy with diet and exercise (n=7), and diet, exercise and Gemfibrozil 1200 mg/day (n=9).

SAMPLE	n	HDL-C	TG	LDL-C	TC/HDL-C
Initial	9	28 + 5	147 + 54	132 + 25	6.8 + 2.5
On Rx, Diet, Exercise		40 + 7§	116 + 66	144 + 27	5.3 + 1.2*
Initial	7	28 + 4	170 + 57	120 + 23	6.5 + 1.2
On Diet, Exercise		29 + 5	183 + 73	122 + 30	6.6 + 1.6

*p<.05; §p<.001, paired t test

Our interventions centered on weight loss towards ideal body weight and supervised aerobic exercise intervention (as tolerated), incorporating at least three periods per week, 25 minutes per session, at approximately 70% of maximum heart rate. None of our probands were cigarette smokers at baseline, and none initiated smoking during followup. In addition to weight loss towards ideal body weight and moderate amounts of aerobic exercise, in nine hypoalphalipoproteinemic probands, Gemfibrozil[15,16] (1200 mg/day in divided doses) was given to determine whether, and to what degree, their HDL-C levels could be further elevated by pharmacologic intervention (Table 2).

As displayed in Table 1, and as previously reported,[1] at baseline, the subjects had (as a group) very low mean HDL-C, 28 mg/dl. Their mean and median triglyceride, total cholesterol and LDL-cholesterol levels were well within the normal range at baseline,[1] so that low HDL-C was (by selection), their sole dyslipoproteinemia. Moreover, it is clear from Table 1 that subjects whose primary dyslipoproteinemia involves solely low levels of HDL-C have an unfavorable ratio of total to HDL-cholesterol, which puts them squarely in the range of increased CHD risk, extrapolated from the Framingham data.[19] It is not surprising that probands and their affected first degree relatives with low levels of HDL-cholesterol, and/or high TC/HDL-C ratios,[19] exhibit rapidly accelerated, premature coronary and cerebrovascular disease.[1-4,8,10]

As displayed in Tables 1 and 2, the hypoalphalipoproteinemic subjects receiving non-pharmacologic interventions alone had no significant changes in HDL-C, LDL-C, TC/HDL-C, or triglycerides on resampling, whereas those receiving Gemfibrozil had a significant increase in HDL-cholesterol (p < .001), and a significant decrement (p < .05) in TC/HDL-C. This suggests that familial hypoalphalipoproteinemia, like familial hypercholesterolemia, may require not only non-pharmacologic intervention, but also pharmacologic intervention to ameliorate the primary lipoprotein abnormality (Table 2).

Figure 1 displays the percent change from baseline in lipids and lipoprotein cholesterols over a nine month period in 11 subjects having primary and familial hypoalphalipoproteinemia and receiving Gemfibrozil 1200 mg/day. The subjects were first stablized for 3 months with non-pharmacological interventions, including weight loss towards ideal body weight and moderate levels of supervised aerobic exercise, 3 bouts per week, 25 minutes per session, at approximately 70% of maximum heart rate.

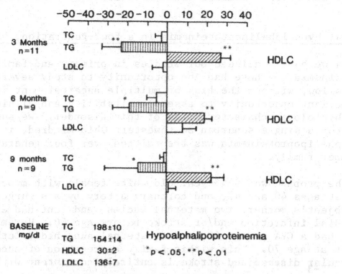

FIGURE 1: Percentage change in lipids and lipo-
        protein cholesterols induced by
        Gemfibrozil 1200 mg/day, after a base-
        line stabilization on diet and aerobic
        exercise, in subjects with familial
        hypoalphalipoproteinemia. Significance
        levels were calculated by paired t tests;
        isocaloric diet, constant exercise,
        non-smoking, and no alcohol intake
        were maintained throughout the study
        period.

They were then given Gemfibrozil 1200 mg/day in divided doses. During the
9 month study period, the diet was held at isocaloric levels, weight was
stable, and the exercise program was maintained stable. None of the subjects smoked or ingested alcoholic beverages during the study. All the
subjects had documented primary and familial hypoalphalipoproteinemia. At
3, 6, and 9 months there were significant increases in HDL-C above baseline of approximately 25% (p<.01). Reductions in triglyceride levels were
most marked at three months (22% below baseline, p<.01), and were also
significant at 6 and 9 months. There were no significant changes in total
or LDL-cholesterol (Figure 1). Thus, as shown in Figure 1 and in Table 2,
combined intervention with diet, exercise, and Gemfibrozil leads to significant increments in HDL-cholesterol and decrements in triglycerides in
familial hypoalphalipoproteinemic subjects. Whether these changes will,
if maintained over long periods of time, reduce the risk of coronary heart
disease and stroke, remains to be determined by controlled clinical
trials.

Familial hypoalphalipoproteinemia in a four-generation, 51-member family

As we have continued our studies in primary and familial hypoalphalipoproteinemia, we have had the opportunity to study several large kindreds
which allow, without the bias of multiple ascertainment in small kindreds,
an excellent opportunity to assess heritability and clinical-
pathophysiologic characteristics of this disorder. We summarize below
data from a single suburban southwestern Ohio kindred, in which familial
hypoalphalipoproteinemia was transmitted over four generations in a
51-member family.

The proband was a 46-year-old white female with myocardial infarc-
tions at ages 40 and 42, and coronary artery bypass surgery at age 44.
The subject's mother, two maternal uncles, and aunt had all sustained
myocardial infarction and/or stroke before age 60. Her maternal grand-
mother had a CVA at age 62, and a maternal aunt had a lethal myocardial
infarct at age 70. This within-family aggregation of accelerated cardiovascular disease and stroke is entirely consistent with our previous
reports.[1-3]

At her initial evaluation, the proband's total cholesterol was 167,
triglyceride 145, HDL-C 30, apo Al 80, and LDL-C 108 mg/dl. The isolated,
severe depression of HDL-C and apo$_1$Al were characteristic of primary and
familial hypoalphalipoproteinemia.[1] There was no evidence for secondary
hypoalphalipoproteinemia, and no evidence of hypertriglyceridemia.
Fifteen of 17 (88%) of the proband's first degree relatives, 21 of 40
(53%) second degree relatives, and 4 of 9 (44%) of third degree relatives
were studied. As displayed in Figure 2, the distribution of
HDL-cholesterol in her first and second degree relatives was shifted horizontally and towards the lower end of HDL-C distribution, with 17 of the
36 relatives (47%) having HDL-cholesterol levels < 39 mg/dl (Figure 2).
As displayed in Figure 3, only 3 of the 36 first and second degree relatives had triglyceride levels > 300, and 4 were in the range of 225-300
md/dl; the majority had triglyceride levels in the low normal range. Of
15 first degree relatives of the proband, 9 had bottom decile HDL-C; of 21
second degree relatives of the proband, 10 had bottom decile HDL-C. Of 36
first and second degree relatives, 27 had triglyceride levels < 90th
percentile; 9 had triglyceride levels > 90th percentile.

There was four generation vertical and horizontal transmission of
bottom decile HDL-C. In 10 critical matings (FHA by other), 27 of 34
offspring were sampled; the ratio of those with HDL-C < the 10th percentile to those with HDL-C > the 10th percentile was 17/$\overline{10}$ (1.7), not significantly different from 1.0, X^2 = 1.81, p > 0.1, the ratio predicted for a

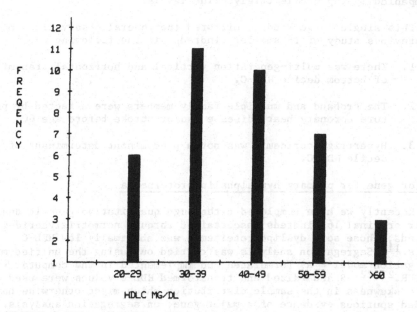

FIGURE 2: Distribution of high density lipoprotein cholesterol
(mg/dl) in 36 first and second degree relatives from
a 51 member kindred with familial hypoalphalipoproteinemia.

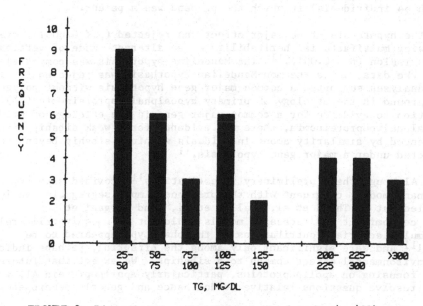

FIGURE 3: Distribution of triglyceride levels (mg/dl) in
36 first and second degree relatives in a 51
member kindred with familial hypoalphalipoproteinemia.

dominant trait. As noted above, bottom decile HDL-C was usually not accompanied by top decile triglyceride levels.

This single large kindred mirrored the general observations made in our previous study of 16 smaller kindreds[1] in the following ways:

1. There was multi-generation vertical and horizontal transmission of bottom decile HDL-C.

2. The proband and multiple family members were affected by premature coronary heart disease and/or stroke before age 60.

3. Hypertriglyceridemia was not a predominant determinant of bottom decile HDL-C.

A major gene for primary hypoalphalipoproteinemia

Recently we have completed a thorough quantitative genetic analysis of our original 16 kindreds[1] ascertained through normotriglyceridemic probands, whose sole dyslipoproteinemia was abnormally low HDL-C levels.[11] Segregation analysis was carried out using the unified mixed segregation model[20] as formulated and implemented in the computer program POINTER.[21,22] Standardized and transformed HDL-C values were used to eliminate skewness in the sample distribution which might otherwise have yielded spurious evidence of a major gene, on segregation analysis.[11]

The unified mixed segregation analysis model[20] postulates that a phenotype is composed of the independent and additive contributions from a major transmissible effect, a multifactorial background, and a residual which is assumed to be unique for every individual.[11] The major effect is assumed to result from segregation at a single bi-allelic locus.

In order to reduce the problems of heterogeneity in ascertainment, we carried out segregation analyses in 14 of the original 16 nuclear families (with 64 individuals) in which the proband was a parent.

The hypothesis of no major effect was rejected ($p < 0.0001$), even when allowing multifactorial heritability as an alternate mode of vertical transmission ($p < 0.001$).[11] The Mendelian hypothesis was compatible with the data, while the non-Mendelian hypothesis was rejected.[11] Thus, our analyses supported a common major gene hypothesis without polygenic background in the etiology of primary hypoalphalipoproteinemia (11). In addition to evidence for a common major gene in the etiology of primary hypoalphalipoproteinemia, there was evidence for a weak sibship effect, evidenced by similarity among individuals within a sibship beyond that expected under a major gene hypothesis.[11]

Although these preliminary investigations[11] provided support for a dominant model, congruent with the previous simple segregation analyses carried out by Third et al,[1] Malloy et al,[4] and Vergani et al,[8] co-dominant and recessive models could not be unequivocally rejected. Any multifactorial contributions to the phenotype appeared to be small[11] and were significant only among the offspring, perhaps indicating an environmental effect common to a sibships. We expect that future studies focusing on apolipoproteins, particularly apolipoprotein A1, will help resolve questions relative to dominance and genetic heterogeneity.

The clinical significance of the identification of a major gene for primary hypoalphalipoproteinemia relates to its potent inverse relationship with premature atherosclerotic, coronary and cerebrovascular disease. Family studies of HDL-C, perhaps in conjunction with other

90

markers such as apolipoprotein Al, should prove to be highly valuable in the identification of individuals at risk for accelerated atherosclerosis.

REFERENCES

1. J.L.H.C. Third, J. Montag, M. Flynn, J. Freidel, P. Laskarzewski, and C.J. Glueck, Primary and familial hypoalphaliporoteinemia. Metabolism 3:16-146, 1984.
2. C.J. Glueck, P. Laskarzewski, D.C. Rao, and J.A. Morrison, Familial aggregations of coronary risk, in "Complications in Coronary Heart Disease", edited by Connor W and Bristow D, Lippincott Co., 1985, pp 173-193.
3. S.R. Daniels, S. Bates, R. Lukin, C. Benton, J.H.L.C. Third, and C.J. Glueck, Cerebrovascular arteriopathy (arteriosclerosis) and ischemic childhood stroke. Stroke 1:360-365, 1982.
4. M.J. Malloy and J.P. Kane, Hypolipidemia. Med. Clin. N. Amer. 66:469-484, 1982.
5. W.P. Castelli, J.T. Doyle, T. Gordon, C.J. Hames, M.C. Hjortland, S.B. Hulley, A. Kagan, and W.J. Zuigi, HDL cholesterol and other lipids in coronary heart disease. The cooperative lipoprotein phenotyping study. Circulation 55:767-772, 1977.
6. G. Heiss, N.J. Johnson, S. Reiland, C.G. Davis, and H.A. Tyroler, The epidemiology of plasma high density lipoprotein cholesterol levels. Circulation 62 (suppl. IV):116-136, 1980.
7. D.C. Rao, P.M. Laskarzewski, J.A. Morrison, P. Khoury, K. Kelly, R. Wette, J. Russell, and C.J. Glueck, The Cincinnati Lipid Research Clinic Family Study: Cultural and biological determinants of lipids and lipoprotein concentrations. Am. J. Hum. Genet. 34:888-903, 1982.
8. C. Vergani and G. Bettale, Familial hypo-alpha-lipoproteinemia. Clin. Chim. Acta 114:45-52, 1981.
9. P.J. Byard, I.B. Borecki, C.J. Glueck, P.M. Laskarzewski, J.L.H.C. Third, and D.C. Rao, A genetic study of hypoalphalipoproteinemia. Genetic Epidemiology 1:43-51, 1984.
10. J. Third and C.J. Glueck, Normotriglyceridemic primary and familial hypoalphalipoproteinemia and coronary heart disease. Circulation 68(4):756, 1983; Arteriosclerosis 3(5):A486, 1983. Presented at the 56th Scientific Session of the American Heart Association Nov., 1983.
11. I.B. Borecki, D.C. Rao, J.H.L.C. Third, P. Laskarzewski, and C.J. Glueck, A major gene for primary hypoalphalipoproteinemia. Am. J. Hum. Genet. in review, 1985.
12. L.H. Myers, N.R. Phillip, and R.J. Havel, Mathematical evaluation of methods for estimation of the concentration of the major lipid components of human lipoproteins. J. Lab. Clin. Med. 88:491-505, 1976.
13. J.A. Morrison, P. Khoury, P.M. Laskarzewski, M.J. Mellies, K. Kelly, and C.J. Glueck, Intrafamilial association of lipids and lipoproteins in kindreds with hypertriglyceridemic probands: The Princeton School Family Study. Circulation 66:67-76, 1982.
14. P.O Kwiterovich, D.S. Fredrickson, and R.I. Levy, Familial hypercholesterolemia (one form of familial type II hyperlipoproteinemia). J. Clin. Invest. 53:1237-1249, 1974.
15. C.J. Glueck, Influence of Gemifbrozil on high-density lipoproteins. Am. J. Cardiology, 52(4):31B-35B, 1983.
16. J. Marks, Dyslipoproteinaemia - Aspects of Gemfibrozil. Research and Clinical Forums. 4(#2):5-133, 1982.
17. C.B. Laurell, Electroimmunoassay of apolipoprotein Al usng high titer monospecific antibodies. Scand. J. Clin. Lab. Inv. 29(Suppl. 124):21, 1972).

18. G. Schonfeld, J.S. Chen, W.F. McDonnel, and I. Jeng, Apolipoprotein-A-II content of human plasma and high density lipoprotens measured by radioimmunoassay. J. Lipid Res. 18:645-655, 1977.
19. T. Gordon and W.B. Cannel, Multiple risk factors for predicting coronary heart disease. The concpt, accuracy, and application. Am. Heart J. 103:1031-1039, 1982.
20. J. M. Lalouel, D.C. Rao, N.E. Morton, and R.C. Elston, A unified model for complex segregation analysis. Am. J. Hum. Genet. 35:816-826, 1983.
21. J.M. Lalouel and N.E. Morton, Complex segregation analysis with pointers. Hum. Heret. 31:312-321, 1981.
22. N.E. Morton, D.C. Rao, and J.M. Lalouel, Methods in Genetic Epidemiology. Basel, Switzerland; S. Karger, 1983.

A FAMILY STUDY OF HYPOALPHALIPOPROTEINEMIA

C. Vergani[x], A. L. Catapano[xx] and A. Sidoli[xxx]

[x]Third Department of Clinical Medicine University of Milan
[xx]Institute of Pharmacology and Pharmacognosy, University of Milan
[xxx]Laboratory of Molecular Biology, Farmitalia Carlo Erba Milan

In recent years the inverse relationship between HDL cholesterol and the risk of coronary artery disease (CAD) has been supported by a lot of clinical and epidemiological data (1, 2).

However, the mechanisms underlying this relationship are not completely clear yet. It is currently believed that HDL might act as a scavenger of cholesterol from the cells and then deliver it to the liver either directly or via a transfer of cholesterol to other lipoproteins (3). Recent studies in vivo have further substantiated this hypothesis (4, 5).

Several factors are involved in the regulation of HDL plasma levels, its synthesis being modulated by exogenous factors (smoking, overweight, alcohol intake, physical activity, etc.) as well as by genetic ones as supported by studies on relatives of subjects suffering from myocardial infarction and by twin studies (6, 7, 8).

Recently the genes coding for the main components of HDL have been located on chromosome 11 (Apo AI) (9) and 1 (Apo AII) (10).

The gene for Apo CIII has also been mapped on chromosome 11, closely linked to the Apo AI gene (9). Alterations in the Apo AI-CIII gene complex resulted in Apo AI deficiency and premature CAD (11).

The study of conditions with low levels of HDL may help in explaining the role of HDL in relation to CAD. Some years ago we described in a three generation family a syndrome (Familial Hypoalphalipoproteinemia) characterized by a primary reduction of plasma HDL (HDL cholesterol below the 10th percentile of normal population, corresponding to 33 mg/dl) which is not associated to any other lipoprotein alterations (12). This syndrome is coupled with a high incidence of premature myocardial infarction and sudden death and is transmitted as an autosomal dominant trait. We summarize here the clinical, biochemical and genetic data obtained in studying this family.

DESCRIPTION OF THE FAMILY

The three generation family originally described affected by hypoalpha lipoproteinemia lives near Milan. The proband in the second generation had myocardial infarction at age of 37. One brother and one cousin had myocardial infarction at 44 and 48 respectively, two cousins died suddenly at 36 and 51 years of age. Sudden death and deaths caused by myocardial infarction are also reported in some first generation relatives (age of death between 49 and 58). All the subjects in the second generation with CAD are hypoalpha. However, 4 hypoalpha subjects, one fertile woman in the second-generation and one girl and two males in the third-generation are apparently healthy. Coronary angiography in one of these males showed a narrowing of more than 70% of the left anterior descending coronary artery (fig. 1) Hypoalpha members of the family were not overweight ($\bar{x}$ =105% + 17 of the ideal body weight), smoked less than 20 cigarettes per day, drank less than 40 g of alcohol per day and had no parenchimal or obstructive liver disease. Neither did they have any of the clinical features observed in other syndromes related to severe HDL deficiency (corneal opacity, planar xantoma, abnormal tonsils, neuropathy, hepatosplenomegaly).

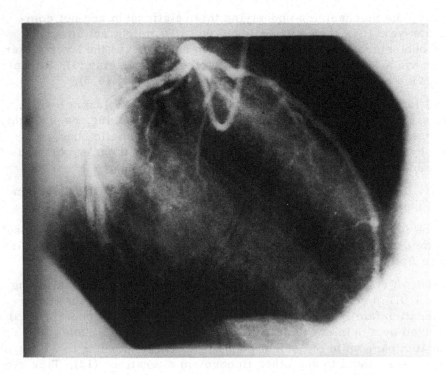

Fig. 1. Coronary arteriogram of a completely asymptomatic 31 year old subject with hypoalphalipoproteinemia (III-5). A severe stenosis of the left anterior descending coronary artery is observed.

BIOCHEMICAL DATA

The hallmark of the disease is a low HDL cholesterol level without changes of plasma levels of VLDL, LDL cholesterol and triglycerides (12). Plasma apolipoprotein levels are also affected. In particular Apo AI , Apo AII and Apo CII are reduced while Apo B, Apo CIII and Apo E are within the normal range (table 1). In one hypoalpha subject bidimensional immunoelectrophoresis shows a reduced Apo A peak (fig. 2). Analysis of the Apo AI isoforms by bidimensional electrophoresis in polyacrylamide gel shows that the isoform distribution is normal (Fig. 3). Study of HDL subfractions by rate zonal ultracentrifugation shows reduced levels of HDL2 (fig. 4). However this cannot be considered peculiar to this disease, since low levels of HDL2 are also found in patients with secondary hypoalphalipoproteinemia. The chemical composition of HDL is normal as reported in table 2. Analysis of HDL by electron microscopy revealed normally shaped particles with an average diameter of about 80 Å (fig. 5). No abnormalities could be detected in VLDL and LDL (data not shown).

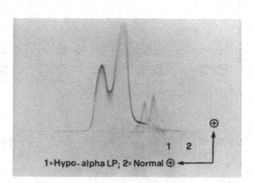

1=Hypo- alpha LP; 2= Normal ⊕ ◄

Fig. 2. Bidimensional immunoelectrophoresis of a hypoalpha serum and pooled normal sera. Antisera to human HDL and LDL were used. A marked reduction of the HDL peak is detected in the hypoalpha as compared to normal. The LDL peak is also slightly reduced in the hypoalpha serum.

Table 1 - Plasma lipid. HDL cholesterol and apolipoprotein levels in hypoalpha and normoalpha blood-related members of the family. Control values are also indicated. Values are means + S. D. (mg/dl). CH = total cholesterol; TG = triglycerides; HDL-C = HDL cholesterol; p = refers to hypoalpha vs controls; NS = not significant

		CH	TG	HDL-C	apoAI	apoAII	apoB	apoCII	apoCIII	apoE
Hypoalpha (n = 6)	x̄	166.8	87.3	31.6	83.5	25.6	86.6	1.6	6.9	3.5
	S.D.	41.5	26.6	2.3	7.8	3.1	13.3	0.4	1.6	1.1
Normoalpha (n = 6)	x̄	177.0	66.6	49.0	116.2	40.6	80.2	2.2	5.6	4.4
	S.D.	23.9	6.4	6.7	9.8	9.5	26.8	0.3	1.2	0.3
Controls										
Males (n= 30)	x̄	179.0	88.0	47.0	135.0	36.0	94.0	2.8	7.3	3.1
	S.D.	27.0	32.0	9.0	21.0	6.0	19.0	1.2	1.5	0.9
Females (n= 30)	x̄	159.0	68.0	52.0	142.0	40.0	79.0	2.2	5.6	3.6
	S.D.	29.0	29.0	8.0	19.0	6.0	15.0	0.8	2.1	0.9
	p	NS	NS	< 0.001	< 0.001	< 0.001	NS	< 0.01	NS	NS

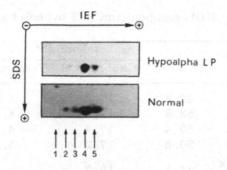

Fig. 3. Bidimensional polyacrylamide gel electrophoresis of Apo AI from
a hypoalphalipoproteinemic subject and from a normal subject.
A normal pattern with the presence of the mature Apo AI isoforms
(Apo AI$_{3-5}$) is observed in the hypoalpha subject.

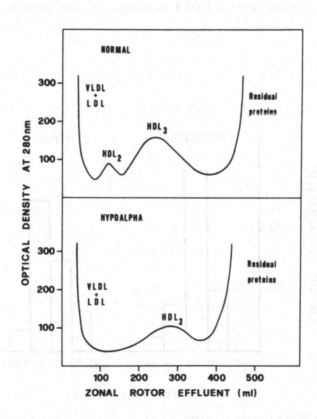

Fig. 4. HDL subfractions in a normal plasma and in the plasma obtained
from a subject with hypoalphalipoproteinemia after rate zonal
ultracentrifugation. HDL2 peak is absent in the hypoalpha subject.

Table 2 – HDL composition in 3 hypoalpha subjects

		% Total weight			
		Protein	Cholesterol	Triglyceride	Phospholipide
Subjects	1	49. 9	17. 2	6	26. 9
	2	52. 3	17. 1	4. 7	26
	3	50. 4	17. 4	4. 9	27. 3
	mean	50. 8	17. 2	5. 2	26. 7
Controls	HDL2[x]	43. 6	19. 7	9. 1	28. 6
	HDL3[x]	52. 4	16. 3	4. 3	27

[x] Values represent the mean of three controls.

In agreement with these observations the activities of hepatic and extra hepatic lipases are normal as well as the LCAT activity (12). To further study the functionality of the HDL obtained from these patients we posed the question whether the HDL interaction with the binding sites on human liver membrane could be increased. The results of these studies are reported in fig. 6. Lipoproteins from one hypoalpha subject are as effective as control HDL3 in competing for these binding sites.

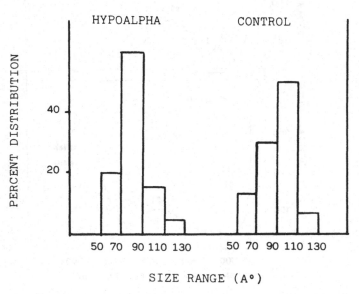

Fig. 5. Size distribution of HDL from a control and a hypoalpha subject. The size distribution was obtained from negatively stained lipo- proteins. 200 free standing particles were measured.

Furthermore Apo AI isolated from a hypoalpha subject and tested in two volunteers is removed from plasma as efficiently as control Apo AI (13). Altogether these findings suggest that the Apo AI from the hypoalpha patients is normal and behaves normally. A question that remains to be answered is whether these patients are hypoalpha because of a defective synthesis or of an increased catabolism of HDL.

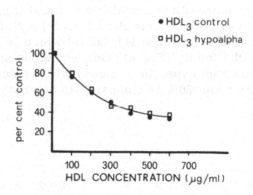

Fig. 6. Competition between control and hypoalpha HDL3 for the binding of control ^{125}I HDL3 to human liver membranes. ^{125}I HDL3 concentration was 10 μg/ml, membrane protein 1 mg/ml. Incubations were carried out at 0°C for 1 hour with or without cold HDL3. Bound lipoproteins were separated from the unbound by centrifugation at 100,000 x g for 30 minutes.

GENETICS

The analysis of 7 critical available matings of hypoalpha with normoalpha subjects, with their 19 offsprings, revealed a ratio of hypoalpha normal of 8/11, not significantly different from a ratio consistent with autosomal dominant transmission (X^2 = 0.46).

The vertical transmission and the delayed appearance of the fenotype are also compatible with this mode of transmission.

Studies on other families suggest that familial hypoalphalipoproteinemia is a hereditary disorder with a pattern of transmission not significantly different from that expected by a hypothesis of Mendelian dominance (14).

RECOMBINANT DNA STUDIES

We performed Southern blotting analysis using different restriction endonucleases on DNA samples from the available members of this family (twelve related members - 6 hypoalpha and 6 normoalpha - and 6 unrelated members) (15).

In particular we focused on the Apo AI and Apo AII genes, looking for the presence of restriction fragment length polymorphisms (RFLPs). When DNA samples were digested with Pst I and hybridized with the Apo AI probe, three types of individuals were present in the family in accordance with Mendelian inheritance : homozygote for 3.3. Kilobase (Kb) band, homozygote for a 2.2 Kb band, and heterozygote for the presence of both bands. After Sac I digestion and hybridization with the Apo AI probe, homozygote for a 3.2 Kb band, homozygote for a 4.2 Kb band, and heterozygote for both bands were observed. We also used the enzyme Msp I to look for RFLPs related to the Apo AII gene. After digestion of DNA samples with this enzyme, followed by hybridization with the Apo AII probe homozygote for a 3.5 Kb band, homozygote for a 3.0 Kb band, and heterozygote for the presence of both bands were found (fig. 7). Analysis of the segregation of these polymorphic Pst I (Pst I+) with a normal Sac I (Sac I-) allele defines a haplotype (Pst I+/ Sac I-) which is present in all hypoalpha members of the family (Fig. 8). One hypoalpha subject (II-9) was homozygous for this haplotype. In this case HDL cholesterol, Apo AI and Apo AII levels were about 80% as compared to heterozygotes.

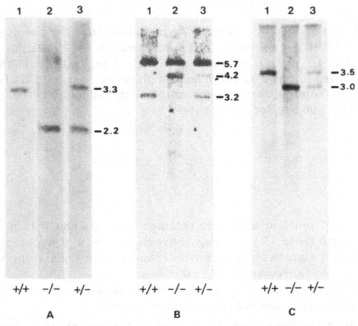

Fig. 7. Autoradiograms showing Southern blot analysis.
A) DNA samples digested with Pst I and hybridized to the Apo AI probe : 1) homozygote for 3.3 (Kb (+/+), 2) homozygote for 2.2 Kb (-/-), 3) heterozygote for both bands (+/-). B) DNA samples digested with Sac I and hybridized to the Apo AI probe : 1) homozygote for 3.2 Kb (+/+), 2) homozygote for 4.2 Kb(-/-),3) heterozygote for both bands (+/-). C) DNA samples digested with Msp I and hybridized to the Apo AII probe: 1) homozygote for 3.5 Kb(+/+), 2) homozygote for 3.0 Kb(-/-), 3) heterozygote for both bands (+/-).

Two carriers of the haplotype (II-5 and III-6) are at present normo-alpha. The allelic frequency of the Pst I+/Sac I- haplotype in our population, defined by analyzing 20 unrelated individuals,was 2.5%.

The 3.5 Kb polymorphic allele observed after Msp I digestion and hybridization with the Apo AII probe is present in 15% of normal subjects in our population and does not segregate with the hypoalpha state in this family.

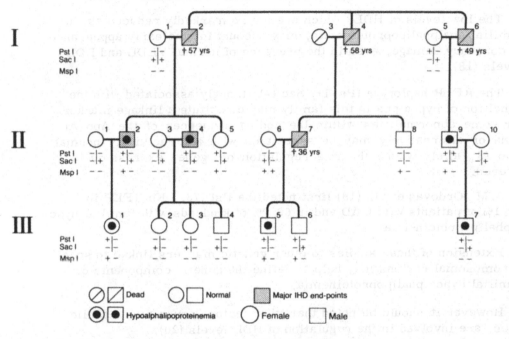

Fig. 8. Kindred with familial hypoalphalipoproteinemia.
Family study showing genotypes resulting from DNA analysis with Pst I and Sac I (Apo AI probe), and Msp I (Apo AII probe). Haplotypes defined by Pst I and Sac I are indicated on the two sides of the vertical line. Hypoalpha phenotype and clinical features are also indicated. Major ischemic heart disease (IHD). end-points mean premature myocardial infarction or sudden death.

Scott et al.(16) reported that this polymorphism is often associated with increased levels of Apo AII and with altered HDL composition. Our data do not seem to be consistent with this observation, moreover Breier et al. (17) found that Apo AII plasma concentration is not primarily related to CAD.

CONCLUSIONS

Familial hypoalphalipoproteinemia is a genetic disorder and the criteria for its definition should include the following :

1) presence of HDL concentration below the 10th percentile of normal population;

2) absence of diseases or factors to which hypoalpha might be secondary;

3) detection of a similar lipoprotein pattern in a first degree relative.

This syndrome is transmitted as an autosomal dominant trait.

The low levels of HDL, which are not so markedly reduced as in familial analphalipoproteinemia, may account for the early appearance of coronary damage, even in the presence of normal VLDL and LDL levels (18).

The RFLP haplotype (Pst I+/ Sac I-) strongly associated with the condition of hypoalpha in this family may constitute a linkage marker for some abnormalities within the coding sequences of the Apo AI gene or alternatively may be associated with a structurally normal Apo AI allele where the fine regulation of gene expression is altered.

J. M. Ordovas et al. (19) first noted the Pst I/3.3 Kb RFLP in 32.1% of patients with CAD and in 66.7% of kindreds with familial hypoalphalipoproteinemia.

Extension of these studies to other genetic markers linked to specific chromosomal regions may help to define the genetic components of familial hypoalphalipoproteinemia.

However it should be noted that other factors, such as metabolic ones, are involved in the regulation of HDL levels (20).

Further studies of the kinetics of Apo AI in the affected members of the family should help to clarify the HDL metabolism in this pathological condition.

REFERENCES

1. G.J. Miller and N.F.Miller. Plasma high density lipoprotein concentration and development of ischaemic heart disease. Lancet 1:16-20(1975)
2. T. Gordon, W.P. Castelli, M.C. Hjortland, W.B. Kannel, and T.R. Dawber. High density lipoprotein as a protective factor against coronary heart disease. The Framingham Study. Am J Med , 62:707-14(1977) .
3. J. A. Glomset. The plasma lecithin; cholesterol acyltransferase reaction. J Lipid Res 9:155-67 (1968).
4. C. Vergani, A.C.Plancher, M. Zuin et al. Bile lipid composition and haemostatic variables in a case of high density lipoprotein deficiency (Tangier disease). Eur J Clin Invest 14:49-54 (1984).

5. N.E. Miller, A. La Ville, and D. Crook. Direct evidence that reverse cholesterol transport is mediated by high-density lipoprotein in rabbit. Nature 314:109-11 (1985).

6. A. Heiberg. The heritability of serum lipoprotein and lipid concentrations. A twin study. Clin Genet 6:307-16 (1974).

7. H. Micheli, D. Pometta, and C. Jornot. High density lipoprotein cholesterol in male relatives of patients with coronary heart disease. Atherosclerosis 32:269-76 (1979).

8. M.S. Nupuf and W.H.F. Sutherland. High density lipoprotein levels in children of young men with ischaemic heart disease. Atherosclerosis 33:365-70 (1979).

9. S.W. Law, G. Gray, H.B. Jr. Brewer, A. Sakaguchi and S. Naylor. Human apolipoprotein A-I and C-III genes reside in p11-q13 region of chromosome 11. Biochem Biophys Res Commun 118:934-42(1984).

10. T.J. Knott, R.L. Eddy, M.E. Robertson, L.M. Priestley, J. Scott and T.B. Schows. Chromosomal localization of the human apoprotein CI gene and of a polymorphic apoprotein AII gene. Biochem Biophys Res Commun 125:299-305 (1984).

11. R.A. Norum, J.B. Lakier, S. Goldstein et al. Familial deficiency of apolipoproteins A-I and C-III and precocious coronary-artery disease. N Engl J Med 306:1513-9 (1982).

12. C. Vergani and G. Bettale. Familial hypo-alpha-lipoproteinemia. Clin Chim Acta 114:45-52 (1981).

13. E.J. Schaefer. Clinical, biochemical and genetic features in familial disorders of high density lipoprotein deficiency. Arteriosclerosis 4:303-22 (1984).

14. J.L. Third, J. Montag, M. Flynn, J. Freidel, P. Laskarzewski and C.J. Gluek. Primary and familial high density lipoproteins : their metabolic properties and response to drug therapy. Biochim Biophys Acta, 751:175-88 (1983).

15. A. Sidoli, G. Giudici, M. Soria and C. Vergani. RFLP haplotype in the AI-CIII gene complex occurring in a family with hypoalphalipoproteinemia. In press.

16. J. Scott, T.J. Knott, L.M. Priestley, M.E. Robertson, D.V. Mann, G. Kostner, G.J. Miller and N.E. Miller. High-density lipoprotein composition is altered by a common DNA polymorphism adjacent to apoprotein AII gene in man. Lancet 1:771-3 (1985).

17. C.H. Breier, V. Mühlberger, E. Knapp et al. Apolipoprotein AII and coronary artery disease. Lancet 1:1339 (1985).

18. S. Eisenberg. High density lipoprotein metabolism. J Lipid Res 25:1017-58 (1984).

19. J.M. Ordovas, E.J. Schaefer, D. Salem et al. Apolipoprotein A-I gene polymorphism in the 3' flanking region associated with familial hypoalphalipoproteinemia and premature coronary artery disease. In press.

20. R.M. Krauss. Regulation of high density lipoprotein levels. Med Clin N Am 66 : 403-30 (1982).

APOPROTEIN A-I AND LECITHIN: CHOLESTEROL ACYLTRANSFERASE

IN A PATIENT WITH TANGIER DISEASE

P. Haydn Pritchard and Jiri Frohlich

Department of Pathology
Shaughnessy Hospital Lipid Research Group
University of British Columbia
Vancouver, B.C. Canada

INTRODUCTION

Tangier disease is a rare autosomal recessive disorder characterized by a virtual absence of high density lipoprotein (HDL) and an accumulation of cholesteryl esters in a number of peripheral tissues (1). The exact genetic defect has not yet been fully elucidated.

During the last 2 years we have had the unique opportunity to investigate the only known patient in Canada with the homozygous form of Tangier disease. Although this patient demonstrates many of the classical features of this disorder, distinct differences from other patients indicate a degree of heterogeneity in Tangier disease. Our studies have been primarily directed towards an understanding of the role of HDL in reverse cholesterol transport in this patient.

CASE REPORT

A 56 year old man presented in 1982 with a progressing neuropathy of the hands and feet. His past history included a tonsillectomy in childhood and a splenectomy in 1964. Absence of HDL on lipoprotein electrophoresis was indicative of Tangier disease. Histological examination indicated the accumulation of a large number of foam cells in the liver, skin, rectal mucosa and in the peripheral nerves. He is the son of first cousins and although he married, he has no children of his own.

The marked decrease in apo A-I (Table 1) was indicative of the Tangier disease. Apo A-II levels were less affected and probably reflects the independent metabolism of apo A-I and A-II. The reduction of apo D levels to 50% of normal is probably due to the absence of normal HDL but it may also be related to abnormalities lecithin: cholesterol acyltransferase (LCAT)/ apo D / apo A-I complex that has been described in normal human plasma. His consistent hypertriglyceridemia underscores the role of HDL in the lipolysis of triglyceride rich particles. However, a decrease in lipoprotein lipase (LPL) activity (table 1), brought about by a decrease in the specific activity of this enzyme (J. Brunzell, unpublished) was also observed. It should be noted that sufficient apo C-II for LPL activation is present in his plasma and that hepatic

Table 1. Apoproteins, lipids and enzyme activities in Tangier plasma.

		Tangier	Controls
Apo A-1	(mg/dl)	2.3*	154±25
Apo A-11	(")	15.7*	65±8
Apo B	(")	85.5*	104±18
Apo C-1	(")	3.9*	8±2
Apo C-11	(")	1.5	3±0.7
Apo C-111	(")	6.8*	7±2
Apo D	(")	6.0	13±3
Apo E	(")	10.3	10±4
Total cholesterol	(")	76*	198±32
Total triglyceride	(")	642*	126±77
LCAT activity	(nmol/h/ml)	13.3*	28±2.3
LCAT mass	(mg/1)	1.2*	6±0.5
LPL activity	(umol/min/ml)	3.7*	16.8±3.1
HTGL activity	(umol/min/ml)	19.7	34.9±15.0

*: greater than 2 SD from mean of control value

triglyceride lipase (HTGL) was not markedly decreased. LCAT activity is decreased in this patient and in other patients with Tangier disease (2). In our patient, this is due to a decrease in LCAT mass (Table 1) rather than a mere decrease in apo A-I activator as addition of pure apo A-I to his plasma did not restore the activity to normal.

REVERSE CHOLESTEROL TRANSPORT IN TANGIER DISEASE

We were interested in the mechanism by which cholesteryl esters accumulate in the foamy cells of peripheral tissues of patients with Tangier disease. Therefore, we have initiated a series of investigations designed to study the reverse cholesterol transport system in this disorder and to elucidate whether the hypercatabolism of HDL in these patients is due to a cellular defect or the result of some plasma modification of HDL. This report describes some of our studies concerning the action of LCAT in Tangier disease.

Our investigation began with the infusion of 31 of normal plasma by plasmapheresis. In this way, we were able to study the fate of a relatively large amount of normal HDL in the plasma compartment of our Tangier disease patient. Figure 1 shows that his LCAT activity prior to he infusion was 22 nmoles/h/ml using an exogenous proteoliposome substrate. The LCAT activity and apo A-1 concentration of the infused plasma were 24.8 nmoles/h/ml and 127 mg/dl respectively.

Following plasmapheresis, the apo A-I level increased but failed to reach a value within the normal range. The level decreased rapidly over the next 3 days suggesting that the infused HDL was rapidly removed from the circulation. This observation supports the hypercatabolism hypothesis for absence of HDL in Tangier disease. The exact fate of the infused HDL cannot be determined from these experiments. However, as we have observed the occurrence of a large amount of immunoreactive apo A-I in the foam cells of a liver biopsy of this patient, we speculate that these hepatic cells may be one of the sites of HDL degradation. Experiments with a rat model suggest that the kidney is an important site of apo A-I degradation

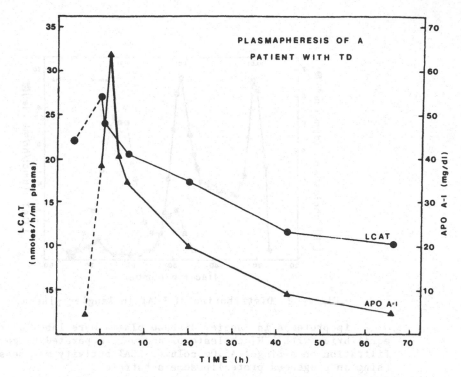

Figure 1. Plasmapheresis of a patient with Tangier disease.

3L of normal plasma were infused over a 4h period. T=0
indicates the end of the infusion period. Initial values for apo A–I
levels and LCAT activity are indicated in the figure and the
corresponding values for the infused plasma are given in the text.

(3). However, neither apo A–I or its precursors forms are excreted in the
urine of this patient (5). Thus, our observations do not indicate that
the kidney is responsible for the hypercatabolism of HDL. Additional
experiments on the binding and degradation of normal HDL by fibroblasts
and peripheral monocytes cultured from our patient with Tangier disease
do not indicate major differences in HDL metabolism. Thus, the site and
mechanism for the hypercatabolism of HDL in Tangier disease remains
elusive.

LCAT ACTIVITY IN TANGIER PLASMA

In the time following the plasmapheresis, we noted that LCAT
activity decreased to approximately 50% of the initial value following
HDL infusion (Figure 1). In an attempt explain this observation, we
continued our study with an investigation of the characterisics of the
LCAT reaction in Tangier disease plasma.

Figure 2 shows the separation of Tangier disease lipoproteins by gel
filtration using equilibrated [^{3}H] cholesterol to estimate the
distribution of lipoprotein cholesterol. VLDL (and other triglyceride-
rich particles) and LDL can be identified as well as a larger
subpopulation of triglyceride- rich LDL. The absence of normal sized HDL
is typical of Tangier disease. Despite this lack of HDL, LCAT activity is
associated with an HDL sized particle with a small amount being recovered
in LDL.

107

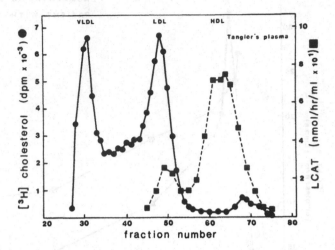

Figure 2. Distribution of LCAT in Tangier plasma.

Lipoproteins in Tangier disease plasma were labelled to equilibrium with [³H] cholesterol and were separated by gel filtration on a Biogel A5.0m column. LCAT activity was measured using an exogenous proteoliposome substrate.

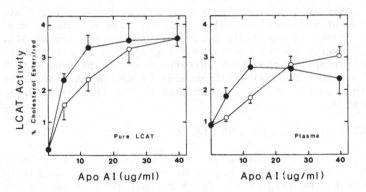

Figure 3. Activation of LCAT by control and Tangier apo A-I.

Apo A-I from control and Tangier plasma was purified by immunoabsorbtion to an anti-A-I affinity column followed by elution with 1M acetic acid. Purified LCAT or plasma were used as a source of enzyme and the results are indicated as follows: Control (●) (mean±SD(n=3)); Tangier (○) (mean±range(n=2)).

In another series of experiments we have shown that LCAT is active in Tangier disease plasma without the addition of exogenous A-I. Furthermore, LCAT also operates in conjunction with cholesteryl ester transfer protein to synthesize esters and transfer them to higher molecular weight lipoproteins (5).

Unlike apo A-I purified from Tangier patients demonstrating abnormal apo A-1 structure (6), apo A-1 isolated from our patients plasma was functionally normal in terms of activation of pure or plasma LCAT (figure 3). Although Tangier disease apo A-I contains a large proportion of isoforms 1 and 2, maximum activation was the same as control A-1. Other workers have reported that isoforms 1 and 2 activate LCAT normally (6).

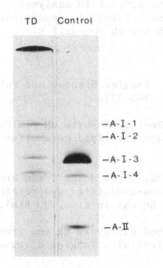

Figure 4 . Isoelectric focussing gel of apoproteins removed plasma by an anti-A-I affinity column.

Aliquots of control and Tangier plasma were applied to an anti A-I affinity column and the bound lipoprotein eluted with 1.0M acetic acid. After deplipidation, the proteins were applied to isoelectric focussing gels (pH 4.0-6.5) and the resulting bands were visualized with Commassie Blue after separation

LIPOPROTEIN LOCATION OF APO A-I AND LCAT IN TANGIER PLASMA

Figure 4 shows the apoproteins associated with apo A-I containing particles isolated from control and Tangier disease plasma by immunoaffinity chromatography. A-I from controls was was primarily isoforms 3 & 4 and was associated with A-II with 90% of the plasma LCAT activity. Thus, we assume that we have isolated a normal HDL particle.

However, Tangier disease apoprotein A-I is not associated with A-II nor is it associated with LCAT activity as 90-100% of LCAT activity was not bound by the anti A-1 affinity column.

Thus, from these experiments we conclude that LCAT in Tangier disease appears to be functionally normal despite the absence of normal HDL and apo A-I. In addition, apo A-I in our patient is functionally normal but is not associated with an apo A-II containing lipoprotein. LCAT appears to be located on a unique particle that is HDL-like in size, but does not contain apoprotein A-1. However, it is associated with a particle of some kind that maintains its activity in vivo. It is possible that a portion of LCAT exists on this particle in normal plasma and that in Tangier disease non LCAT containing HDL has been removed. The loss of LCAT activity following plasmapheresis suggests that the LCAT activity equilibrated with this rapidly turning over pool of HDL. This assumedly resulted in the net removal of LCAT protein during this hypercatabolism of the normal HDL.

ACKNOWLEDGEMENTS

The help and encouragement of Drs. J. Frohlich, M. Hayden, A. Lacko and R. McLeod are gratefully appreciated. Dr. P. Alaupovic, Oklahoma, kindly carried out the apoprotein analyses and Dr. J. Albers, Seattle, measured the LCAT mass. Financial support for this research was provided by B.C. Health Care Research Foundation.

REFERENCES

1. Schaefer EJ (1984) Tangier disease and related disorders Atherosclerosis 4, 303-322.

2. Clifton-Bligh P, Nestel PJ Whyte HM (1972) Tangier disease: report of a case and studies on lipid metabolism N Engl J Med 286, 567-571.

3. Class C, Pittman RC, Civen M and Steinberg D (1985) Uptake of High Density Lipoproteins-associated apoprotein A-I and cholesterol esters by 16 tissues in the rat in vivo. J. Biol. Chem. 260, 744-750.

4. Pritchard PH, Mcleod R, Hayden MR and Frohlich J (1985) Urinary proteins in a patient with Tangier disease. Clin. Biochem. 18, 98-101.

5. Pritchard PH, Mah E, McLeod R, Frohlich J and Lacko AG (1985) Lecithin: cholesterol acyltransferase in a patient with Tangier disease. Fed. Proc. 44, 1786

6. Rall SC, Weisengraber KH, Mahley RW, Ogawa Y, Fielding CJ, Utermann G, Haas J, Steinmetz, Menzel H-J, Assmann G. (1984) Abnormal lecithin:cholesterol acyltransferase activation by a human apolipoprotein A-I variant in which a single lysine residue is deleted. J. Biol. Chem. 259, 10063-10070.

SEVERE HYPOALPHALIPOPROTEINEMIA INDUCED BY A COMBINATION OF PROBUCOL AND CLOFIBRATE

Jean Davignon[1], A. Christine Nestruck[1], Petar Alaupovic[2],
and Daniel Bouthillier[1]

[1] Department of Lipid Metabolism and Atherosclerosis
Research, Clinical Research Institute of Montreal
Montreal, Canada
[2] Lipoprotein and Atherosclerosis Research Program
Oklahoma Medical Research Foundation
Oklahoma City

SUMMARY

Observation of a markedly depressed HDL-cholesterol (5 mg/dL) in a
patient with familial hypercholesterolemia (FH) receiving probucol
(1 g/day) and clofibrate (2 g/day) prompted a review of all cases
treated by this combination at our lipid clinic. Hypoalphalipoproteinemia (HDL-C <15 mg/dL) developed in 19 of 28 (70%) hyperlipidemic
subjects who received this combination for an average of 1.5 years.
This effect was sustained and reversible; it did not occur on either
drug alone and was manifested on average 17 weeks after the combination
was started. Plasma triglycerides increased significantly in most of
those patients susceptible to this reduction in HDL-C. Plasma
apolipoprotein A-I was decreased 82%, in proportion to the HDL-C fall,
whereas apo A-II was lowered 65%. Since apo C-III concentrations tended
to be high, the apo A-I/C-III ratio was markedly depressed. Apo E
levels were unchanged and apo B levels reflected the high LDL
concentrations of the underlying disease. An intermediate response was
observed in subjects whose HDL-C remained well above 15 mg/dL on the
combination. No deleterious side-effects could be attributed directly
to the administration of the combined drugs in this high-risk group.
One patient actually showed complete regression of xanthelasma and
extensor tendon xanthomas of the finger on the combination. A parallel
is drawn with Fish-Eye disease and the presence of the apolipoprotein
A-I Milano variant, where similar HDL-C levels are observed in the
absence of an increased atherogenic risk. It is mandatory to monitor
plasma HDL-C in hypercholesterolemic patients treated with this
combination, otherwise the pronounced HDL-deficiency could go unnoticed.

Key words: High density lipoprotein cholesterol, Apolipoprotein A-I,
Probucol, Clofibrate, Hyperlipidemia treatment

INTRODUCTION

Probucol is a well-tolerated cholesterol lowering drug used in the treatment of familial hypercholesterolemia (FH) and familial combined hyperlipidemia (FCH)[1],[2]. Although its major effect is to lower low density lipoprotein-cholesterol (LDL-C), it also lowers plasma high density lipoprotein-cholesterol (HDL-C). In view of the protective role currently ascribed to HDL, the usefulness of probucol in the treatment of hyperlipidemia has been a matter of controversy[3],[4]. However, renewed interest in this drug as a therapeutic agent stems from studies showing regression of atherosclerosis in hyperlipidemic monkeys[5] and discovery of a new mechanism for its cholesterol lowering effect[6]. While probucol has little effect on plasma triglycerides and very low density lipoprotein-cholesterol (VLDL-C), clofibrate has a marked effect on both[7]. The major indication for administration of clofibrate is familial dysbetalipoproteinemia (type III). It has also been found effective in some patients with familial hypertriglyceridemia (type IV) as well as in some FH patients with a IIa lipoprotein electrophoretic phenotype and tendon xanthomas, although a paradoxical LDL-raising effect has also been observed[7],[8],[9]. In FH patients who are responsive to the cholesterol-lowering effect of clofibrate, a rise in plasma HDL-C has been reported[10]. An HDL-raising effect has also been observed in type IV[11]. Thus, it would appear that a combination of both drugs might have a beneficial effect on the plasma lipoprotein profile: the potential LDL-raising effect of clofibrate being opposed by probucol and the HDL-lowering effect of probucol being countered by clofibrate.

The unexpected observation of a fall in HDL-C to 5 mg/dL in a FH patient receiving both clofibrate (2 g/day) and probucol (1 g/day) prompted a review of our lipid clinic records and of the lipoprotein patterns occurring on this combination.

PATIENTS AND METHODS

Subjects. The clinical features and baseline lipoprotein profile of the 28 patients included in this study are given in Table 1. Since there was a bimodal distribution of the HDL-C response to the combination, the patients were divided into two groups; those whose HDL-C remained above 20 mg/dL were entered in group I (n=9), while those whose HDL-C dropped to or below 15 mg/dL were included in group II (n=19). The subgroups did not differ in age or body mass index (BMI). There was one mildly obese subject in each group: two women with BMI's of 29 and 30 during the combination period. Hypertension was equally distributed between the two groups. Twenty three of 28 patients had at least one manifestation of atherosclerosis; these being almost equally distributed except for peripheral vascular disease which tended to be more frequent in group I. Manifestations of hyperlipidemia were also frequent, 23 patients having xanthelasma, arcus corneae or tendon xanthomas, singly or in combination. Xanthelasma tended to be less frequent in group II and there were proportionately more FH in group I. Two patients with mixed hyperlipidemia (type V) and one with type IV had also received the combination of drugs because, after they were seen to respond to clofibrate in terms of triglyceride reduction, probucol was added to achieve a better control of cholesterol and LDL-C. Although plasma lipid determinations were available in all subjects during baseline, lipoprotein data was available in only 22 of the 28 subjects. There were no significant differences between groups I and II but HDL-C was higher in group II, the subjects which later developed the HDL

Table 1. Characteristics of the patients

Variable	Total sample	Group I HDL-C >20 mg/dL	Group II HDL-C <15 mg/dL
Number of subjects (males)	28 (11)	9 (3)	19 (8)
Age at combination	55.3 ± 8.4[b]	53.1 ± 6.9	56.8 ± 8.9
Body mass index[a] females:	25.2 ± 2.9	24.7 ± 3.1	25.4 ± 2.9
males:	24.2 ± 3.4	24.1 ± 4.5	24.2 ± 3.3
Obesity	2 [c]	1	1
Hypertension	7 (25.0)	2 (22.2)	5 (26.3)
Coronary heart disease	12 (42.9)	4 (44.4)	8 (42.1)
Peripheral vascular disease	12 (42.9)	6 (66.6)	6 (31.6)
Cerebrovascular disease	2 (7.0)	1	1
Xanthelasma	11 (39.3)	5 (55.5)	6 (31.6)
Arcus corneae	14 (50.0)	4 (44.4)	10 (52.6)
Tendon xanthomas	14 (50.0)	6 (66.6)	8 (42.1)
FH	15 (53.6)	7 (77.7)	8 (42.1)
FCH	10 (35.7)	2 (22.2)	8 (42.1)
Other hyperlipidemia	3 (10.7)	0	3
Cholesterol (mg/dL)	368 ± 81[b]	394 ± 73	359 ± 85
Triglycerides (mg/dL)	310 ± 375	230 ± 154	252 ± 133[e]
VLDL-C (mg/dL)[d]	54 ± 30	75 ± 27	55 ± 40
LDL-C (mg/dL)	275 ± 72	283 ± 91	295 ± 113
HDL-C (mg/dL)	39 ± 9	35 ± 8	41 ± 9
Number of subjects with:			
Apo E phenotype E3/3	20	6	14
E3/2	3	0	3
E4/3	4	3	1
E4/2	1	0	1

a: BMI = weight /height^2, upper limit of normal is 30 for males
        and 28.6 for females.
b: Values are $\bar{x}$ ± S.D.
c: n (%)
d: Number of subjects in each group for lipoprotein measurements were
        22, 5 and 17 for total sample, group I and group II respectively.
e: Excludes one at 2095 mg/dL.

deficiency. The numbers are too small to ascertain whether either
"response group" was enriched in a specific apo E phenotype; it is
noticeable, however, that the ε2 allele was not represented in group I.

Record review. Most of the patients had been studied for several
years and their hyperlipidemia had been well characterized at the
initial visits. The records were reviewed for a total period of
observation which varied a great deal from patient to patient.

Chronologically the first period considered was a baseline period
(B) on a lipid-lowering diet but off lipid-lowering drugs (13 ± 5 weeks,
range 5-26 weeks). The diet was maintained throughout the various

treatment periods and monitored by changes in body weight and regular visits with the dietician.

During the "first drug period", the patients received either clofibrate (2 g/day) or probucol (1 g/day). To achieve further lipid reduction, the complementary drug was then added during the combination (C) period. A "post combination drug period" in which the initially prescribed drug was discontinued followed in some patients. These periods were sequential in all instances except for 5 subjects (2 in group I and 3 in group II) where the baseline period occurred 5 months to 3 years prior to the administration of the first drug. Analysis of the data showed that the position of the one-drug-period relative to the combination did not affect the lipid-lowering effect: these data were therefore pooled to form a clofibrate period and a probucol period (Figure 1).

The baseline plasma lipid and lipoprotein values are given in Table 2 for the various classes of hyperlipidemia. As expected, plasma cholesterol and LDL-C were highest and VLDL-C and triglycerides lowest in FH. The mean HDL-C was similar for FH and FCH patients but was lower for the hypertriglyceridemics.

<u>Laboratory methods</u>. All blood samples were obtained after 12-14 h of fasting and drawn into Vacutainer tubes containing EDTA (1 mg/mL). Plasma were stored at 4°C until analysis within 3-4 days. Total plasma cholesterol, triglycerides and lipoprotein cholesterol were measured enzymatically[12],[13] using an autoanalyzer (ABA-100, Abbott Laboratories, Pasadena, CA). Lipoproteins were isolated by a combination of ultracentrifugation (d=1.006 g.mL^{-1}) and heparin manganese precipitation

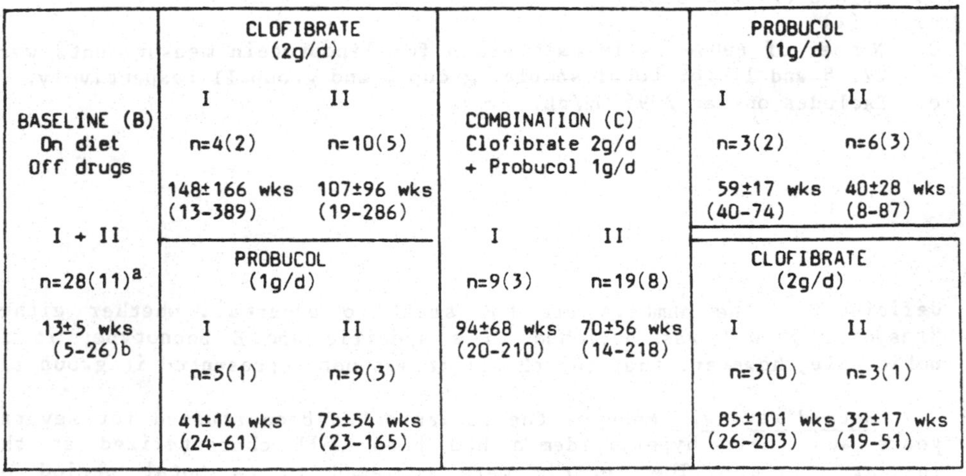

BASELINE (B) On diet Off drugs I + II n=28(11)[a] 13±5 wks (5-26)[b]	CLOFIBRATE (2g/d)		COMBINATION (C) Clofibrate 2g/d + Probucol 1g/d		PROBUCOL (1g/d)	
	I	II			I	II
	n=4(2)	n=10(5)			n=3(2)	n=6(3)
	148±166 wks (13-389)	107±96 wks (19-286)			59±17 wks (40-74)	40±28 wks (8-87)
	PROBUCOL (1g/d)		I	II	CLOFIBRATE (2g/d)	
	I	II	n=9(3)	n=19(8)	I	II
	n=5(1)	n=9(3)	94±68 wks (20-210)	70±56 wks (14-218)	n=3(0)	n=3(1)
	41±14 wks (24-61)	75±54 wks (23-165)			85±101 wks (26-203)	32±17 wks (19-51)

a: number of subjects (no. of males); b: range

Clofibrate periods were pooled: Group I n=7, Group II n=13
Probucol periods were pooled: Group I n=8, Group II n=15

Figure 1. Characteristics of the study periods for groups I and II.

Table 2. Baseline plasma lipid and lipoprotein concentrations (mg/dL, mean±SD) in the classes of hyperlipidemia.

| | Hyperlipidemia | | |
	FH	FCH	Others
Number of subjects	15	10	3
Cholesterol	418 ± 85	306 ± 42	324 ± 43
Triglycerides	166 ± 74	298 ± 115	1057 ± 915
Lipoprotein cholesterol:			
Number of subjects	11	8	3
VLDL-C	40 ± 18	62 ± 29	123 ± 53
LDL-C	360 ± 35	266 ± 37	176 ± 28
HDL-C	42 ± 9	38 ± 8	31 ± 8

by Lipid Research Clinic methods[14]. In a very few instances, retrospective HDL-C measurements were carried out on plasma which had been frozen (-35°C) for less than 6 months. Repeat determinations on plasma of known cholesterol and HDL-C, which had been frozen in the same batch, were simultaneously carried out as controls. Apolipoprotein E phenotypes were determined after isoelectric focusing[15]. Apolipoprotein quantitation by electroimmunoassay (A-I, A-II, B, E and C-III) was performed on fresh plasma aliquots from a subset sent overnight by air express to Oklahoma City[16].

RESULTS

Effects on plasma lipids and lipoproteins. Table 3 presents plasma lipids and lipoprotein cholesterol for each group during the four study periods. In group I, both clofibrate and probucol had a cholesterol-lowering effect of about the same magnitude; clofibrate being more effective for reduction of triglycerides, and elevation of HDL-C. The combination of the two drugs brought about a 24% fall in cholesterol, 7% more than on either drug alone, but did not have an additive effect on triglycerides. The lowest levels of HDL-C and LDL-C occurred on the combination and the lowest absolute value observed at any time in this group for HDL-C was 21 mg/dL. In group II, response to probucol in terms of plasma lipids was similar to group I, with the exception that HDL-C was strikingly reduced (to 26 mg/dL). Clofibrate appeared to be slightly less effective for reduction of triglycerides in group II, but this was due to the 3 non-type II patients whose triglycerides were reduced from an average 1038 ± 915 to 343 ± 50 mg/dL, if these are excluded, the clofibrate effect in the remaining 10 is a 33% fall (to 150 ± 58 mg/dL); an effect similar to that in group I. The combination of the two drugs resulted in a further 19% reduction in cholesterol, 6.5% greater than that observed on either drug alone. Triglycerides significantly increased in group II during the combination in 16 of 19 patients (paired t test, p <0.001). In contrast, only 3 patients (1 FCH IIb, 1 FH IIa, 1 FH IIb) in group I, showed a mild elevation. The increase in triglycerides occurred independently of the drug preceeding the combination period. In group II, HDL-C was practically unchanged by clofibrate, significantly lowered by probucol and severely reduced by the combination; 17 of the 19 subjects reached at one time during follow-up a value of 8 mg/dL or less. Changes in plasma lipids and lipoproteins with time are illustrated in a typical case from group I (Figure 2) and from group II (Figure 3).

Table 3. Plasma lipids and lipoprotein cholesterol (mg/dL) during the study periods.

	Baseline	Probucol	Clofibrate	Combination
Group I (HDL >20 mg/dL)				
n	9	8	7	9
Cholesterol	394 ± 73[a]	323 ± 71	326 ± 34	301 ± 33
Triglycerides	230 ± 154	175 ± 110	148 ± 107	147 ± 71
VLDL-C	75 ± 27[b]	46 ± 20	55 ± 47	43 ± 20
LDL-C	283 ± 91[b]	261 ± 43	234 ± 28	230 ± 44
HDL-C	35 ± 8[b]	34 ± 10	46 ± 8	30 ± 9
Group II (HDL <15 mg/dL)				
n	19	15	13	19
Cholesterol	357 ± 85	310 ± 63	330 ± 60	290 ± 68
Triglycerides	252 ± 133[c]	196 ± 87	193 ± 96	266 ± 113
VLDL-C	55 ± 40[d]	49 ± 24	48 ± 25	79 ± 111
LDL-C	295 ± 113[d]	238 ± 68	245 ± 63	219 ± 55
HDL-C	41 ± 9[d]	26 ± 9	39 ± 13	8 ± 2

a: $\bar{x}$ ± S.D.; b: no. of subjects = 5; c: excludes one value at 2095 mg/dL;
d: no. of subjects = 17

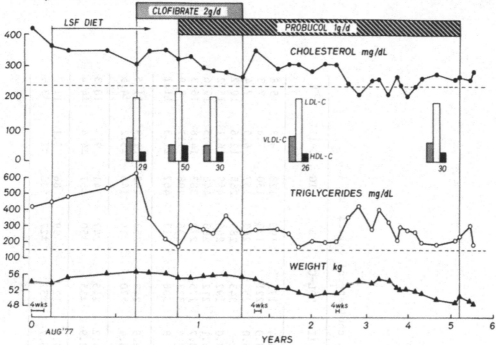

Figure 2. Response of plasma lipids and lipoproteins in a patient from group I affected with FCH. A low-saturated fat, low-cholesterol diet, restricted in simple sugars (LSF) failed to affect plasma TG. Clofibrate caused a striking reduction in plasma TG (and VLDL-C) but raised LDL-C. Probucol was added to oppose the latter effect. HDL-C, raised by Clofibrate, returned to initial levels on the combination. Later, probucol alone was able to control the hyperlipidemia. A period of weight gain was associated with an increase in TG and a reciprocal decrease in cholesterol.

Effect on apolipoproteins. Because this study was not designed a priori and indeed some patients had already been taken off the combination, plasma was obtained from only 9 subjects with severely reduced HDL-C (group A) for apolipoprotein determinations. These results are presented in Table 4 and for comparison, the mean results of 4 patients whose HDL-C was >20 mg/dL at the time of sampling (group B) and a group of normals (group C) are given.

Both apo A-I and apo A-II were very low in group A. The A-I:A-II ratio was 0.86 ± 0.42, about half the mean normal ratio of 1.96. The A-I:HDL-C ratio was close to normal since the depression of HDL-C (82.4%) was as marked as that of apo A-I (82.9%). The A-II:HDL-C ratio (2.58) was however, strikingly higher than normal (1.29). For the subjects in group B, the apo A-I and apo A-II levels, although reduced, yielded A-I:A-II, A-I:HDL-C and A-II:HDL-C ratios close to normal.

Table 4. Apolipoprotein and lipoprotein cholesterol values (mg/dL)

Group	Patient	HLP[a]	Weeks on C	HDL-C	A-I	A-II	LDL-C	B	TG	C-III	E
II	AJ 65F	FCH/IIb	168	17	49.7	38.4	175	196	128	6.3	5.8
II	RC 44M	FCH/IIb	44	14	53.2	33.0	221	180	130	7.6	9.3
II	BR 53F	FH/IIa	57	12	25.4	25.0	257	263	157	6.5	7.4
II	AM 47M	FH/IIb	76	9	22.5	-	343	289	265	12.6	10.2
II	RG 51M	FCH/IIb	16	9	18.9	27.0	192	244	459	16.1	15.2
II	BA 64F	FH/IIb	75	7	10.5	16.1	212	283	446	14.5	19.6
II	DA 56M	V	67	6	14.8	18.7	175	272	383	14.6	16.3
II	GG 65F	FH/IIa	26	6	8.7	17.4	328	236	208	7.7	8.9
II	AA 45M	FH/IIb	15	6	5.4	15.8	336	313	462	21.3	11.1
A	HDL-C ≤15 mg/dL: (n=9)		60[b] ±47	9.5 ±3.9	23.2 ±17.2	23.9 ±8.4	249 ±70	253 ±44	293 ±145	11.9 ±5.2	11.5 ±4.5
B	HDL-C >20 mg/dL: (n=4)		70 ±68	31.5 ±8.9	75.6 ±28.8	44.2 ±12.9	245 ±55	247 ±50	141 ±57	9.4 ±4.4	10.3 ±4.4
C	Normals: (n=108-150)			54.1 ±14.3	136.1 ±27.6	69.3 ±15.7	123 ±32	99 ±27	86 ±33	8.1 ±2.1	11.0 ±3.9

a: Hyperlipidemia

b: Values are $\bar{x}$ ± S.D.

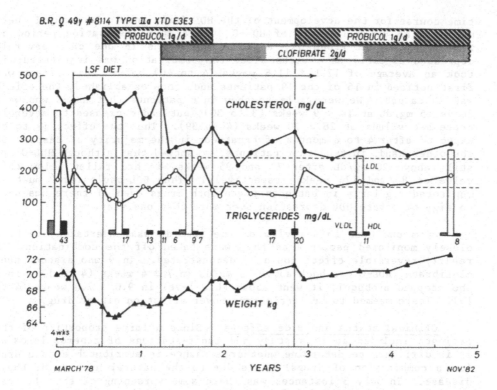

Figure 3. Response of plasma lipids and lipoproteins in a patient from
group II affected with FH and tendon xanthomatosis. The
lipid-lowering (LSF) diet caused a weight loss and a fall in
plasma TG but cholesterol was unaffected. Probucol added
little to the effect of the diet on cholesterol but lowered
HDL-C to less than half the pretreatment values and TG were
raised as body weight increased. Clofibrate was added to
enhance the cholesterol lowering effect of probucol. This
was successful but HDL-C was further decreased to levels as
low as 6 mg/dL, which gradually increased when probucol was
stopped. A second challenge with the drug combination again
resulted in severely depressed HDL-C levels. It was during
the latter period that xanthoma regression was observed in
this subject (see text).

The high levels of apo B reflect the increased LDL-C of all
patients on the combination. The concentration of apo C-III was
significantly elevated (p <0.001) in the patients of group A and the
concentrations of apo C-III and apo E were positively correlated with
plasma triglycerides. Although the highest levels of triglycerides, apo
C-III and apo E were found in those patients with the lowest HDL-C (<10
mg/dL), there were some exceptions: patient GG, HDL-C at 6 mg/dL with
normal apo C-III and low apo E. The apo E levels of the three subjects
with normal triglycerides also appeared to be normal or slightly
reduced. The A-I/C-III ratio was significantly lower in group A than in
controls and intermediate values were found in group B.

Time scale of the changes in HDL-C. Since this study was partially
retrospective and sampling frequency varied from patient to patient,
ranging for most from 1 to 3 months it is difficult to determine the

time course for the development of the HDL-deficiency. In group I there had been no undue lowering of HDL-C over a mean observation period of 21.2 months. In group II, although HDL-C was in one case severely depressed as early as 4 weeks after the combination was instituted; it took an average of 17.1 ± 11.6 weeks (4 to 40) before this effect was first noticed in 16 of the 19 patients and, once established, the effect was sustained. We were able to show in 6 patients that HDL-C was well above 15 mg/dL at 14 ± 9 weeks (3 to 30), but had decreased to severely depressed values at 28 ± 10 weeks (4 to 39). Thus the effect is to be expected after 4 to 6 months of treatment in the majority of cases. On the other hand, in two other patients, we showed that although HDL-C was still above 20 mg/dL after 72 and 95 weeks, it had fallen below this value at 86 and 124 weeks respectively. Thus HDL-cholesterol must be monitored regularly in patients given both drugs even if there has been no sign of severe HDL depression over more than one year.

From the time this review of the records was undertaken, we have closely monitored patients as they were taken off the combination. A readily reversible effect could be demonstrated. In 9 who discontinued clofibrate, HDL-C was back above 15 mg/dL in 7 ± 4 weeks (4 to 17); in 5 who stopped probucol, it went above this level in 9.0 ± 7.0 weeks (4 to 19). There seemed to be little carry-over effect on either drug.

Clinical status and side effects. Since a large proportion of the patients included in this study had manifestations of atherosclerosis, it is difficult to determine whether a change is attributable to a drug (or a combination of drugs) or is due to the natural history of their disease. In only 3 instances was there some worsening of the clinical condition.

One patient of group II, a 50y old male with FCH IIb, was on the combination for 18 weeks when he noticed an increased frequency of constrictive chest pain which led us to discontinue the combination by removing clofibrate. At this time his HDL-C was 9 mg/dL and had been at about this level for at least 16 weeks. His plasma cholesterol and triglycerides were rather resistant to the lipid-lowering drugs either singly or in combination (the biggest change in cholesterol, -13.2%, occurred during the combination; probucol tended to raise his triglycerides). Although he had had atypical chest pain for several years, this high-strung individual developed attacks which were more typical of angina, yielding to nitroglycerin and were considered to be of ischemic origin by a cardiologist. An ECG stress test was negative however, and the cardiologist felt that a coronary angiogram was unnecessary. A duodenal ulcer was found and his angina was controlled by a beta-blocker. He is still on probucol and appears to maintain low HDL-C concentrations on this drug alone (mean of 16.8 ± 6.3 over 53 weeks). He is currently asymptomatic.

Another patient of group II, a 60y old male with familial type IV and severe atherosclerosis, was on the combination for 14 weeks. Probucol was discontinued because of rising triglycerides. A lipoprotein analysis made on that day disclosed an unsuspected severe decrease in HDL-C to 7 mg/dL. Six months later (one year after cessation of the combination) on clofibrate (2 g/day), he had a slight stroke with left hemiparesis which regressed completely on a combination of dypiridamole and ASA; his HDL-C averaged 24 mg/dL during this period. Six months later, he developed unstable angina and was submitted to aorto-coronary bypass and correction of a left ventricular aneurysm, he was at the time being treated by diet plus nicotinic acid and his HDL-C averaged 41 mg/dL. According to the time sequence of the various cardiovascular events, it is very doubtful that the combination was involved; the natural history of the disease would appear to have followed its course, in spite of the various attempts at therapy.

A patient from group I, a 43y old male with FH IIb and tendon xanthomas has been monitored for more than 14 years at our clinic. The combination was begun at a time when he complained of atypical chest pain and has been continued for 154 weeks. His HDL-C averaged 26 ± 6 (n=8). At the 130th week of treatment with the combination, he developed unstable angina. An ECG stress test and a coronary angiogram confirmed the diagnosis of ischemic heart disease. His symptoms were successfully controlled by verapamil and metoprolol. Later, a duodenal ulcer was diagnosed. At week 172 of treatment with the combination, his cholesterol was 293 mg/dL and his triglycerides 81 mg/dL, as compared to values of 421 ± 43 and 73 ± 3 (n=3) on diet alone during the baseline period.

Interestingly, regression of xanthomas was observed in a patient of group II (Figure 3) during treatment with the combination in spite of a severe depression of plasma HDL-C (11 ± 2 mg/dL, n=7). This 49y old woman with FH IIa, harboring xanthelasma and tendon xanthomas, was studied over 3.5 years. During the second combination there was a gradual regression of her xanthelasma and of the xanthomas at the extensor tendons of the fingers, by the 14th month there was no trace of these lesions. She had also herself noticed some thinning of the Achilles tendon xanthomas but no objective measurement of this change was available in her record. The following differences from the baseline period were observed in the second combination period: -36.0% for TC, -31.0% for LDL-C, -14.0% for TG and -74% for HDL-C. The LDL-C/HDL-C ratio went from 9.1 to 24.7, a 2.7 fold increase. Similar changes had occurred during the first combination. Clofibrate or probucol alone had not been as effective as the combination.

DISCUSSION

Except at initial evaluation, measurement of HDL-cholesterol has not been a routine procedure in the monitoring of patients with hyperlipidemia. In the past few years, we have begun to evaluate the lipoprotein profile 3 to 4 months after a change in the therapeutic regimen, preferably when the plasma lipid levels and body weight have stabilized. This led us to the discovery of a very pronounced drug-induced HDL deficiency which could have gone unnoticed, since in several subjects the cholesterol and triglyceride response had been very satisfactory.

Although Miettinen et al.[17] have reported lower HDL-C values on a combination of probucol and clofibrate than on probucol alone, this is the first report of very marked HDL deficiency occurring on this combination. Our retrospective study of the clinical records of 28 patients who received this combination disclosed that in 70% of the subjects, HDL-C concentrations decreased to 15 mg/dL or less. This effect occurred on average after 17 weeks but there were wide individual variations. It was sustained and reversible. Few conditions have been associated with such low levels of HDL-cholesterol. Concentrations of 3 mg/dL or less have been reported in analphalipoproteinemia[18], apo A-I absence[19] and apo A-I, C-III deficiency[20]. Levels between 4 and 14 mg/dL are associated with fish-eye disease[21] and the A-I-Milano mutation[22], while mean concentrations of 14 mg/dL are found in cholestanolosis[23] and familial lipoprotein lipase deficiency (type I)[24]. The smaller group of subjects whose HDL-C fell only slightly on the combination had low levels during the baseline period and reached plasma concentrations (30 ± 9) which are in the range of what is considered to be hypoalphalipoproteinemia in an adult sample. Indeed, in two studies of familial hypoalphalipoproteinemia, mean HDL-C values of 26 ± 5[25] and 28 ± 4 mg/dL[26] have been reported in affected individuals.

Measurements of the major plasma apolipoproteins in subsets of subjects did not on average disclose abnormal levels of apo E. However, some bimodality in apo E and apo C-III response was observed in group II; those subjects with normal or only slightly elevated triglycerides having low or reduced levels of apo E and apo C-III. The significance of this remains to be determined. Apo C-III was higher on average and the A-I/C-III ratio was markedly lower than control values. This is of interest in view of the recently reported association of peripheral vascular disease and a lowered A-I/C-III ratio[27]. Unfortunately, baseline values for this ratio are not available. Three of the Group A subjects had peripheral atherosclerosis. Apo B tended to be high, reflecting the high levels in plasma LDL-C. The plasma apo A-I concentrations were 80% lower than those of normal controls and in proportion to the very low HDL-C concentrations. The mean apo A-I of 23.2 mg/dL is in the range of the values estimated for fish-eye disease[21]. Even in familial lipoprotein lipase deficiency with hyperchylomicronemia (type I), where HDL-C of 15 mg/dL or less is frequently seen, the apo A-I concentrations are 2 to 3 times higher than those reported for group A[28]. Because of a lesser fall in plasma apo A-II (65%), the A-I:A-II molar ratio was markedly reduced, a situation reported only in the severest forms of HDL deficiency[18,19,20].

There is an inverse relationship between the incidence of coronary heart disease (CHD) and HDL-cholesterol concentrations[29]. Premature atherosclerosis has been reported in apo A-I absence[19] and apo A-I/C-III deficiency[20]. Pediatric stroke victims tend to have lower HDL-C[30] and there is an increased incidence of CHD in families with hypoalphalipoproteinemia[25,26]. Hence our observation raises the possibility of potential deleterious effects of a severe drug-induced HDL-deficiency. The clinical finding over an average of 17.7 months, representing a total observation period of 44.2 man-years on the combination does not seem to support this notion. All of these subjects were at high risk for cardiovascular complications and 82% of them had at least one manifestation of coronary, peripheral or cerebral atherosclerosis (Table 1). During treatment with the combination of drugs, only two subjects (one in each group) showed signs of clinical deterioration which could be ascribed equally probably to the natural history of their disease as to the drugs they were receiving. Furthermore, in one instance, there was regression of xanthelasma and tendon xanthomas in a patient who received the combination for 2.5 years with a mean HDL-C of 11.2 ± 2.3 mg/dL and an LDL/HDL ratio of 24.7. As pointed out elsewhere[4,31,32], there are exceptions in whom the inverse relationship between HDL-C levels and CHD incidence does not seem to apply. CHD may develop even in patients with high levels of HDL-C[33], and Stamler has reminded us that oestrogen treatment in men surviving a myocardial infarction was associated with a considerable rise in plasma HDL levels but proved ineffective in preventing recurrent infarction and prolonging life[34]. Increased cardiovascular mortality has also been found associated with estrogen therapy, despite the reported increase in HDL[35]. Conversely, the very low concentrations of HDL-C and apo A-I associated with the A-I Milano apoprotein variant are not associated with an increased incidence of atherosclerotic vascular disease[22]. The same appears to be true in Tangier disease[36] and fish-eye disease[21]. It is possible that the probucol/clofibrate-induced HDL deficiency is another such exception.

Until the mechanism whereby this occurs and until the lipoprotein and lipoprotein kinetics are understood in this situation, patients given the combination should be closely monitored for changes in HDL-C concentrations and cardiovascular status.

ACKNOWLEDGEMENTS

This study was supported by grants from the Medical Research Council of Canada, the Quebec Heart Foundation, the U.S. Public Health Service, and by the resources of the Oklahoma Medical Research Foundation.

The collaboration of Denise Dubreuil, R.N. is gratefully acknowledged especially for her help with review of the records, and of Michel Tremblay, Lucie Boulet, Louis-Jacques Fortin and Josée Gingras for technical assistance.

REFERENCES

1. J. LeLorier, S. Dubreuil-Quidoz, S. Lussier-Cacan, Y. S. Huang and J. Davignon, Diet and probucol in lowering cholesterol concentrations, Additive effects on plasma cholesterol concentrations in patients with familial type II hyperlipoproteinemia, Arch. Int. Med. 137:1429 (1977).
2. J. Davignon and J. LeLorier, Utilité du probucol dans le traitement à long-terme des hypercholestérolémies héréditaires, in: "Symposium sur le Probucol", J. L. Beaumont, ed., Excerpta Medica, Amsterdam, (1981).
3. C. Gagné, P. J. Lupien, M. Brun, S. Moorjani and M. Toussaint, Probucol and high density lipoprotein cholesterol, Can. Med. Assoc. J. 123:356 (1980).
4. J. Davignon and D. Bouthillier, Probucol and familial hypercholesterolemia, Can. Med. Assoc. J. 126:1024 (1982).
5. R. W. Wissler and D. Vesselinovitch, Combined effects of cholestyramine and probucol on regression of atherosclerosis in rhesus monkey aortas, Appl. Pathol. 1:89 (1983).
6. M. Naruszewicz, T. E. Carew, R. C. Pittman, J. L. Witztum and D. Steinberg, A novel mechanism by which probucol lowers low density lipoprotein levels demonstrated in the LDL receptor-deficient rabbit, J. Lipid Res. 25:1206 (1984).
7. J. Davignon, The hyperlipoproteinemias, in: "Current Therapy", H. F. Conn, ed., W. B. Saunders, Philadelphia, (1978).
8. R. Pichardo, L. Boulet and J. Davignon, Pharmacokinetics of clofibrate in familial hypercholesterolemia, Atherosclerosis 26:573 (1977).
9. J. Davignon, A. Sniderman, M. N. Cayen, W. T. Robinson, M. Kraml, E. Reichel and S. Lussier-Cacan, Sensitivity and resistance to the cholesterol-lowering effect of clofibrate - pharmacokinetic studies, plasma apolipoprotein B levels and therapeutic implications, in: "Drugs Affecting Lipid Metabolism", R. Fumagalli, D. Kritchevsky and R. Paoletti, eds., Elsevier/North Holland, New York, (1980).
10. D. B. Hunninghake, Drug treatment of type II hyperlipoproteinemia. Effects on plasma lipid and lipoprotein levels, in: "Atherosclerosis V", A. M. Gotto, L. C. Smith and B. Allen, eds., Springer-Verlag, New York, (1980).

11. J. Davignon, Les médicaments dans le traitement des hyperlipidémies - nécessités et limites, in: "Symposium sur le Probucol", J. L. Beaumont, ed., Excerpta Medica, Amsterdam, (1981).

12. C. C. Allain, L. S. Poon, F. C. S. Chan, W. Richmond and P. C. Fu, Enzymatic determination of total serum cholesterol, Clin. Chem. 20:470 (1974).

13. E. J. Sampson, L. M. Demers and A. F. Krieg, Faster enzymatic procedure for serum triglycerides, Clin. Chem. 21:1983 (1975).

14. Manual of Laboratory Operations, Lipid Research Clinics, Program I: National Heart and Lung Institute, NIH, DHEW, Publ. no (NFH) 75-628, (1974).

15. D. Bouthillier, C. F. Sing and J. Davignon, Apolipoprotein E phenotyping with a single gel method - Application to the study of informative matings, J. Lipid Res. 24:1060 (1983).

16. P. Alaupovic, M. D. Curry and W. J. McConnathy, Quantitative determination of human plasma apolipoproteins by electroimmunoassays, in: "International Conference on Atherosclerosis", L. A. Carlson, R. Paoletti, C. R. Sirtori and G. Weber, eds., Raven Press, New York, (1978).

17. T. A. Miettinen, J. K. Huttunen, V. Naukkarinen, S. Mattila, T. Strandberg and T. Kumlin, Multifactorial primary prevention of cardiovascular diseases, Circulation 64:IV-80 (1981).

18. E. J. Schaefer, L. L. Kay, L. A. Zech, H. B. Brewer, Tangier disease, high density lipoprotein deficiency due to defective metabolism of an abnormal apolipoprotein AI, J. Clin. Invest. 70:934 (1982).

19. E. J. Schaefer, W. H. Heaton, M. G. Wetzel and H. B. Brewer, Plasma apolipoprotein AI absence associated with a marked reduction of high density lipoproteins and premature coronary artery disease, Arteriosclerosis 2:16 (1982).

20. R. A. Norum, J. B. Lakier, S. Goldstein, A. Angel, R. B. Goldberg, W. D. Block, D. K. Noffze, P. J. Dolphin, J. Edelglass, D. D. Bogorad and P Alaupovic, Familial deficiency of apolipoproteins AI and CIII and precocious coronary artery disease, New Engl. J. Med. 306:1513 (1982).

21. L. A. Carlson and B. Philipson, Fish-eye disease - a new familial condition with massive corneal opacities and dyslipoproteinemia, Lancet II:921 (1979).

22. G. Franceschini and C. R. Sirtori, AI Milano apoprotein - Decreased high density lipoprotein cholesterol levels with significant lipoprotein modifications and without clinical atherosclerosis in an Italian family, J. Clin. Invest. 66:892 (1980).

23. V. Shore, G. Salen, T. Cheng, T. Forte, S. Shefer, G. S. Tint, F. T. Lindgren, Abnormal high density lipoproteins in cerebrotendinous xanthomatosis, J. Clin. Invest. 68:1295 (1981).

24. D. S. Fredrickson, J. L. Goldstein and M. S. Brown, The familial hyperlipoproteinemias, in: "The Metabolic Basis of Inherited Disease", J. B. Stanbury, J. B. Wyngaarden and D. S. Fredrickson eds., McGraw-Hill, New York, (1978).

25. C. Vergani and G. Bettale, Familial hypo-alpha-lipoproteinemia, Clin. Chim. Acta. 114:45 (1981).

26. J. L. H. C. Third, J. Montag, M. Flynn, J. Freidel, P. Laskarzewski and C. J. Glueck, Primary and familial hypoalphalipoproteinemia, Metabolism 33:136 (1984).

27. W. J. McConathy, R. M. Greenhalgh, P. Alaupovic, N. E. Woolcock, S. P. Laing, V. Lund, E. T. Lee and G. W. Taylor, Plasma lipid and apolipoprotein profiles of women with two types of peripheral arterial disease, Atherosclerosis 50:295 (1984).

28. J. D. Brunzell, N. E. Miller, P. Alaupovic, R. J. St-Hilaire, C. S. Wang, D. L. Sarson, S. R. Bloom and B. Lewis, Familial chylomicronemia due to a circulating inhibitor of lipoprotein lipase activity, J. Lipid Res. 24:12 (1983).

29. G. J. Miller and N. E. Miller, Plasma high density lipoprotein concentration and development of ischemic heart diseade, Lancet 1:16 (1975).

30. C. J. Glueck, S. R. Daniels, S. Bates, C. Benton, T. Tracy and J. L. H. C. Third, Pediatric victims of unexplained stroke and their families: familial lipid and lipoprotein abnormalities, Pediatrics 69:308 (1982).

31. P. K. Kinnunen, High-density lipoprotein may not be antiatherogenic after all, Lancet 2:34 (1979).

32. J. Davignon, R. Dufour and M. Cantin, Atherosclerosis and hypertension, in: "Hypertension: Physiopathology and Treatment", J. Genest, O. Küchel, P. Hamet and M. Cantin, eds., McGraw-Hill, New York, (1983).

33. A. Rotsztain, Risk factors and HDL, Circulation 57:1032 (1978).

34. J. Stamler, Dietary and serum lipids in the multifactorial etiology of atherosclerosis, Arch. Surg. 113:21 (1978).

35. L. Wallentin and E. Varenhorst, Changes of plasma lipid metabolism in males during estrogen treatment for prostatic carcinoma, J. Clin. Endocr. Metab. 47:596 (1975).

36. G. Assmann, The metabolic role of high density lipoproteins: perspective from Tangier disease, in: "High Density Lipoproteins and Atherosclerosis", A. M. Gotto, N. E. Miller and M. F. Oliver eds., Elsevier, North Holland, (1978).

CONTROL OF PLASMA HDL LEVELS AFTER PLASMAPHERESIS

G.R. Thompson and A. Jadhav

MRC Lipoprotein Team
Hammersmith Hospital
London, U.K.

INTRODUCTION

Plasma exchange, or plasmapheresis, was first introduced for the
treatment of homozygous familial hypercholesterolaemia more than 10 years
ago (1). Since then it has become increasingly clear that twice monthly
exchanges with plasma protein fraction (PPF) provide a safe and effective
means of treating this potentially fatal condition. Several patients have
undergone this regimen for over 9 years without any significant acute complications or long-term side-effects. Mean plasma cholesterol levels have
been reduced to 50% of untreated values (2) and this has been accompanied
by gradual disappearance of cutaneous and tendon xanthomata. However,
there has been little evidence of regression of atheromatous lesions in the
root of the aorta or coronary arteries, although the impression is that
progression of such lesions has been arrested or slowed (3). Analysis of
the life-span of 5 homozygous sibling pairs, one member of each being
treated by plasma exchange and the other not, shows that this form of
treatment improves the chances of survival (unpublished data).

Despite its safety, however, plasma exchange has one major disadvantage,
at least in theory, which is that both low density lipoprotein (LDL) and
high density lipoprotein (HDL) are removed to an equivalent extent from
their respective intravascular pools. An inverse association between HDL
and coronary heart disease (CHD) was first noted by Barr (4). Subsequently
Miller and Miller (5) suggested that the protective effect of HDL was
related to its capacity to act as an acceptor of tissue cholesterol,
including that which accumulates in the arterial wall. The chief apolipoprotein of HDL is apoA-I and several studies have shown an inverse association between plasma apoA-I levels and CHD (6, 7) similar to that which has
been documented for HDL cholesterol and CHD (8, 9, 10). In addition both
HDL cholesterol (11, 12) and apoA-I (13) levels are inversely correlated
with the severity of angiographically-demonstrable coronary arteriosclerotic
lesions. Thus it can be argued that high concentrations of HDL in plasma
influence CHD by retarding atherogenesis or promoting regression and that
reducing HDL levels would be deleterious in these respects.

Mobilisation of tissue cholesterol

As has already been stated plasma exchange promotes mobilisation of
cholesterol from xanthomata, notwithstanding the fact that reverse

cholesterol transport is commonly regarded as being dependent upon HDL. One way of attempting to quantitate this process is to administer ^{14}C-cholesterol several weeks prior to plasma exchange, so as to achieve higher levels of specific activity in the tissues than in plasma by the time the exchange is undertaken. Using this approach it can be shown that the combined specific activity of very low density lipoprotein (VLDL) and LDL shows only slight changes after plasma exchange whereas the specific activity of HDL rises steeply, falls and then rises again during the subsequent 10 days (Fig. 1). The initial rise in specific activity occurred at a time when the concentration of HDL was still subnormal and possibly represents movement of extravascular HDL, in equilibrium with tissue cholesterol and having a higher specific activity than plasma, into the intravascular compartment whereas the subsequent fall and rise in specific activity could represent rapid synthesis of new, unlabelled HDL, which in turn re-equilibrates with tissue cholesterol. Thus despite removing over 50% of the intravascular pool of HDL plasma exchange appears to promote movement of cholesterol out of tissues, perhaps by stimulating the flux of HDL from the extravascular into the intravascular compartment. HDL differs from LDL in being distributed fairly evenly between these two compartments whereas over two-thirds of LDL is intravascular. In contrast to these human data, however, a similar study in a cholesterol-fed pig showed parallel rises in the specific activity of both LDL and HDL following plasma exchange (Fig. 2), without evidence of any differential effect (14).

Effects of LCAT and cholesterol esterification

In view of the evidence that a significant proportion of the enzyme lecithin cholesterol acyl transferase (LCAT) is bound to HDL and is activated by its major apoprotein, apoA-I, it is not surprising that plasma exchange causes a short-term reduction in LCAT activity, which is closely correlated with changes in HDL cholesterol (Fig. 3). However, it is also apparent that although net cholesterol esterification is reduced this occurs to a lesser extent than the reduction in LCAT and is correlated more closely with changes in plasma triglyceride (15). As a result net cholesterol esterification is reduced more transiently than is LCAT activity, due to the rapid return to normal of triglyceride levels following plasma exchange. Similar changes were observed following plasma exchange in a cholesterol-fed pig (Fig. 4).

Addition of a phospholipid/bile salt suspension to PPF

In an attempt to offset the temporary decrease in HDL which results from plasma exchange it was decided to investigate the possible benefits of adding Lipostabil, a 5% suspension of soyabean phospholipid in 4% sodium deoxycholate, to the PPF used as an exchange medium. Theoretical advantages of so doing include observations that Lipostabil coalesces with HDL_3 to form HDL_2 (16), promotes efflux of cholesterol from cells in tissue culture (17), dissolves cholesterol crystals *in vitro* when combined with HDL (18) and induces regression of atheroma in experimental animals when given intravenously (19).

Preliminary experiments in Rhesus monkeys showed no evidence of serious haemolysis nor any hepato- or nephrotoxicity when 0.5 - 2 g Lipostabil were infused together with 100 ml monkey plasma. Further studies were then undertaken in 2 monkeys, each pre-labelled with ^{14}C-cholesterol. Both underwent plasma exchange under general anaesthesia using a Haemonetics Model 30 Cell Separator fitted with a paediatric bowl, one monkey receiving fresh frozen plasma mixed with 1.25 g Lipostabil, the other plasma to which an equivalent volume (25 ml) of sterile saline had been added. Changes in the concentration of plasma cholesterol, HDL_2 and HDL_3 cholesterol and in the specific activity of cholesterol in VLDL plus LDL as well as HDL_2 and

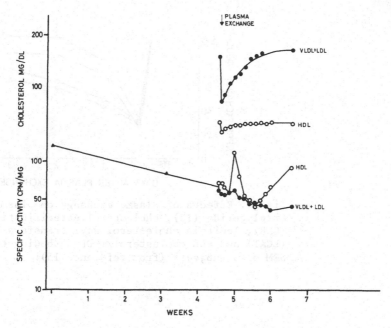

Fig. 1. Changes in the concentration and specific
activity of cholesterol in combined VLDL + LDL and HDL
(using heparin manganese precipitation of plasma as
soon as possible after sample had been taken) following
plasma exchange in a patient with type IV hyperlipoproteinaemia, who had been injected previously with
[14]C-cholesterol (from reference 2).

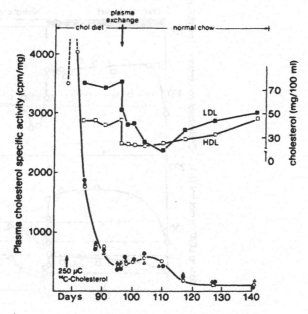

Fig. 2. Effect of plasma exchange on the
specific activity of plasma total cholesterol
(O), LDL cholesterol (●) and HDL cholesterol
(△) in a cholesterol-fed pig labelled with
[14]C-cholesterol (from reference 14).

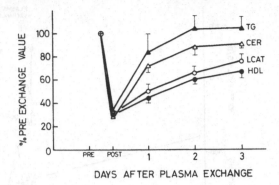

Fig. 3. Effects of plasma exchange on plasma
triglyceride (TG), cholesterol esterification rate
(CER), lecithin cholesterol acyl transferase activity
(LCAT) and HDL cholesterol (HDL), showing the mean ±
SEM of 5 subjects (from reference 15).

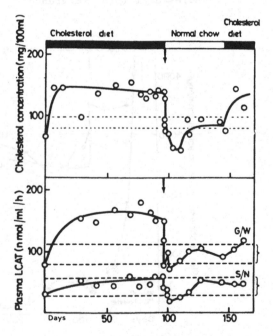

Fig. 4. Effect of plasma exchange (arrow) on
plasma cholesterol concentration, LCAT
activity (G/W) and net esterification rate
(S/N) in a cholesterol-fed pig (from
reference 14).

130

HDL$_3$ are shown in Figs 5 and 6. HDL$_3$ cholesterol levels fell and HDL$_2$ cholesterol levels rose in both monkeys during the week following each procedure. After plasma exchange without Lipostabil the specific activities of both VLDL and LDL and HDL$_3$ cholesterol rose, the latter after an initial fall, whereas that of HDL$_2$ cholesterol fell (Fig. 5). Somewhat similar albeit more erratic changes occurred following plasma exchange in the monkey which received Lipostabil except that HDL$_2$ cholesterol specific activity had nearly regained its pre-exchange level within 2 days.

Although rather difficult to interpret these studies in monkeys confirmed the safety if not the efficacy of Lipostabil. In view of this 2 FH homozygotes who were regularly undergoing plasma exchange with 4 litres of PPF every 2 weeks were each given 2.5 – 5 g Lipostabil mixed with PPF on several occasions. Changes in the concentration of cholesterol and phospholipid in both plasma and HDL are shown in Fig. 7. Decreases in plasma and HDL phospholipid were less marked when Lipostabil was infused than when PPF was given alone; note the relatively high concentrations of phospholipid in HDL following 5 g Lipostabil. However, decreases in plasma or HDL cholesterol were similar irrespective of whether Lipostabil was or was not added to PPF.

Changes in serum bile acids and HDL$_2$ and HDL$_3$ cholesterol during these experiments are shown in the Table. Serum bile acids were markedly elevated immediately post-plasma exchange with PPF plus Lipostabil but declined to normal within 3 hours. There was a concomitant increase in the ratio of HDL$_2$:HDL$_3$ cholesterol but this was transient.

These results do not suggest that addition of Lipostabil to PPF had any significant effect in counteracting the reduction in HDL cholesterol which follows plasma exchange. One possible explanation for this apparent lack of effect is the rapidity with which Lipostabil is removed from plasma after intravenous administration (20). Alternatively the high concentration of bile salts in this preparation may have had a disruptive effect on interaction between Lipostabil and residual HDL in plasma.

Other approaches to conservation of HDL

The major constituents of HDL are apoA-I and phosphatidyl choline (lecithin). Complexes of HDL apoproteins, mainly apoA-I, and lecithin promote the efflux of free cholesterol from cells *in vitro*, including cultured arterial smooth muscle cells (21, 22). Intravenous infusion of phospholipids into animals with experimentally-induced atheroma favourably influences the latter (23, 19) but the effectiveness of this approach may be limited by the rapidity with which lecithin vesicles are cleared from plasma (20), as discussed above. Recently, however, it was shown that apoA-I/lecithin complexes are cleared from the plasma at a much slower rate than lecithin alone, similar to that of native HDL particles (24). This suggests that such complexes might be more effective than Lipostabil alone and this possibility is currently being explored.

An alternative approach is that of LDL-apheresis which was initiated by Lupien *et al* (25) and subsequently adopted and modified by Stoffel *et al* (26). This involves selectively removing LDL from the plasma of FH homozygotes, thus conserving their HDL. A recent development in this area has been the utilisation of WHHL rabbits as a means of assessing the efficiency and safety of a dextran sulphate affinity column (27), which may eventually prove to be cheaper and simpler to use in man than the immunoadsorbent columns (26) or heparin-agarose beads (25) currently in use. Replacement of plasma exchange by LDL-apheresis should probably be determined more by economic factors than by scientific arguments, at least on the basis of the evidence currently available.

131

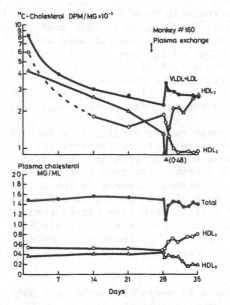

Fig. 5. Changes in concentration of plasma
total and HDL₂ and HDL₃ cholesterol and specific
activity of cholesterol in VLDL + LDL and HDL₂
and HDL₃ before and after plasma exchange in a
Rhesus monkey.

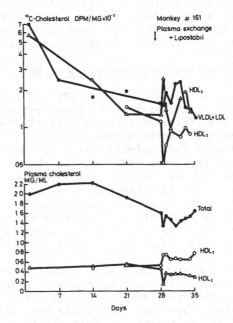

Fig. 6. Changes in concentration of plasma
total and HDL₂ and HDL₃ cholesterol and specific
activity of cholesterol in VLDL + LDL and HDL₂
and HDL₃ before and after plasma exchange combined with infusion of 1.25 g Lipostabil in a
Rhesus monkey.

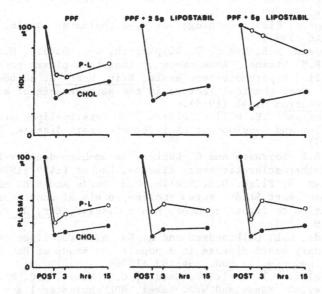

Fig. 7. Changes in concentration of cholesterol
and phospholipid of plasma and HDL at various
times after plasma exchange alone (PPF) or combined with 2.5 g or 5 g Lipostabil. Results
represent mean of 2 FH homozygotes and are
expressed as % of pre-exchange values.

Table 1. Effects of plasma exchange ± Lipostabil 5 g on HDL subfractions and serum bile acids (mean values of 2 FH
homozygotes).

	HDL₂	HDL₃	HDL₂:HDL₃	Bile acids
	cholesterol, mg/dl			μmol/l
Basal	15	28	0.5	32
Post-PE	3	12	0.25	20
3 h	4	9	0.4	–
15 h	5	11	0.45	–
Post-PE + L-S	5	5	1.0	170
3 h	4	9	0.4	36
15 h	4	13	0.3	21

REFERENCES

1. G.R. Thompson, R. Lowenthal and N.B. Myant, Plasma exchange in the management of homozygous familial hypercholesterolaemia, Lancet 1:1208 (1975).
2. G.R. Thompson, Plasma exchange for hypercholesterolaemia, Lancet 1:1246 (1981).
3. G.R. Thompson, N.B. Myant, D. Kilpatrick, C.M. Oakely, M.J. Raphael and R.E. Steiner, Assessment of long-term plasma exchange for familial hypercholesterolaemia, Brit. Heart J. 43:680 (1980).
4. D.P. Barr, Some chemical factors in the pathogenesis of atherosclerosis, Circulation 8:641 (1953).
5. G.S. Miller and N.E. Miller, Plasma-high-density-lipoprotein concentration and development of ischaemic heart disease, Lancet 1:16 (1975).
6. K. Berg, A.L. Borresen and G. Dahlen, Serum-high-density-lipoprotein and atherosclerotic heart disease, Lancet 1:499 (1976).
7. T. Ishikawa, N. Fidge, D.S. Thelle, O.H. Forde and N.E. Miller, The Tromso Heart Study: serum apolipoprotein AI concentration in relation to future coronary heart disease, Eur. J. Clin. Invest. 8:179 (1978).
8. G.G. Rhoads, C.L. Gulbrandsen and A. Kagan, Serum lipoproteins and coronary heart disease in a population study of Hawaii Japanese men, New Eng. J. Med. 294:293 (1976).
9. W.P. Castelli, J.T. Doyle, T. Gordon, C.G. Hames, M.C. Hjortland, S.B. Hulley, A. Kagan and W.J. Zukel, HDL cholesterol and other lipids in coronary heart disease. The cooperative lipoprotein phenotyping study, Circulation 55:767 (1977).
10. S. Yaari, U. Goldbourt, Z. Even-Zohar and H.N. Neufeld, Associations of serum high density lipoprotein and total cholesterol with total, cardiovascular and cancer mortality in a 7-year prospective study of 10,000 men, Lancet 1:1011 (1981).
11. P.J. Jenkins, R.W. Harper and P.J. Nestel, Severity of coronary atherosclerosis related to lipoprotein concentration, Brit. Med. J. 2: 388 (1978).
12. T.A. Pearson, B.H. Bulkley, S.C. Achuff, P.O. Kwiterovich and L. Gordis, The association of low levels of HDL cholesterol and arteriographically defined coronary artery disease, Amer. J. Epidem. 109:285 (1979).
13. J.J. Maciejko, D.R. Holmes, B.A. Kottke, A.R. Zinsmeister, D.M. Dinh and S.J.T. Mao, Apolipoprotein A-I as a marker of angiographically assessed coronary artery disease, New Eng. J. Med. 309:385 (1983).
14. A. Angel, S. Thanabalasingham, D. Reichl, J.J. Pflug, G.R. Thompson and N.B. Myant, Effects of starvation and plasma exchange on lecithin: cholesterol acyltransferase activity and cholesterol efflux in cholesterol-fed pigs, Res. Exp. Med. 184:231 (1984).
15. S. Thanabalasingham, G.R. Thompson, I. Trayner, N.B. Myant and A.K. Soutar, Effect of lipoprotein concentration and lecithin: cholesterol acyltransferase activity on cholesterol esterification in human plasma after plasma exchange, Eur. J. Clin. Invest. 10: 45 (1980).
16. O. Zierenberg, G. Assmann, G. Schmitz and M. Rosseneu, Effect of polyenephosphatidylcholine on cholesterol uptake by human high density lipoprotein, Atherosclerosis 39:527 (1981).
17. A. Postiglione, B.L. Knight, A.K. Soutar and G.R. Thompson, Comparison of cholesterol efflux induced by exogenous phospholipid *in vitro* and during plasma exchange, in press.
18. C.W.M. Adams and Y.J. Abdulla, The action of human high density lipoprotein on cholesterol crystals. Part 1. Light-microscopic observations, Atherosclerosis 31:465 (1978).

19. C.W.M. Adams, Y.H. Abdulla, O.B. Bayliss and R.S. Morgan, Modification of aortic atheroma and fatty liver in cholesterol-fed rabbits by intravenous injection of saturated and polyunsaturated lecithins, J. Path. Bact. 94:77 (1967).

20. P. Dewailly, E. Decoopman, C. Desreumaux and J.C. Fruchart, Plasma removal of intravenous essential phospholipids in man, in "Phosphatidylcholine," H. Peeters, ed., Springer Verlag, Berlin, pp 80-86 (1976).

21. O. Stein and Y. Stein, The removal of cholesterol from Landschütz ascites cells by high-density apolipoprotein, Biochem. Biophys. Acta 326:232 (1973).

22. Y. Stein, M.C. Glangeaud, M. Fainaru and O. Stein, The removal of cholesterol from aortic smooth muscle cells in culture and Landschütz ascites cells by fractions of human high-density apo-lipoprotein, Biochem. Biophys. Acta 380:106 (1975).

23. M. Friedman, S.O. Byers and R.H. Rosenman, Resolution of aortic athero-sclerotic infiltration in the rabbit by phosphatide infusion, Proc. Sco. Exp. Biol. Med. 95:586 (1957).

24. C.L. Malmendier, C. Delcroix and J.P. Ameryckx, In vivo metabolism of human apoprotein A-I-phospholipid complexes. Comparison with human high density lipoprotein - apoprotein A-I metabolism, Clin. Chem. Acta 131:201 (1983).

25. P-J. Lupien, S. Moorjani and J. Awad, A new approach to the management of familial hypercholesterolaemia: removal of plasma cholesterol based on the principle of affinity chromatography, Lancet 1:1261 (1976).

26. W. Stoffel, H. Borberg and V. Greve, Application of specific extra-corporeal removal of low density lipoprotein in familial hyper-cholesterolaemia, Lancet 2:1005 (1981).

27. S. Yokoyama, R. Hayashi, T. Kikkawa, N. Tani, S. Takada, K. Hatanaka and A. Yamamoto, Specific sorbent of apolipoprotein B-containing lipoproteins for plasmapheresis. Characterization and experimental use in hypercholesterolemic rabbits, Arteriosclerosis 4:276 (1984).

CLINICAL SYNDROME AND LIPID METABOLISM IN HEREDITARY DEFICIENCY OF

APOLIPOPROTEINS A-I AND C-III, VARIANT 1

Robert A. Norum,[1,4] Trudy M. Forte,[2] Petar Alaupovic[3] and
Henry N. Ginsberg[4]

[1]Department of Internal Medicine, Henry Ford Hospital,
Detroit, MI 48202 [2]Donner Laboratory, University of Cali-
fornia, Berkeley, Berkeley, CA 94720 [3]Laboratory of Lipid and
Lipoprotein Studies, Oklahoma Medical Research Foundation,
Oklahoma City, OK 73104 [4]Department of Medicine, Mt. Sinai
School of Medicine, New York, NY 10029

The DNA alteration in hereditary deficiency of apoA-I and C-III,
variant 1, is one of the best characterized of all the dyslipoproteinemias.
Since that discovery lipid metabolism has been examined in some detail in
the two patients with this disorder. This report will review the clinical
findings in the two patients and present results of studies on the
physico-chemical properties and metabolism of their lipoproteins.

The first manifestations of this disorder were skin xanthomas (Fig. 1),
noted by the patients in adolescence. In the past 4 years, even though
the patients were on a low cholesterol diet, the deposits have become
slightly more extensive. The corneal clouding as seen on slit lamp examination extends diffusely throughout the corneas, although it is more concentrated in the periphery (Fig. 2). This corneal clouding became noticeable when the patients were in their 30's and is similar to that found in
Tangier disease.

Patient 1 presented at age 31 with heart failure, and was found to
have extensive atherosclerosis of the coronary arteries. She underwent
coronary artery bypass grafting, and her heart failure improved. The other
patient also had extensive atherosclerosis of the coronary arteries on
artereogram but has been unable to cope with medical management. Neither
patient has had evidence of atherosclerosis of the carotid arteries on
physical examination. In May, 1985 patient 1 was found to have atherosclerosis of the abdominal aorta on a radiograph done for low back pain.

Plasma lipid measurements in these patients are shown in Fig. 3. The
levels of HDL cholesterol and VLDL triglyceride were very low. Note also
that apoA-I was undetectable in most assays used and was only minimally
detectable in one assay. Fig. 4 shows additional apolipoprotein levels
determined by electroimmunoassay of whole plasma. In addition to apoA-I,
apoC-III was not detectable in the patients. Levels of apolipoproteins
A-II, C-I, C-II and D were also low. Not shown here are levels of the
patients' apoA-IV, which according to Dr. Bisgaier are about 1/2 the
normal mean.

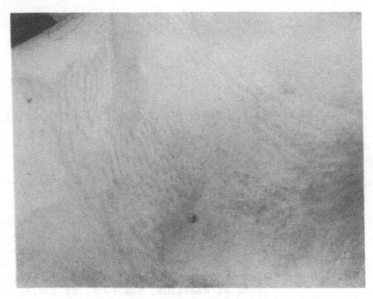

Fig. 1. Skin xanthoma. The neck in patient 1 at age 31.

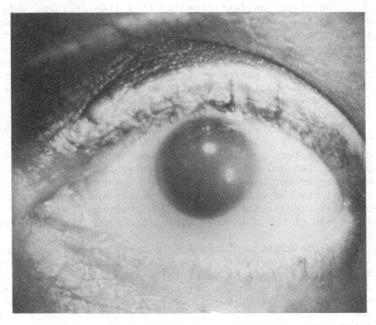

Fig. 2. Corneal clouding.   Patient 2 at age 34.

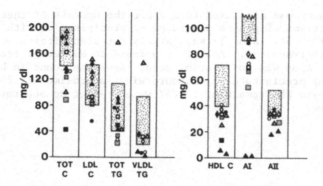

Fig. 3. Plasma lipid and serum apolipoprotein concentrations in the patients and their families. Closed triangles are the patients. Closed and open circles, crossed, closed and double boxes are children of the patients. Open boxes and open diamonds are parents of the patients. Closed diamonds are the patients' brother. Open triangles are the patients' husbands. The dotted area is the normal range. Reprinted, by permission of The New England Journal of Medicine, 306;1513, 1982.

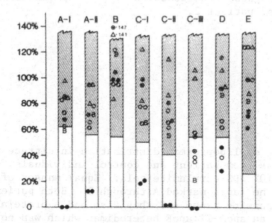

Fig. 4. Apolipoprotein concentrations in whole plasma. Vertical axis shows percent of mean of normal subjects. The hashed area extends down to 2 standard deviations below the normal mean. ● patients, ⊙ parents, ○ children, B brother, Δ husbands. Reprinted from Norum, R. A., Dolphin, P. J., and Alaupovic, P., 1984, Familial deficiency of apolipo-proteins A-I and C-III and precocious coronary artery disease: Family Study, in: "Latent Dyslipoproteinemias and Atherosclerosis," J. L. De Gennes, J. Polonovski, and R. Paoletti, eds., by permission of Raven Press, New York.

An immunoassay was also used to measure the amounts of these apolipoproteins in fractions of plasma separated by precipitation with heparin
and manganese (Fig. 5). ApoA-II was largely not precipitated while apoC-1
was found in both fractions. ApoC-II was found in different fractions in
the 2 patients. ApoD was not precipitated. ApoE was found in both
fractions. Among precipitated apolipoproteins, approximately 25-30% were
bound to apoB while the remaining 70-75% were present on separate lipoprotein particles.

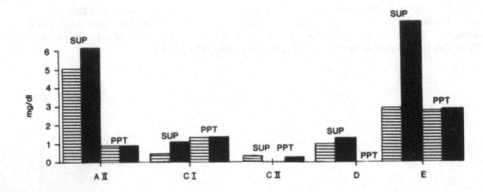

Fig. 5. Apolipoproteins in supernatant (SUP) and precipitate (PPT) from
heparin-managnese fractionation of plasma. Hashed bars, patient 1.
Closed bars, patient 2.

Fig. 6 compares apolipoprotein distribution in ultracentrifugally
isolated lipoproteins from patient and control individuals following
electrophoresis in 10% SDS acrylamide gels. Equal masses of protein were
applied to each of the lanes marked A through D. Both patients had
apoA-II in the HDL density range and they also had a protein whose size
was compatible with an apoA-II:apoE heterodimer which was not seen in the
control. This identification was confirmed when the sample was treated
with beta mercaptoethanol, as seen in lane B. The heterodimer nearly
disappeared, and the apoE band is relatively increased in intensity.

In contrast, HDL from the control (lane C & D) has no detectable apoE
in either the reduced or unreduced condition. Lanes E and F contain
protein from the LDL density range. ApoB stays at the top of the gel.
The control proteins, in lane F contain apoA-I in the LDL density range,
while the patients' proteins do not.

Fig. 7 shows electron micrographs of the particles found in the VLDL, LDL, and HDL from a normal subject, and one of the patients. The VLDL and the LDL particles in the patients were similar to those of controls in both size distribution and morphology. The round particles in the HDL density range in the patients were morphologically similar to controls. However, particle size distributions obtained from micrographs show (Fig. 8) that two distinct populations of particles are present; one large,

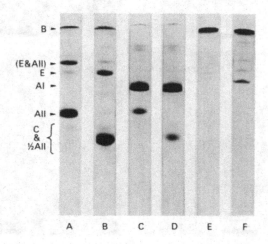

Fig. 6. 10% SDS-PAGE gels of patient and control HDL and LDL. A. Patient HDL, unreduced. B. Patient HDL reduced with beta mercaptoethanol. C. Control HDL, unreduced. D. Control HDL reduced with beta mercaptoethanol. E. Patient LDL unreduced. F. Control LDL unreduced. Positions of the apolipoproteins are indicated on the left side of figure. Proteins applied to gels consisted of 25-30 ug for HDL and 60 ug for LDL. Reprinted from Journal of Clinical Investigation 74:1601, 1984.

9.5-12 mm diameter and the other small, 5.5-7.5 mm. Both subjects are deficient in particles with diameters of 8 to 9 nanometers. The shape of the distributions suggests that the missing class may be merely an artifact of measurement. However, the existence of two populations of particles in patients is supported by analysis of their HDL by non-denaturing gradient gel electrophoresis, which separates particles on the basis of size (Fig. 9).

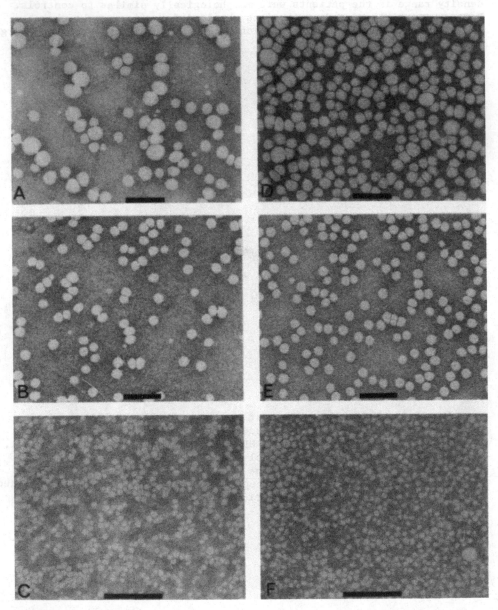

Fig. 7. Electron micrographs of negatively stained lipoproteins from control subject and patient 1. A, B, C are VLDL, LDL and HDL, respectively, of the control subject; and D, E and F are VLDL, LDL and HDL, respectively, of the patient. Bar markers represent 100 nm. Reprinted from Journal of Clinical Investigation 74:1601, 1984.

The gradient gel electrophoretic tracings in Fig. 9 indicate the density of coomassie-staining protein in these gels. The patients had a dominant population of particles whose mean size was somewhat larger than that found in normal subjects. In the patients, there was also a population of particles smaller than found in the controls. Overall, the shape of the gradient gel curves is similar to that obtained by electron microscopy including a deficiency of particles in the (HDL 3a) gge region which corresponds to particles 8-9 mm in diameter.

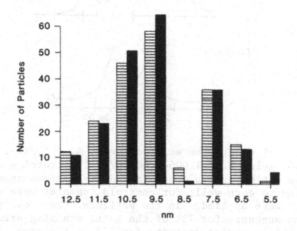

Fig. 8. HDL particle diameter distribution. Diameters were measured on electron micrographs. Hashed bars are patient 1. Closed bars are patient 2.

To analyze the protein constituents within the broad spectrum of HDL particle sizes noted in gradient gel electrophoretograms subfractions of patients' HDL were obtained by density gradient ultracentrifugation and analyzed on 10% SDS gels. Fig. 10 shows the distribution of unreduced proteins in each density fraction. The less dense particles contain the apoA-II : apoE heterodimer as well as apoE which is not covalently bound to apoA-II. In contrast apoE in the more dense fractions is present primarily as a heterodimer. The C apolipoproteins appear to be associated with the less dense particles.

For each of these density gradient fractions, 1 thru 5, the protein concentration, and the protein and lipid compositions were analyzed (Fig. 11). The fractions of greater density had most of the protein found in the HDL region and also had a somewhat higher protein to lipid ratio. The lipid compositions of the subfractions are presented as percent of the total mass. The phospholipid concentration, as well as the unesterified and esterified cholesterol concentrations all decreased with increasing density of the subfraction. Paucity of material in fraction 2 precluded accurate compositional measurements.

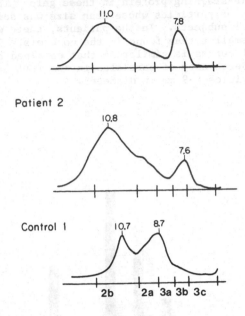

Fig. 9. Gradient gel electrophoresis patterns: A. d 1.063-1.21 g/ml lipoproteins on 4-30% gels; B. d 1.006-1.063 g/ml lipoproteins on 2-16% gels.
The HDL from the patients were concentrated 24-fold and approximately 25 ul
of sample applied to each well; for controls, samples were concentrated 2-
fold, and 6-8 ul were applied. In the patients, the broad peak in the
(HDL2)gge region accounts for 73% of the total staining area, and the sharp
peak in the (HDL3)gge region accounts for 27% of the area. All gels were
stained with Coomassie G250. Spectrophotometric scans of the gels show the
relative distribution of the various lipoprotein subpopulations. The
numbers above the peaks indicate particle diameters in nm. The designations
at the bottom indicate size classes. Reprinted from Journal of Clinical
Investigation 74:1601, 1984.

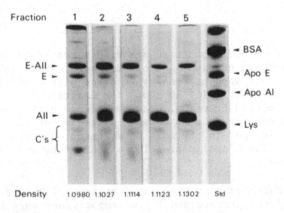

Fig. 10. 10% SDS-PAGE gel patterns of HDL gradient subfractions of patient
1. The average density of each of the subfractions is indicated below the
gel. 30 ug of protein were applied to each gel, and samples were unreduced.
The protein standard (std) are: bovine serum albumin, BSA; apolipoprotein
E, apoE, apolipoprotein A-I, apoA-I; and egg lysozyme, lys. The apolipoprotein bands in the patients' subfractions are identified on the left-
hand side.

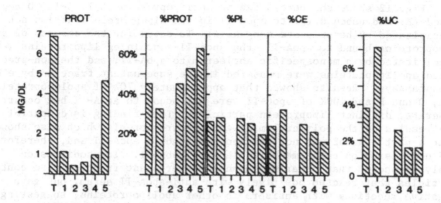

Fig. 11. HDL subfraction composition. T means total HDL. Numbers at
bottom correspond to density subfraction in Fig. 10. Prot is protein, PL
is phospholipid, CE is cholesterol ester, UC is unesterified cholesterol.
% is percent of total mass of protein and lipid in the subfraction.

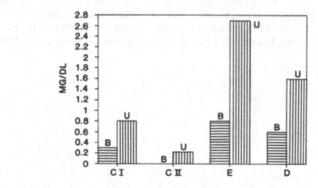

Fig. 12. HDL apolipoproteins bound (B) and unbound (U) to apoA-II in
patient 1. Lipoproteins not precipitated by heparin and
manganese were treated with antibody to apoA-II to precipitate apoA-II and apolipoproteins bound to apoA-II.
Apolipoproteins indicated at the bottom were measured by
electroimmunossay before and after precipitation by
antibody to apoA-II.

To summarize, the patients with deficiency of apoA-I and C-III have 2 sizes of particles in the HDL density range although these are present in extremely low concentration. The smaller of these 2 is smaller than found in controls, and contains relatively less lipid than the larger particles.

Fig. 12 shows the percentage of apolipoproteins C-I, C-II, D and E bound (B) and unbound (U) to apoA-II in the lipoprotein fraction not precipitated by heparin and manganese. To determine the amounts of apo-lipoproteins bound to apoA-II, the apoA-II-containing lipoproteins were precipitated by a monospecific antiserum to apoA-II, and the non-precipitated apolipoproteins were measured in the supernatant fraction by electro-immunoassays. Results showed that approximately 70% of apolipoproteins C-I, D and E and 100% of apoC-II were not bound to apoA-II but occurred as separate, distinct lipoprotein particles. This finding is compatible with the results of the polyacrylamide gel electrophoresis which also showed that a part of apoE was not covalently bound to apoA-II and, therefore, did not reside on the same particle with apoA-II. Immunodiffusion analysis showed that heparin-manganese supernatant fraction also contained particles that reacted positively with anti-apoA-II serum, but gave negative reactions with antisera to other apolipoproteins, suggesting the existence of particles with apoA-II as the sole protein constituent.

Kinetic studies of fat metabolism have been performed in these 2 patients. One of the patients was given a meal of 100 grams of fat. The other patient has gallstones and so could not participate in this study. Fig. 13 shows that the total cholesterol level rose substantially and was still high at 8 hrs. after the meal. The triglyceride level rose to a peak at 4 hrs. and then declined. The HDL cholesterol level was unde-tectable in the fasting state and rose to a few mg/dl before declining.

Fig. 14 shows the composition of VLDL at various time points. The percent of VLDL mass composed of protein decreased slightly over the course of this experiment. The percent VLDL cholesterol rose slightly, and the percent VLDL triglyceride rose rather more. Fig. 15 shows the composition of lipoproteins in the HDL density range after the fat meal. The proportion of HDL triglyceride rose slightly at four hours and then declined. The composition of LDL showed little change over the course of the experiment.

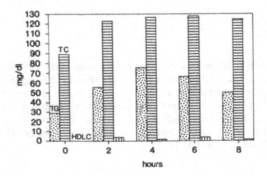

Fig. 13. Plasma lipids in patient 1 after meal of 100 g of fat.
The dotted bars are triglycerides. The horizontally
hashed bars are total cholesterol. The vertically
hashed bars are HDL cholesterol.

146

The removal of ... VLDL and LDL ... was established in this ...
... The particle ... VLDL ... LDL ... ... removed by attenuated ...
... and the particles ... and ... similarities ... The lateral ...
... these ... cholesterol transfer ... the physical ... distribution of cholesterol ...
... the lipoprotein ... VLDL ... observed in patient ... and larger ...
... cholesterol ... of ... VLDL. In some clinical cases more than normal ...
... Cholesterol of VLDL was analyzed in mass ...
... converted ... only as candidate glyceride ...

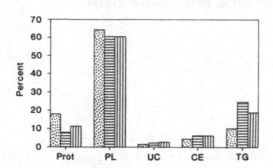

Fig. 14. VLDL composition in patient 1 after meal of 100g fat.
The dotted bars are at prior to the meal. The horizontally
hashed bars are at 4 hours after the meal. The vertically
hashed bars are from combined 6 and 8 hour samples.

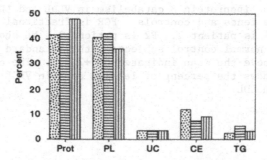

Fig. 15. HDL composition in patient 1 after meal of 100g fat.
The dotted bars are at prior to the meal. The horizontally
hashed bars are at 4 hours after the meal. The vertically
hashed bars are from combined 6 and 8 hour samples.

The turnover of apoB in VLDL and LDL has been examined in this disorder. The patient's own VLDL and LDL were isolated by ultracentrifugation, and the proteins labeled with radioactive iodine. The labeled lipoproteins were infused into the patients, and distribution of apoB among the lipoprotein density classes was observed (Fig. 16). The fractional catabolic rate of apoB in VLDL is about 7 times faster than normal. This rapid removal of VLDL was confirmed by assessment of triglyceride turnover in VLDL using radiolabeled glycerol.

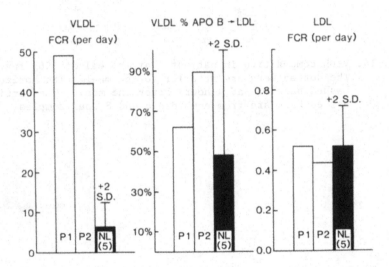

Fig. 16. Apolipoprotein B catabolism in VLDL and LDL in the patients and controls. FCR is fractional catabolic rate. P1 is patient 1. P2 is patient 2. NL shows the mean of 5 normal control subjects with 2 standard deviations above the mean indicated by +2 S.D. The center panel shows the percent of labeled apoB in VLDL that appeared in LDL.

The apoB in VLDL was converted to intermediate density lipoprotein, and then to LDL. The proportion of VLDL apoB that was converted to LDL was 65% and 90% in patients 1 and 2, respectively. This proportion is toward the high end of the very wide range found in normal subjects.

The result is remarkable in that it accounts for much of VLDL apoB, and indicates that a large fraction of VLDL apoB does not go directly into peripheral tissues. The fractional catabolic rate of apoB in LDL in the patients was similar to normal controls.

The two patients are homozygotes; both have the same two abnormal genes involving the loci for apolipoproteins A-I and C-III, and the two abnormal genes are identical on molecular analysis thus far. Some of the patients' relatives are heterozygotes as indicated by a single dose of the abnormal gene on DNA analysis.

Family records indicate that the parents of the patients are third cousins. The parents each have a single dose of this gene as assessed by DNA analyses. The father's father has been found to also have one dose of the gene. We infer that the mother's mother had the gene since the mother's father has been tested with DNA analysis and he was normal.

Only one path of consanguinity has been identified thus far. Since this gene seems to be rare, as indicated by the rarity of such low HDL's, it is likely that the two doses of the gene came from the same source in their ancestry. We presume that one of the two great, great, great grandparents of the patients had a single dose of this gene.

Among the proven heterozygotes, neither parent of the patients shows evidence of atherosclerosis, though their mother has xanthelasma. The father's father is alive at 83 years of age. The mother's mother died at the age of 72, while her mother lived to the age of 92. Hence, a single dose of this gene does not seem to be an impediment to longevity.

We plan to study more heterozygotes for evidence of atherosclerosis and to assess further the effects of a single dose of this gene on lipid metabolism.

Fig. 3 includes results of lipid measurements in the heterozygotes as well as those for the patients. Everyone except the probands and their spouses is a proven heterozygote for this gene by DNA analysis. All but 1 of the husbands have VLDL triglyceride that is less than the normal mean. The levels of HDL cholesterol of all family members, including husbands, are remarkably low, as are apoA-I and apoA-II levels. As seen in Fig. 4, plasma levels of apoC's, especially apoC-III, are somewhat low in the heterozygotes. If heterozygotes have no prominent atherosclerosis then low levels of A-I and C-III seem to be adequate.

Severe deficiency of apolipoproteins A-I and C-III has complex effects on lipid metabolism that result in lipid accumulation in the skin, corneas, and coronary and peripheral arteries, and perhaps other places not yet known.

Part of the mechanism leading to this result appears to be very rapid conversion of VLDL to LDL. The very small amount of lipoprotein in the HDL density range is highly abnormal in morphology and apolipoprotein composition. Further assessment of the pathogenesis of lipid accumulation may yield information relevant to more common conditions. Perhaps the combined deficiency of apoA-I and C-III has special significance.

THE MOLECULAR BIOLOGY OF HUMAN ApoA-I,

ApoA-II, ApoC-II and ApoB

Simon W. Law, Karl J. Lackner, Silvia S. Fojo,
Ashok Hospattankar, Juan C. Monge, and H. Bryan Brewer, Jr.

Molecular Disease Branch
National Heart, Lung and Blood Institute
National Institutes of Health
Building 10, Room 7N112
Bethesda, MD 20892

Introduction

Plasma lipoproteins are the major vehicles for lipid transport in man. Plasma levels of cholesterol, lipoproteins and apolipoproteins are important parameters in the evaluation of patients for the potential risk of development of premature cardiovascular disease. Thirteen apolipoproteins have been isolated and characterized. These proteins function as enzyme cofactors, ligands for interaction with specific high affinity cellular receptors, exchange proteins for the transfer of lipid constituents between lipoprotein particles, and structural components of lipoprotein particles. Thus plasma apolipoproteins plays a key role in lipid and lipoprotein metabolism in man. Major advances in our understanding of the molecular biology of the human apolipoproteins have occurred over the last several years. The objective of this report is a brief review of our current knowledge of the molecular biology of human apolipoprotein A-I, A-II, C-II and B.

I) Molecular Biology of Human apoA-I

a) Normal apoA-I.

Human apolipoprotein (apo)A-I is a single polypeptide chain composed of 243 amino acids[1,2] and is a major protein constituent of plasma high density lipoproteins (HDL). HDL has been of particular interest since it has been inversely associated with the development of premature cardiovascular diseases[3-5].

ApoA-I mRNA has been cloned and the nucleotide sequence determined[6-10]. ApoA-I mRNA encodes a precursor apolipoprotein, pre-proapoA-I which contains a 24 amino acids peptide extension at the NH_2-terminal end of the mature plasma apoA-I. Eighteen amino acids are contained within the hydrophobic prepeptide followed by a 6 amino acid propeptide (Arg-His-Phe-Trp-Gln-Gln)[6]. Our results on human apoA-I are in agreement with the partial sequence of the precursor of rat apoA-I[11] with respect to the length of the precursor sequence, location of the prepeptide, and the presence of an unusual propeptide sequence terminating in a neutral dipeptide Gln-Gln instead of the basic dipeptide which is characteristic of many secretory proteins.

Human apoA-I is polymorphic in plasma[2,12,13] and of particular interest is the secretion into plasma of the proapoA-I, and its increase following fat feeding[14]. These results indicate that there are two sites of processing of the A-I apolipoprotein (Figure 1).

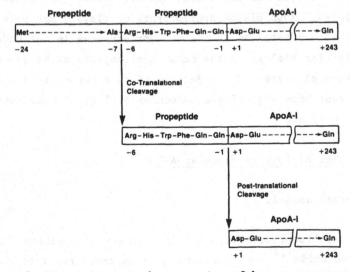

Figure 1. Biosynthesis and processing of human preproapoA-I.

Initially, the apolipoprotein is synthesized as a preproapoA-I precursor. Co-translational cleavage of the prepeptide occurs predominantly intra-cellularly, and the proapoA-I is secreted into plasma and lymph. The mature form of plasma apoA-I is produced by a second post-translational cleavage of proapoA-I to yield the mature A-I apolipoprotein[15].

b). ApoA-I in Tangier Disease

Tangier disease is a rare familial disorder of lipid transport characterized by hypocholesterolemia, moderate hypertriglyceridemia, low levels of low density lipoproteins, and a marked deficiency of HDL[16-18] Clinically, these patients have enlarged orange tonsils, hepatosplenomeg-aly, lymphadenopathy, and recurrent transient sensory-motor peripheral neuropathies. The tissue lipid deposits are composed principally of cholesteryl esters and are confined primarily to the macrophages, Schwann cells, and nonvascular smooth muscle cells[19]. In normal subjects, the predominant components of HDL are cholesterol, apoA-I and apoA-II. In plasma from Tangier homozygotes, HDL cholesterol concentrations are 5% of normal and apoA-I and apoA-II levels are approximately 1% and 7% of normal, respectively[18,20,21]. Previous studies in our laboratory have shown an increased fractional catabolic rate of apoA-I isolated from Tangier disease subjects[22,23], demonstrating kinetic abnormality in apoA-I isolated from these patients. In addition, plasma apoA-I isolated from a Tangier disease patient was found to be defective in recombining with HDL[24]. Using the technique of two-dimensional gel analysis, we and others have also observed that plasma apoA-I from Tangier patients has an increased amount of proapoA-I when compared to normal subjects[25-27], suggesting altered proapoA-I to mature apoA-I conversion in Tangier disease. However, amino acid sequence analysis of the proapoA-I isolated from a Tangier disease subject showed the same amino-terminal sequence as normal proapoA-I[27]. Our data thus ruled out the possibility of a defec-tive conversion of proapoA-I to mature plasma apoA-I due to a structural mutation at the propeptide cleavage site in Tangier disease as was pro-posed by Zannis et al[25]. In vitro analysis of the conversion of pro-apoA-I to mature apoA-I in plasma from Tangier subjects revealed a normal conversion of the proisoprotein to the mature isoprotein[28,29]. Further-more, detailed study of the in vivo conversion of radiolabeled proapoA-I to mature apoA-I in Tangier patients definitively established that the conversion was normal and the relative increase in the proisoprotein was due to markedly accelerated catabolism of the mature apoA-I isoproteins with no change in the conversion rate of proapoA-I to mature apoA-I[30]. We have previously presented results of Southern blot hybridization analysis of normal and Tangier disease DNA showing no major gene deletions or insertions of the Tangier apoA-I gene[7]. We now have cloned Tangier liver apoA-I cDNA and determined the complete coding sequence of preproapoA-I mRNA.

A cDNA library of Tangier liver mRNA was established in pBR322[31]. Sequence analysis of the longest apoA-I cDNA clone established the derived amino acid sequence of residue 116-243 of plasma apoA-I. The remaining portion of the sequence of Tangier preproapoA-I mRNA was established by sequence analysis of specific primer extensions of synthetic oligonucleotides on Tangier liver mRNA. This later technique provided the derived amino acid sequence of residue -24 to 116, thus completing the entire preproapoA-I structure. The sequence of Tangier preproapoA-I was identical to normal preproapoA-I except for a single base substitution (G T) which resulted in the isosteric replacement of a glutamic acid residue at position 120 to aspartic acid. Our current understanding of the molecular defect of Tangier disease is as follows: 1) Southern blot hybridization of genomic DNA fragments showed no major gene deletions or insertions. 2) The apoA-I gene in Tangier disease is functional and its mRNA of the same size as normal apoA-I mRNA. 3) The sequence of Tangier apoA-I mRNA is identical to that of normal apoA-I. The molecular defect in Tangier disease is due to post-translational or receptor abnormality in which normal apoA-I undergoes processing which results in rapid catabolism of the protein.

II) Molecular Biology of Human ApoA-II

Human apoA-II is the second most abundant protein constituent of HDL[32]. ApoA-II in human plasma is a 154 amino acid protein composed of two identical chains of 77 amino acid residue linked by a single disulfide bridge at position 6 in the sequence[33]. The physiological role of apoA-II in lipoprotein metabolism is not as yet clearly defined. To understand better the biosynthesis and processing of apoA-II and to evaluate the regulation of the apoA-I and apoA-II genes in patients with dyslipoproteinemias, we have cloned the cDNA and genomic DNA of human apoA-II[34,37].

154

a)   ApoA-II mRNA

ApoA-II mRNA codes for a 100 amino acid protein, preproapoA-II that has an 18 amino acid prepeptide and a 5 amino acid propeptide terminating with a basic dipeptide (Arg-Arg) at the cleavage site to mature apo-A-II[34]. Our data confirms the previous in vitro translation studies on the precursor forms of apoA-II[35,36]. We have now also resolved the discrepancy of the structural organization of the A-II prepropeptide[34]. The biosynthesis and processing of apoA-II is summarized in Figure 2.

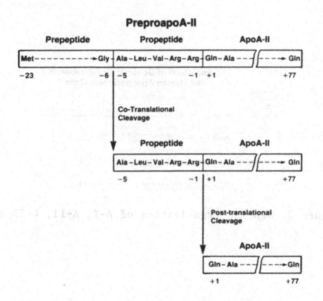

Figure 2. Biosynthesis and processing of human preproapoA-II

Thus the post-translational processing of apoA-I and apoA-II are different and appear to be under separate control. The regulation of the processing of the two major apolipoproteins of HDL may play an important role in HDL metabolism.

b). Genomic structure organization of apoA-II

ApoA-II genomic clones were isolated from a human DNA library
packaged in charon 28 phage (This library was kindly provided to us by
Dr. P. Leder). The nucleic acid and derived amino acid sequence of the
apoA-II gene has been determined[37]. Typical eukaryotic promoter sequences such as the (CAAT) and (TATA) boxes are found at appropriate
locations. Six bases upstream from the 3' splice site of the second
intron is a stretch of 33 base sequence composed of 16 pairs of GT

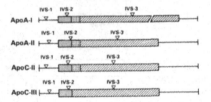

Figure 3. Genomic organization of A-I, A-II, C-II and C-III,

dinucleotide which can potentially form Z-DNA conformation. Similar to
apoA-I and apoC-III[38], the apoA-II gene is also interupted by three
introns at similar sites in the gene. An intron in the 5' untranslated
region, an intron in the prepeptide sequence, and one in the coding
sequence of the mRNA. Thus the concept of a common origin of apolipoproteins[39] is further substantiated by detailed analysis of these genes.
The genome organization of apoA-I, apoA-II, apoC-II and apoC-III is
summarized in Figure 3.

III). Molecular Biology of Human ApoC-II.

The cDNA, and genomic DNA of apoC-II has been cloned[40] and localized to chromosome 19[41]. Similar to apoA-I and apoA-II, the apoC-II mRNA encodes a precursor protein preproapoC-II. A schematic diagram summarizing the biosynthesis and processing of apoC-II is illustrated in Figure 4.

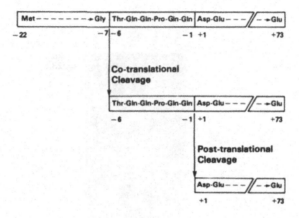

Figure 4. Biosynthesis and processing of human preproapoC-II.

The molecular defects in patients with apoC-II deficiencies are currently being investigated[42].

IV). Molecular Biology of Human Hepatic apoB-100:

a). Cloning and Analysis of Hepatic ApoB-100 mRNA in Normal Subjects

Human apoB-100 is the major apolipoprotein of low density lipoproteins (LDL) and the principal ligand for interaction with the LDL receptor[43]. Human apoB-100 has been cloned in a λgt-11 expression vector, and the apoB-100 cDNA clones identified by screening with a monospecific apoB-100 antiserum, synthetic oligonucleotides based on the amino acid sequence of peptides isolated from apoB-100, and by immunoblot analysis of the expressed protein with a monoclonal antibody to apoB-100[44,45].

The nucleic acid derived 560 amino acid residues of apoB-100 [45] contain no unique linear or repeating sequence of amino acids. The computer predicted conformation of the apoB-100 protein include segments of helical structure, however a significant portion of the protein is organized into β-structure. The β-structure may be important in lipid-apoB-100 interactions in LDL and contributed to the insolubility of delipidated apoB-100 in aqueous buffers.

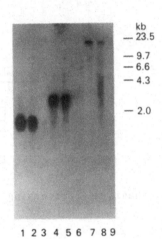

Figure 5. Northern blot filter hybridization analysis of human liver RNA with apoA-I cDNA probe (lanes 1-3); albumin cDNA probe (lanes 4-6) and apoB-100 cDNA probe (lanes 7-9), lanes 1, 4, 7 are 1 ug each of poly(A)$^+$ RNA; lanes 2, 5, 8 are 10 ug each of total RNA; lanes 3, 6, 9 are 2 ug of poly(A)$^-$ RNA.

Northern blot analysis of liver mRNA utilizing an NcoI/Hind III apoB-100 cDNA probe revealed that the apoB-100 mRNA is of enormous size (Figure 5).

158

Accurate determination of the apoB-100 mRNA size is not possible due to lack of single stranded RNA size markers. Our estimation is 15-18 kb long which is of sufficient size to code for an apolipoprotein of 250,000 to 387,000 daltons, the proposed molecular weight of delipidated plasma apoB-100[46,47]. Contrary to the liver, intestinal mRNA contains both apoB-100 and apoB-48 species.

The gene for human apoB-100 has been localized to the region of p23 →pter of chromosome 2 by filter hybridization of human-mouse somatic cell hybrids utilizing nick-translated cDNA probes[44,45]. It is interesting to note the location of the apoB-100 gene, the major ligand for the LDL receptor, is on chromosome 2 and not in synteny with the apoE and the LDL receptor genes.

b)   Analysis of apoB-100 gene expression in Abetalipoproteinemia.

High molecular weight DNA was isolated from white cells of normal lipidemic and abetalipoproteinemic subjects as described by Bell et al[48]. DNA's were cleaved by a panel of restriction endonucleases and analyzed by the technique of Southern[49]. In all cases, DNA fragments of similar size were observed (Lackner et al unpublished observations). Extensive analysis with probes corresponding to the amino terminal, middle, and carboxyl terminal region of apoB-100 mRNA did not reveal any major deletions and or rearrangement of the apoB gene. However, due to the enormous size of apoB-100 mRNA, subtle but important structural defects could escape this kind of analysis. Is there any structural defect in the apoB gene that would result in the absence or the formation of an abnormal apoB mRNA? To address this aspect of expression of the apoB gene, we have isolated mRNA from an abetalipoproteinemic subject and analyzed the mRNA by northern blot hybridization. Our results as illustrated in Figure 6, clearly demonstrated the presence of apoB-100 mRNA in the liver tissue from a patient with abetalipoproteinemia.

Thus the lack of the B apolipoprotein in the plasma of an abetalipoprotein subject is not due to a non-functioning apoB gene. We are currently studying additional patients to establish that the molecular defect in abetalipoproteinemia is at the post-translational level.

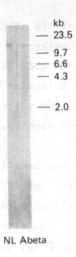

kb
— 23.5
— 9.7
— 6.6
— 4.3

— 2.0

NL Abeta

Figure 6. Northern Blot Analysis of liver RNA from a normal subject and a patient with abetalipoproteinemia. 1 g poly A$^+$ RNA was applied to each lane. NL, Normal Liver RNA. abeta, abetalipoproteinemia.

Summary

The application of molecular biology techniques has enabled us to determine the gene sequence, organization, transcription and processing of apolipoprotein genes. Consequently, new insights have been gained in the biosynthesis and processing of these proteins. In addition to apoA-I, apoA-II and apoC-III reported here, other apolipoprotein genes such as apoC-II and apoE genes were found to share common intron-exon organizations. The results suggest that these genes most probably arise from a common ancestral gene.

Utilizing cDNA as hybridization probes, we have localized apoA-I, apoA-II, apoC-II, apoC-III, apoE and apoB to specific locations of individual chromosomes (for review, see ref. 6). There is no clear relationship between currently known physiological function and the organization of the apolipoproteins in the chromosomes with the exception of the LDL receptor and its ligand, apoE which are localized to chromosome 19. However, apoB-100, the major ligand for the LDL receptor is on chromosome 2 and not in synteny with the apoE and the LDL receptor genes. The cloning of the major human apolipoprotein genes have also allowed us to inititate studies on the molecular defects leading to various dyslipoproteinemias including Tangier disease and abetalipoproteinemia. Undoubtedly, information derived from these studies will provide the basis for future _in vitro_ and _in vivo_ studies on patients with dyslipoproteinemia and premature atherosclerosis.

Acknowledgement

The authors would like to acknowledge Drs. R.E. Gregg and J.H. Hoeg for helpful discussion.

REFERENCES

1. Brewer, H.B., Jr., Fairwell, T., LaRue, A., Ronan, R., Houser, A. & Bronzert, T.J. (1978) Biochem. Biophys. Res. Commun. 80, 623-630.
2. Osborne, J.C. Jr., & Brewer, H.B., Jr. (1977) Adv. Protein. Chem. 31, 253-337.
3. Rhoades, G.G., Gillbrandsen, C.L. & Kagan, A. (1976) N. Engl. J. Med. 294, 293-295.
4. Castelli, W.P., Doyle, J.Y., Gordon, T.G., Hames, C.G., Hjortland, M.C., Hylley, S.B., Kajahn, A. & Zukel, W.J. (1977) Circulation 55, 767-772.
5. Avogaro, P., Bittolo Bon, G., Cazzolato, G., Quinci, G.B. & Belussi, F. (1978) Artery 4, 385-394.
6. Law, S.W., Gray, G., and Brewer, H.B. Jr., (1983) Biochem. Biophys. Res. Commun. 112, 257-264.
7. Law, S.W., Gray, G., and Brewer, H.B. Jr., (1984) Proc. Natl. Acad. Sci. USA. 81, 66-70.
8. Shoulders, C.C., and Baralle, F.E. (1982) Nucl. Acids. Res. 10, 4872-4882.
9. Breslow, J.L., Ross, D., McPherson, J. (1982) Proc. Natl. Acad. Sci. USA. 79, 6861-6865.
10. Cheung, P., Chan, L. (1983) Nucl. Acids. Res. 11, 3703-3715.
11. Gordon, J.I., Smith, D.P., Andy, R., Alpers, D.H., Schonfeld, G., and Strauss, A.W. (1982) J. Biol. Chem. 257, 971-978.
12. Nestruck, A.C., Suzue, G., and Marcel, Y.L. (1980) Biochem. Biophys. Acta. 617, 110-121.
13. Zannis, V.I., Breslow, J.L., and Katz, A.J. (1980) J. Biol. Chem. 255, 8612-8617.
14. Ghiselli, G., Schaefer, E.J., Light, J.A., and Brewer, H.B., Jr. (1983) J. Lipid Res. 24, 731-736.
15. Bojanovski, D., Gregg, R.E., Ghiselli, G., Schaefer, E.J., Light, J.A., and Brewer, H.B. Jr. (1985) J. Lipid Res. 26, 185-193, 1985.
16. Fredrickson, D.S., Altrocchi, P.H., Avioli, L.V., Goodman, D.S., and Goodman, H.C. (1961) Ann. Int. Med. 55, 1016-1031.
17. Herbert, P.N., Gotto, A.M., Jr., and Fredrickson, D.S. (1978) in The Metabolic Basis of Inherited Disease (Stanbury, J.B., Wyngaarden, J.B., and Fredrickson, D.S., eds) pp. 544-588, McGraw-Hill, New York.
18. Schaefer, E.J., Zech, L.A., Schwartz, D.E., and Brewer, H.B., Jr. (1980) Ann. Int. Med. 93, 261-266.
19. Ferrans, V.J., and Fredrickson, D.S. (1975) Am. J. Pathol. 78, 101-158.
20. Assmann, G. (1979) in Atherosclerosis Reviews (Gotto, A.M., and Paoletti, R., eds) Vol. 6, pp. 1-28, Raven Press, New York.
21. Schaefer, E.J., Blum, C.B., Levy, R.I., Jenkins, L.L., Alaupovic, P., Foster, D.M., and Brewer, H.B., Jr. (1978) N. Engl. J. Med. 299, 905-910.
22. Kay, L.L., Ronan, R., Schaefer, E.J., and Brewer, H.B. Jr., (1982) Proc. Natl. Acad. Sci. USA. 79, 2485-2489.
23. Schaefer, E.J., Kay, L.L., Zech, L.A., and Brewer, H.B., Jr. (1982) J. Clin. Invest. 70, 934-945.

24. Schmitz, G., Assmann, G., Rall, S.C., and Mahley, R.W. (1983) Proc. Natl. Acad. Sci. USA. 80, 2574-2578.

25. Zannis, V.I., Lees, A.M., Lees, R.S., and Breslow, J.L. (1982) J. Biol. Chem. 257, 4978-4986.

26. Gordon, J.I., Sims, H.F., Lentz, S.R., Edelstein, C., Scanu, A.M., and Strauss, A.W. (1983) J. Biol. Chem. 258, 4307-4344.

27. Brewer, H.B. Jr., Fairwell, T., Meng, M., Kay, L., and Ronan, R. (1983) Biochem. Biophys. Res. Commun. 113, 934-940.

28. Bojanovski, D., Gregg, R.E., and Brewer, H.B., Jr. (1984) J. Biol. Chem. 259, 6049-6051.

29. Edelstein, G., Gordon, J.I., Vergani, C.A., Catapano, A.J., Peitrini, V., and Scanu, A.M. (1984) J. Clin. Invest. 74, 1098-1103.

30. Bojanovski, D., Gregg, R.E., Zech, L.A., Meng, M.S., Ronan, R., and Brewer, H.B. Jr., (1984) Clin. Res. 309A.

31. Law, S.W., and Brewer, H.B., Jr. (1985) J. Biol. Chem. 260, 12810-12814.

32. Kostner, G.M., Patsch, J.R., Sailer, S., Braunsteiner, H. and Holasek, A. (1974) Eur. J. Biochem. 45, 611-621.

33. Brewer, H.B. Jr., Lux, S.E., Ronan, R. and Jahn, K.M. (1972) Proc. Natl. Acad. Sci. USA. 69, 1304-1308.

34. Lackner, K.J., Law, S.W. and Brewer, H.B., Jr. (1984) FEBS Lett. 175, 159-164.

35. Stoffel, W., Krueger, E. and Deutzmann, R. (1983) Hoppe-Seyler's Z. Physiol. Chem. 364, 227-237.

36. Gordon, J.L., Budelier, K.A., Sims, H.F., Edelstein, C., Scanu, A.M. and Strauss, A.W. (1983) J. Biol. Chem. 258, 14054-14059.

37. Lackner, K.J., Law, S.W. and Brewer, H.B. Jr., (1985) Nucl. Acids Res. 13, 4597-4608.

38. Protter, A.A., Levy-Wilson, B., Miller, J., Bencen, G. White, T., and Seilhamer, J.J. (1984) DNA 3, 449-456.

39. Barker, W.C., and Dayhoff, M.O. (1977) Comp. Biochem. Physiol. 57B, 309-315.

40. Fojo, S.S., Law, S.W., Brewer, H.B., Jr. (1984) Proc. Natl. Acad. Sci. USA. 81, 6354-6357.

41. Fojo, S.S., Law, S.W., Brewer, H.B., Jr., Sakaguchi, A.Y., and Naylor, S.L., (1984) Biochem. Biophys. Res. Commun. 122, 687-693.

42. Fojo, S.S., Law, S.W., Sprecher, D.L., Gregg, R.E., Baggio, G., Brewer, H.B., Jr., (1984) Biochem. Biophys. Res. Commun. 124, 308-313.

43. Goldstein, J.L., and Brown, M.S. (1977) Ann. Rev. Biochem. 46, 897-930.

44. Law, S.W., Lee, N., Monge, J.C., Brewer, H.B., Jr. (1985) Biochem. Biophys. Res. Commun. 131, 1003-1012.

45. Law, S.W., Lackner, K.J., Hospattankar, A.V., Anchors, J.M., Sakaguchi, A., Naylor, S.L. and Brewer, H.B. Jr., (1985) Proc. Natl. Acad. Sci. USA 82, 8340-8344.

46. Steele, J.C.H. Jr., and Reynolds, J.A. (1979) Biol. Chem. 254, 1639-1643.

47. Elovson, J., Jacobs, J.C., Schumaker, V.N., and Puppione, D.L. (1985) Biochemistry 24, 1569-1578.

48. Bell, G.I., Kazam, J.H. and Rutter, W.J. (1979) Proc. Natl. Acad. Sci. USA 78, 5759-5763.

49. Southern, E.M. (1975) J. Mol. Biol. 98, 503-517.

LECITHIN: CHOLESTEROL ACYLTRANSFERASE,

A REVIEW AND IMMUNOCHEMICAL STUDIES

Yves L. Marcel, Camilla A. Vézina, Philip K. Weech,
François Tercé and Ross W. Milne

Laboratory of Lipoprotein Metabolism,
Clinical Research Institute of Montreal
110, Pine Avenue West, Montreal, Quebec H2W 1R7, Canada

Lecithin:cholesterol acyltransferase (LCAT; EC 2.3.1.43) is the major source of cholesteryl esters in human plasma[1]. Despite the considerable progress made in the characterization of this enzyme, the precise mechanism for the formation of cholesteryl esters and their distribution in the various plasma lipoproteins remains unknown and is still actively studied. We will summarize here some of the most recent investigations made in this area and describe some pertinent results from our laboratory.

The mechanism of activation of LCAT by apo AI has been extensively studied and was reviewed recently[2]. Additional information was gained following the discovery of several apo AI mutants: in apo AI-Marburg, the deletion of Lys-107 decreases LCAT activation by 40 to 60%[3]; in apo AI-Giessen, the substitution of Pro-143 by Arg decreases activation by 30 to 40%[4]. The latter mutation lowers the ß-turn probability of the sequence in a portion of the apo AI molecule which is most potent for LCAT activation[5]. In contrast the lysine-107 is located in a region that is inactive by itself in LCAT activation but the deletion alters the nature and orientation of the hydrophilic and hydrophobic faces of α-helices which probably affects the binding of apo AI to lipids[3]. In contrast, it is noteworthy that apo AI from a variety of mammalian species which most likely differ in several positions of their sequence, are functionally similar in LCAT activation[6].

Apolipoproteins other than apo AI such as apo CI[7,8], CII[8,9], CIII[8,9], D[8], AII[9] and AIV[10] have also been reported to activate the LCAT reaction. However the levels of activation achieved by the different apolipoproteins vary greatly and a range of 400 to 6 fold has been reported for apo AI and apo AII respectively in one study[9]. These studies which are carried out with purified enzyme and artificial substrate recombinants, must be interpreted with caution and cannot be extended directly to the physiological situation. For example the relative activating efficiency of apo AI and apo AIV varies extensively as a function of the type of phospholipid substrate used and apo AIV becomes more efficient than apo AI with saturated diacyl lecithins[10]. In addition complexes of saturated lecithins and cholesterol can react with LCAT in the absence of any activator protein[10]. Therefore the protein activators which are essential with most lipid substrates are apolipoproteins (that is with a requirement for lipid binding properties as shown with apo AI mutants [3,4] and with apo AI fragments [5]) which possess additional specific but not unique properties that are shared to different extents by apo AI, AI fragments, AII, AIV, CI, CI fragments, CII, CIII and D.

LCAT binds to lipid vesicles at low ionic strength [11], and, in addition, can react with lipid vesicles made with saturated lecithins in the absence of apolipoproteins [10]. Therefore LCAT binding and reaction with its lipid substrates may depend on a defined configuration of the lipids mediated by ionic strength and the saturation of the acyl chains. Such configuration may not exist in vivo, but it remains that LCAT can bind and interact directly with the lipid surface and that activation may occur as a secondary phenomenon which results from the change in lipid packing upon binding to apolipoproteins characterized by amphipathic helices and specially those with tandem repetitions of 22-amino acid segments [12]. It has also been observed that apo AI activation of LCAT is inhibited by apo AII [13] as well as apo CII, CIII$_1$ and CIII$_2$ [8]. In contrast apo CI activation of LCAT is enhanced by addition of apo CII, CIII and D [8]. These effects are attributable in the case of apo AI to the displacement of an optimum activator by less potent activators or in the case of apo CI by cumulative effects of suboptimal activators .

To summarize the observations made to date on the interaction of LCAT with artificial recombinant substrates, it is clear that purified LCAT can bind directly to lecithin-cholesterol vesicles, although the enzyme may have a higher binding affinity for these vesicles when they contain apo AI. Some controversy exists on this point and certain investigators have concluded that enzyme activation occurs upon binding to complexes via apoproteins [9]. LCAT binding affinity to the lipid vesicles is analogous to that of apo AI and low compared to apo AII and CIII. Therefore we can propose that while LCAT associates preferentially with its substrates, such as complexes rich in lecithin - cholesterol and apo AI, its weak affinity could facilitate the transfer of the enzyme between substrate lipoproteins.

The association of LCAT with plasma lipoproteins has also been studied by determinations of either enzymatic activity or antigenic activity. LCAT activity has been found associated with small HDL-like particles isolated from plasma by selective precipitation followed by hydrophobic and ion-exchange chromatographies [14]. Yamazaki et al [15] have observed that variation in ionic strength could modify the affinity of LCAT for lipoproteins: while LCAT activity is associated exclusively with HDL$_2$ and HDL$_3$ in a medium of ionic strength 0.16, it can also associate with LDL and VLDL upon an increase of ionic strength to 0.5. This effect is attributed to a greater unfolding of the enzyme molecule with greater exposure of the binding sites [15]. Other investigators have also found a small but consistent fraction of LCAT mass and activity that is associated with LDL independently of the isolation methods [16]. In the same study, it has also been noted that LCAT mass and activity is associated more with large than with small HDL-like particles isolated by gel filtration whereas this association is more with small and dense HDL than with large and light HDL after single spin gradient centrifugation [15]. Some conflicting results have also been obtained with lipoproteins isolated by immunoadsorption. Fielding and Fielding first reported that 100% of plasma LCAT activity is bound to apo AI while none is associated with apo AII [17]. In contrast, Cheung and Albers [18] have found about that about 75% of plasma LCAT immunoreactivity is bound to apo AI and that this LCAT mass can be found associated with lipoproteins containing apo AI and AII (48%) and with lipoproteins containing apo AI but no apo AII (27%). It is presumed that these discrepancies are attributable to the different parameters monitored such as enzyme activity [17] versus enzyme immunoreactivity [18], and also perhaps to the different specificities of the antisera used.

We can conclude from this brief review that there is a concensus for the association of LCAT with HDL particles or according to another definition to apo AI-containing lipoproteins. However a variable proportion of LCAT may be associated with large versus small HDL-type particles and with lipoproteins with apo AI versus lipoproteins with apo AI and AII or other lipoproteins. It is also likely that LCAT is associated with both substrate-lipoproteins and product-lipoproteins.

Further progress in the definition of LCAT mechanisms of reaction will require both the development of specific antibodies against this enzyme and the application of non-disruptive methods for the separation of plasma lipoproteins and the evaluation of lipoprotein-associated LCAT activity. In the remainder of this paper we shall describe the development and characterization of monoclonal antibodies directed against LCAT and present a preliminary report on LCAT activity and immunoreactivity associated with plasma lipoproteins separated by non-denaturating gradient gel electrophoresis.

MATERIALS AND METHODS

Plasma used for the purification of LCAT was obtained by centrifugation of blood obtained from the Canadian Red Cross. Lipoproteins used in the different assays were prepared by sequential preparative ultracentrifugation [19] of plasma isolated from normolipemic blood donors. For electrophoresis of plasma on polyacrylamide gradient gels, blood was freshly obtained from normal volunteers and immediately chilled on ice. All plasmas contained 0.02% sodium azide and, except for those which were subsequently assayed for LCAT activity, 1 mM phenylmethylsulfonylchloride (Sigma, St.Louis, USA). The plasma samples were applied to the gradient gels within 6h of blood collection.

LCAT Purification

Lipoprotein-deficient plasma (1500 ml) was prepared by dextran-sulfate $MnCl_2$ precipitation, adjusted to 4M NaCl and incubated with phenyl-Sepharose CL-4B (300 ml; Pharmacia) for 1.5 h at $4^{\circ}C$ as described previously by others [20] for cholesteryl ester transfer protein (CETP). After extensive washing with 10 mM Tris-HCl, 150 mM NaCl, 1 mM EDTA, 0.02% NaN_3 pH 7.4, the LCAT active fraction was eluted with H_2O and dialysed against 1 mM Tris-HCl, 50 mM sodium acetate pH 7.4. CETP was removed from the LCAT preparation by chromatography of the phenyl-Sepharose fraction on CM-Sepharose (40 ml; Pharmacia) in 50mM NaCl, 10 mM sodium acetate pH 4.5 [20]. The unretained fraction from CM-Sepharose which contained LCAT activity but no CETP activity was dialysed against at least 7 l of 10 mM Tris-HCl, 150 mM NaCl, 1 mM EDTA, 0.02% NaN_3 pH 7.4 and applied to a 20 ml column of wheat-germ agglutinin-Sepharose (WGA-Sepharose; Pharmacia) conditionned with the dialysis buffer [21]. After extensive washing, the LCAT bound to WGA-Sepharose was eluted in two peaks by a sequential elution with 0.08 M and 0.5 M N-acetyl-D-glucosamine (Sigma). The two purified LCAT fractions thus obtained were dialysed exhaustively against 10 mM Tris-HCl, 150 mM NaCl, 1 mM EDTA, 0.02% NaN_3 pH 7.4 and stored in aliquots at $4^{\circ}C$ or $-20^{\circ}C$. LCAT activity was stable for at least one month at $4^{\circ}C$.

Assays for LCAT and cholesterol ester transfer activities

To monitor LCAT activity during the purification of the enzyme, apo AI-proteoliposomes labelled with (1,2-^{3}H)-cholesterol were used in the assay[22]. LCAT activity in plasma samples was assayed as described by others[23]. CETP activity in partially purified fractions was measured by determination of cholesterol ester transfer between 4-(^{14}C) cholesterol ester labelled LDL and HDL[24].

Production of monoclonal antibodies

Female 8-week old BALB/c mice were immunized with the fraction eluted from WGA-Sepharose with 0.5 M N-acetyl-D-glucosamine by the procedures described before[25]. Cells from the spleens of these mice were fused with SP2-0 cells and the resulting hybridomas were cloned by limiting dilution[25]. Antibody-producing clones were selected by immunoassay[26] using the 0.5 M N-acetyl glucosamine fraction (about 5 ug protein/ml) as antigen coated onto Immulon 2 Removawells (Dynatech Laboratories Inc.). Further selection of potential anti-LCAT producing clones was made in immunoassays using as antigen pure LCAT preparations made in our laboratory[21] or received as a gift from Dr. A. G. Lacko (Denton, TX) and pure apo D given to us by Dr. W. McConathy (Oklahoma City, OK) and Dr. C. J. Fielding (San Francisco, CA). The immunoblogulin class and subclass of the antibodies were determined in redissolved 40% ammonium sulfate precipitates of the cell culture fluid[27]. Cell clones were injected intraperitoneally into mice, and their ascites was used as the source of antibodies for experiments.

Radioimmunoassay of LCAT

Plastic wells (Removawells, Dynatech Laboratories), were coated overnight with 200 ul of the fraction eluted from WGA-Sepharose with 0.5 M N-acetylglucosamine (1 ug/ml in 5 mM glycine pH 9.2), and saturated by incubation for 30 min with 300 ul of 1% bovine serum albumin (BSA) in phosphate buffered saline (PBS). Each antibody, appropriately diluted in the same buffer, was incubated overnight with dilutions of the reference LCAT. 200 ul of these mixtures were added to the coated wells which had been washed with 0.15 M NaCl containing 0.025% Tween 20. The wells containing the mixtures were then incubated overnight and washed with the Tween-saline solution as above. Labelled goat anti-mouse IgG[28,29], diluted in PBS-BSA was added and incubated overnight. The wells were finally washed with the Tween-PBS and counted for radioactivity.

Nitrocellulose replicas of electrophoretically separated proteins and their immunoreactions

Fractions eluted from WGA-Sepharose containing LCAT were solubilized in 9.5 M urea containing 1% Nonidet and 2% ampholytes pH 4-6 and separated by 2-dimensional electrophoresis[30]. The resolved proteins were transferred electrophoretically to nitrocellulose paper which, after saturation with 3% BSA in Tris buffered saline, was incubated with the different antibodies as described earlier[29]. The nitrocellulose print was first reacted with one antibody (5D4 or 2H11), washed with saturation buffer, reacted with ^{125}I goat antimouse IgG, and washed extensively. After sealing in a plastic envelope the print was submitted to contact autoradiography on XAR-5 Kodak films with an intensifier screen (Cronex, Dupont) at -70°C. This nitrocellulose print was then reacted with a second antibody (4E11) and submitted to a second series of treatments leading to a second autoradiography.

Neuraminidase treatment of LCAT preparations

Neuraminidase-agarose (20-30 units/g of agarose; Sigma), was equilibrated with the incubation buffer just before use. Purified LCAT preparations from WGA-Sepharose eluted with 0.5 M N-acetyl-D-glucosamine (7 ug protein) or from phenyl-Sepharose (2 mg protein) were incubated at 37°C with the insoluble neuraminidase (0.02 to 0.04 units) in 0.025 M phosphate, 0.005% EDTA pH 6.0 (final volume 1.0 ml). Incubations with the WGA-Sepharose fraction contained 1% human serum albumin to stabilize the LCAT activity. At various time intervals from 0 to 3 h, the insoluble neuraminidase was centrifuged and aliquots of the supernatant were removed for measurement of LCAT activity and for isoelectric focusing on 5% polyacrylamide gel slabs in a pH gradient of 3.5 to 6.0 as described by others[31]. After focusing, the proteins were transferred to nitrocellulose paper which was incubated with monoclonal antibodies and processed for autoradiography as above.

Incubations of LCAT with antibodies and immunoadsorbers

The IgG containing the 2H11 and 5D4 antibodies, purified from ascitic fluid by affinity chromatography on protein-A Sepharose, were concentrated by precipitation with 40% saturated ammonium sulfate and dialysed against PBS. Purified LCAT fractions were incubated overnight at 4° with the purified IgG at various concentrations up to five-fold molar excess of IgG with respect to LCAT concentration. The purified LCAT fractions were stabilized during the incubations by the presence of 1% ovalbumin. LCAT substrate was then added and LCAT activity was measured[22,23].

Immunoadsorbers were prepared by coupling purified IgG to activated Sepharose 4B (3 mg protein/ml Sepharose 4B) as described elsewhere[26]. After equilibration of 2.5 ml columns of 2H11-Sepharose, 5D4-Sepharose, or 5E11-Sepharose (an anti apo B immunoadsorber used as control) with 10 mM Tris-HCl, 150 mM NaCl, 0.02% NaN$_3$ pH 7.4, purified LCAT eluted from WGA-Sepharose with 0.5 M N-acetyl-D-glucosamine (4 to 8 ug protein) was applied to each immunoadsorber at 25°C at a flow rate of less than 5 ml/h. In some experiments the equilibration and washing buffers contained 0.1% BSA, in others only the sample contained BSA. Aliquots of unbound fractions were assayed for LCAT activity and for 2H11 and 5D4 antigen levels. Protein bound to the immunoadsorber columns was eluted with 3.0 M sodium thiocyanate pH 6.8, and analysed after dialysis.

Gradient gel electrophoresis of plasma

Aliquots of plasma were pre-stained with an equal volume of acetylated Sudan Black B (Serva, Heidelberg, Germany) dissolved in ethylene glycol (0.5%) for 30 min at 4°C according to Gambert et al[32]. Electrophoresis was carried out on polyacrylamide vertical slab gels 3 mm thick with a linear concentration gradient of 2 to 15% (Pharmacia, Uppsala, Sweden), 2.3 to 18% (Spiral, Dijon, France) or 2.5 to 27% (Isolab, Akron, USA). After pre-electroporesis of the gels at 50v for 30min, a prestained plasma sample (50 μl) was applied in a small well and unstained plasma containing 12% sucrose (200 to 500 ul) was applied over the remaining surface of the gel. Electrophoresis was started at 30v for 30 min, the voltage was then increased to 60v for 60 min and finally to 80v for 6 or 18 h. The shorter time allowed detection of LCAT activity in the protein fraction separated on the 2.3 to 18% gels which migrated almost out of the gel after 18 h. The migration buffer was 14 mM Tris, 110 mM glycine 0.01% azide pH 8.3.

After electrophoresis the separated lipoproteins and proteins were transferred electrophoretically from the gel to nitrocellulose paper (Millipore, Bedford, USA) in the same Tris buffer at 4°C for 24 to 72 h. In some cases, slices of the gradient gel containing the separated lipoproteins were incubated in 1% SDS buffer with or without 5% -mercaptoethanol prior to transfer to nitrocellulose paper in Tris-glycine pH 8.3 with 20% methanol. In other experiments, the plasma proteins were migrated in the 2.3 - 18% gradient gels, a slice of the gel was treated with SDS and applied to a 10% polyacrylamide gel and the proteins were separated in a second dimension in the presence of SDS. The proteins were then transerred to nitrocellulose paper in Tris-glycine pH 8.3 with 20% methanol. Immunoblots of the nitrocellulose prints were obtained by the standard incubation procedures described earlier.

Determination of LCAT activity after gradient gel electrophoresis

For measurement of LCAT activity, fresh plasma (500 ul) was separated on 2.3 - 18% gradient gels, the separated plasma lipoproteins were then electroeluted into 1.5% agarose to facilitate equilibration with the LCAT substrate. The 1.5% agarose slab gel was sandwiched between the gradient gel and a 20% polyacrylamide gel. After electroelution at 100 mAmp in Tris-glycine 5 mM -mercaptoethanol pH 8.3 for 24 h at 4°C, the agarose gel was then sliced into sixteen 5 mm fractions. The slices were incubated overnight in 1.0 ml of 10 mM Tris, 150 mM NaCl, 1 mM EDTA, 5 mM -mercaptoethanol pH 7.4 at 4°C and then with LCAT substrate, either apo A-I proteoliposomes or ^{14}C-cholesterol- albumin, for 4 h at 37°C in the presence of 0.5% human serum albumin. After incubation the lipids were extracted and radioactivity associated with unesterified and esterified cholesterol determined as described previously[21].

Analytical methods

SDS polyacrylamide gel electrophoresis was carried out in 10% gels according to Weber and Osborne[33]. Protein was measured by the method of Lowry et al[34] or in the case of purified LCAT according to Bensadoun and Weinstein[35].

RESULTS

LCAT purification

Elution of protein bound to WGA-Sepharose with two concentrations of N-acetyl-D-glucosamine resulted in two peaks of LCAT activity. The fraction eluted at 0.08 M N-acetyl-D-glucosamine was more concentrated in protein while the fraction eluted at 0.5 M had higher LCAT specific activity. On SDS-polyacrylamide gel electrophoresis the 0.08 M fraction exhibited at least 3 protein bands with major bands at 60 to 65,000 daltons and at 35 to 40,000; the 0.5 M fraction consisted of a major band at 60 to 65,000 daltons, a minor band at 35 to 40,000 and in some preparations one or two trace bands. The LCAT eluted at 0.08 M was purified approximately 9,500 fold; that at 0.5 M was purified 44,000 fold.

Cell fusion and identification of the clones

Growth occurred in all the wells that were seeded and 75 positive wells were identified upon screening with an LCAT preparation eluted from WGA Sepharose with 0.5 M N-acetyl-D-glucosamine. The cells in 55 wells proved to be stable secretors of antibody and were cloned by

168

Table 1: Screening for the specificity of monoclonal antibodies against LCAT

Antigens		Antibodies screened			
Source	Protein dilution (ug/ml)	2H11	5D4	4E11	5G10
LCAT-1	0.5	54845[a]	58195	14080	17195
	0.1	35090	47220	2100	4360
	0.05	24655	40125	1100	2365
	0.025	15230	34225	420	1025
LCAT-2	2.5	42035	50990	22265	17665
	0.5	17625	41485	10160	6605
	0.1	11665	29955	2320	1500
	0.05	5555	21260	770	485
	0.025	2915	11320	560	300
LCAT-3	10	52340	56420	43200	33245
	2	32805	45690	23415	12485
	0.4	7495	25395	3720	3135
	0.2	2855	14325	1660	1490
	0.1	1005	9300	680	700
APO D*	10	160	500	5945	10310
APO D**	10	730	1155	46750	54190

LEGEND TO TABLE 1

Ther antigens used in this screening are identified as follows: LCAT-1 and LCAT-2, two separate preparations [21] with hydroxylapatite chromatography at the last stage of purification and in which apo D could not be visualized after electrophoresis in the presence of SDS or in alkaline Tris-urea; LCAT-3 [39], apo D* [40] and apo D** [41] were generous gifts from Drs. A.G. Lacko, W. McConathy and C.J. Fielding respectively. The antigen at the indicated concentration was coated to the plastic wells which were then saturated with BSA. The appropriate antibody was added, incubated and washed away. The antibody specifically bound was then quantified with ^{125}I-goat anti mouse IgG as described under Methods.

a. cpm bound to the well.

limiting dilutions. Sixteen of the clones whose culture supernatants had the highest apparent titres of antibody were submitted to a double screening using either the LCAT preparation or pure apo D preparations which were gifts from Dr. W.J. McConathy and Dr. C.J. Fielding. Of these 8 were positive and 8 were negative in the tests with the apo D preparations. Therefore 8 of these hybridomas appeared to secrete antibodies which may be directed against an antigen in the apo D preparations[36]. Two hybridomas from each series were recloned by limiting dilution while the others were stored frozen. The reactivities of the 4 antibodies with serial dilutions of several LCAT or apo D preparations has been summarized in Table 1.

Each of the four antibodies, at their lowest dilutions, reacted with our purest LCAT preparations obtained by a previously described method[21]. However as the antigens were diluted, only antibodies 2H11 and 5D4 maintained a significant reacton with LCAT preparations. Similar results were obtained with a LCAT preparation which was a generous gift from Dr. A. G. Lacko. In contrast, when the 4 antibodies were tested against two different preparations of apo D, only antibodies 4E11 and 5G10 reacted significantly (Table I). From these results we tentatively concluded that antibodies 2H11 and 5D4 reacted with LCAT and not with apo D.

For radioimmunoassay of antigen recognized by antibodies 5D4 and 2H11, purified LCAT from WGA-Sepharose was used as the immobilized antigen coated on the Removawells and also as the competing antigen for the standard displacement curve (Fig. 1). Although both 5D4 and 2H11 had good affinity for the immobilized LCAT fraction on the wells, in the case of the 2H11 assay, the LCAT fraction in solution only competed with the immobilized fraction at high concentration. To optimize the sensitivity of the 2H11 assay, the effects of different carrier proteins in the competition assay were assessed. BSA, ovalbumin or gelatin were added to the dilution and saturation buffers at concentrations of 0.1% or1.0%. The best displacement curve was obtained with 0.1% BSA as carrier protein with a working range from 4 to 0.250 ug protein/ml similar to that for the 5D4 assay.

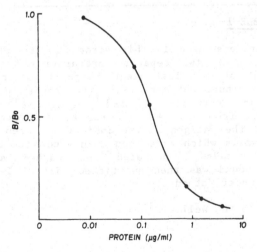

Fig. 1: Radioimmunoassay of LCAT with antibody 5D4. Dilutions of the LCAT preparation eluted from WGA-Sepharose with 0.5 M N-acetylglucosamine were tested in their ability to compete with the same immobilized antigen.

170

Characterization of the antigens recognized by the antibodies

In order to further define the antigens recognized by each antibody, purified LCAT was submitted to 2-dimensional electrophoresis, transferred to nitrocellulose paper and reacted sequentially with the different antibodies as described under Methods. A typical autoradiograph of the immunoreactions observed with antibody 5D4 is shown on Fig. 2A. Antibody 5D4 reacted with a protein of 65,000 daltons characterized by multiple isoforms with pI ranging from 4.4 to 5.0. The same specificity was obtained with antibody 2H11 which also reacted with a protein of 65,000 daltons and with the same range of pI (not illustrated).

When the nitrocellulose print already incubated with antibody 5D4, was further reacted with antibody 4E11 another series of spots appeared on the autoradiograph (Fig. 2B). These corresponded to a protein which migrated as a broad band on SDS gel electrophoresis with a median apparent molecular weight of 35,000. This protein also presented as multiple isoforms with pI ranging from 4.0 to 5.0. The autoradiograph of Fig. 2B indicated that at least 6 different isoproteins were present, which reacted with antibody 4E11.

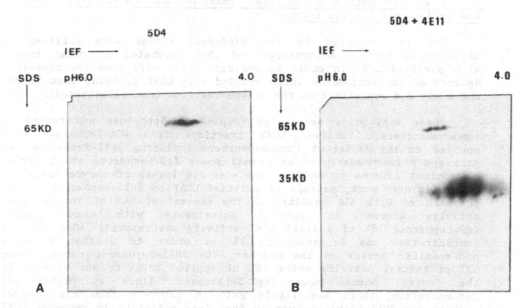

Fig. 2: Electrophoretic blot of the 2-dimensional electrophoresis of the fraction eluted from WGA-Sepharose with 0.5 M N-acetylglucosamine. Fig. 2A is the autoradiograph of the nitrocellulose replica incubated with antibody 5D4 and Fig. 2B is the autoradiograph of the same replica incubated first with antibody 5D4 and then with antibody 4E11.

171

Therefore antibodies 5D4 and 2H11 reacted with a protein which has a molecular weight and isoproteins of pI analogous to those described for purified LCAT while antibody 4E11 reacted with a different protein, present as a contaminant in the LCAT fraction and now identified as apo D [36].

Effect of neuraminidase treatment on the immunoreactivity and enzymatic activity of LCAT

After treatment with insoluble neuraminidase for varying periods of time, aliquots of LCAT preparations were subjected to isoelectricfocusing, transfer to nitrocellulose paper and immunoreaction with antibodies 5D4 or 2H11. Figure 3 illustrates the resulting autoradiograph of a typical experiment with an LCAT preparation from phenyl-Sepharose. The untreated LCAT, (lane 1), exhibited a number of bands which react with antibody 2H11, and with isoelectric points in the pH 4.0 region. After 30 min of neuraminidase treatment, (lane 3), the immunoreactive isoproteins showed a shift of the bands towards a slightly higher pH region and the intensity of the immunoreaction decreased. After two hours of treatment (lane 2), the immunoreaction was barely visible. In contrast there was no effect of neuraminidase treatment on LCAT enzyme activity measured in aliquots of the same neuraminidase treated samples. An LCAT fraction from WGA-Sepharose eluted with 0.5 M N-acetyl-D-glucosamine exhibited the same loss of immunoreaction after treatment with neuraminidase for 3 hours whereas the LCAT activity in the sample remained unchanged. It seems, therefore, that sialic acid residues are important in the antigenic determinants recognized by antibodies 5D4 or 2H11 either as part of the primary structure or in the conformation of the protein, but sialic acids are not essential for expression of LCAT activity in agreement with others [31].

Effect of the antibodies on LCAT reaction and their properties as immunoadsorbers for the enzyme

The IgG secreted by the different clones were purified by affinity on Protein A-Sepharose and then incubated in molar excess with purified LCAT in order to evaluate their effect on the reaction. Neither of the antibodies which reacted with LCAT protein, that is 5D4 and 2H11, had any effect on the enzyme reaction (not illustrated).

These antibodies were also coupled to Sepharose and tested as immunoadsorbers. Purified LCAT fractions from WGA-Sepharose were applied to the different immunoadsorbers including 5E11-Sepharose (an anti apo B immunoadsorber) or normal mouse IgG-Sepharose which served as control columns to monitor non specific losses of enzyme activity. In preliminary work, passage of purified LCAT on 2H11-Sepharose in the presence of 0.1% BSA resulted in the removal of 94% of initial LCAT activity whereas in control experiments with normal mouse IgG-Sepharose 18% of initial LCAT activity was removed. When the BSA concentration was increased to 1% in order to further minimize non-specific losses on the columns, the 2H11-Sepharose removed only 53% of initial activity while 12% of applied activity was removed by the control normal mouse IgG-Sepharose. Since at higher BSA concentration, there was little gain in recovery from the control columns yet 2H11-Sepharose was in fact less effective in removing LCAT activity, these results indicated an interference of the higher albumin concentration on the antigen-antibody interaction. These observations were corroborated in the 2H11 radioimmunoassay where better competition of soluble antigen was obtained in the presence of 0.1% BSA rather than with 1% BSA.

Therefore in subsequent experiments, the LCAT preparations were diluted in 0.1% BSA prior to application on the immunoadsorber columns (Table 2). In experiment 1, the immunoadsorbers and the elution buffer were also equilibrated with 0.1% BSA. Under these conditions, a single passage on 2H11 Sepharose and 5D4 Sepharose removed respectively 71% and 38% of initial LCAT activity while 9% of applied activity was lost on the control 5E11-Sepharose. In experiment 2, where only the applied LCAT preparation contained 0.1% BSA, a

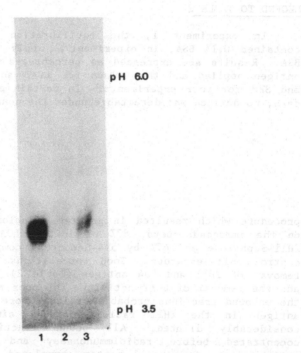

Fig. 3: Effect of neuraminidase treatment on the immunoreactivity of LCAT with antibody 2H11. A partially purified LCAT preparation from phenyl-Sepharose (2 mg protein) was incubated with neuraminidase (0.04 units) in 1.0 ml centrifuge tubes. Aliquots (175 g protein) were withdrawn at intervals and focused on a 5% polyacrylamide gel slab for 18 h at 4°C. A nitrocellulose paper blot of the focused protein was incubated with antibody 2H11, washed and incubated with ^{125}I-goat anti-mouse IgG to produce the autoradiograph in the figure. Samples applied were: untreated LCAT (lane 1); LCAT incubated with neuraminidase for 2 h (lane 2), and 30 min (lane 3).

Table 2: Immunoaffinity chromatography of LCAT

Immuno-adsorber	LCAT activity		2H11 antigen		5D4 antigen	
	unbound		unbound	bound	unbound	bound
Exp. 1						
2H11	29	± 4.4	-	45	-	53
5D4	62	± 4.1	-	81	13	58
5E11	91	± 7.6	50	11	73	10
Exp. 2						
2H11	17	± 6.9	-	68	22	38
5D4	53	± 5.0	-	82	41	46
5E11	90	± 4.1	43	28	83	8

LEGEND TO TABLE 2

In experiment 1, the equilibration and washing buffer also contained 0.1% BSA, in experiment 2, only the sample contained 0.1% BSA. Results are expressed as percentages of total activity or total antigen applied and the values for LCAT activity represent the mean and SEM for four experiments. In certain experiments, indicated by a dash, no antigen was detectable under the conditions of assay.

procedure which resulted in greater dilution of the BSA upon passage on the immunoadsorbers, 83% of initial LCAT activity was removed by 2H11-Sepharose and 47% by 5D4-Sepharose compared to only 10% by the control 5E11-Sepharose. Good correlations were observed between the removal of 2H11 and 5D4 antigen (Table 2) by the respective columns and the removal of LCAT activity. The poor recovery of 2H11 antigen in the unbound fractions probably reflects more the limit of detection of antigen in the 2H11 radioimmunoassay since these fractions were considerably diluted. All bound fractions were dialysed and cocentrated before radioimmunoassay and their concentrations of antigens were in general better correlated with the amounts of LCAT activity bound by the immunoadsorbers.

Analysis of LCAT associated with plasma lipoproteins separated by gradient gel electrophoresis

When plasma was electrophoresed on non-denaturating gradient gels as described under Methods, the lipoproteins were fractionated by size as illustrated by the prestained sample (Fig. 4A). LCAT activity associated with the fractions of different size was analyzed by 2 methods: one measured the net LCAT activity of each fraction and

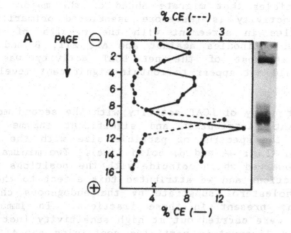

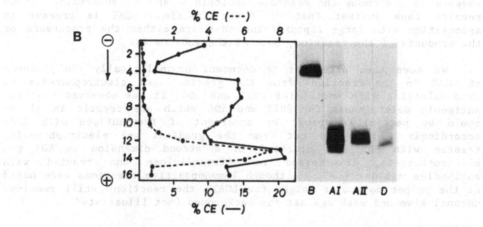

Fig. 4: Evaluation of LCAT activity associated with plasma lipoproteins separated by gradient gel electrophoresis. Panel A and B are two representative experiments. After electrophoresis, LCAT activity present in the separated fractions was assayed by two methods, one relying on the endogenous substrate present in each fraction[23] (o--o) and the other using an exogenous substrate[22] (o—o). The position of the different fractions is indicated by the number on the vertical axis and the position of the lipoproteins by the illustration of the prestained lipoproteins after electrophoresis (Panel A, right)). The position of different apolipoproteins in the same electrophoresis is obtained by immunoblots with antibodies against apo B, AI, AII and D (Panel B, right).

reflected both enzyme level and endogenous substrate availability[23] (Fig. 4A and B, dotted line), while the other[22] measured mostly the enzyme level but was affected by the dilution of labelled cholesterol due to the plasma lipoproteins (Fig. 4A an B, solid line). Net LCAT activity was only observed in the fractions containing the small lipoprotein particles that migrate ahead of the major HDL fraction. This net LCAT activity is therefore associated primarily with small HDL-like particles in agreement with the results of others. The immunoblots with antibodies against apo AI, AII, B and D (Fig. 4B) also showed that most of the net LCAT activity was present in fractions that did not appear to contain significant levels of apo AII and D.

In contrast assay of LCAT activity with the second method yielded a very different distribution and significant enzyme activity was present accross the spectrum of particle size with the exception of the LDL fraction (Fig. 4A and B, solid line). Two minima of enzymatic activity were observed which coincide with the positions of the major LDL and HDL fractions and we attributed this effect to the dilution of the labelled cholesterol substrate by the endogenous cholesterol of the lipoproteins present in these fractions. In immunoblots with antiapo AI that were carried out at high sensitivity (not illustrated) we observed that lipoprotein particles containing apo AI were present in all the fractions that expressed LCAT activity, including those that are larger than LDL, but which are inactive as substrates in the absence of exogenous cholesterol- lecithin - apo AI substrate. These results thus suggest that in these fractions LCAT is present in association with large lipoproteins that are either the precursors or the products of the reaction, but not the substrates.

We have also attempted to document immunologically the presence of LCAT in the fractions from the gradient gel electrophoresis by immunoblotting with antibodies 2H11 and 5D4. It was observed that the antigenic determinants for 2H11 and 5D4 which are cryptic in plasma could be partially exposed by treatment of the antigen with SDS. Accordingly a lane was cut from the gradient gel electrophoresis, treated with SDS and migrated in a second dimension in SDS gel electrophoresis, transferred to nitrocellulose and reacted with antibodies against LCAT. Although immunopositive reactions were noted at the proper molecular weight for LCAT, the reactions still remained unconclusive and weak against the background (not illustrated).

DISCUSSION

Immunization of mice with a purified LCAT preparation followed by differential screening with a series of purified LCAT and apo D preparations have generated a series of clones that secrete antibodies against LCAT, two of which have been characterized in detail. The evidence for the specificity of antibodies 2H11 and 5D4 can be summarized as follows. These antibodies 5D4 and 2H11 react with a protein of 65,000 daltons as evaluated by SDS-gel electrophoresis; such a molecular weight corresponds well with the published values[2]. The published pI for LCAT isoproteins differ somewhat as Doi and Nishida[31] have found a pI ranging from 3.9 to 4.2 and Utermann et al[37] have reported a pI from 4.2 to 4.4. Albers and colleagues who have measured the pI in the presence of 8 M urea, have observed more alkaline values from 5.1 to 5.5. In our experiments, antibodies 5D4 and 2H11 react with multiple isoproteins of pI ranging from 4.4 to 5.0. While these values do not correspond directly to the published pI, we do not think that they are contradictory, since apparent pI are directly influenced by the techniques used and especially by the

176

purity of urea which may often shift the pH gradient toward more alkaline values. When the LCAT preparations were treated with neuraminidase (Fig. 3), we observed a shift of the proteins recognized by antibodies 5D4 and 2H11 toward more alkaline pI. This confirms the presence of sialic acid residues in the proteins recognized by these antibodies and is in agreement with the reported presence of sialic acid in LCAT [31]. Although the resolution afforded by the autoradiographs did not allow us to determine exactly the number of isoproteins reacting with these antibodies, we evaluated these to range from 4 to 6, a value in agreement with that observed by Doi and Nishida [31]. It should be pointed out that the double antibody reaction and autoradiography yield a much higher sensitivity than Coomassie blue staining; however since the antigenic reaction disappeared as the sialic acid residues were removed, other isoproteins could have been present which would be undetected. From these experiments, we concluded that antibodies 5D4 and 2H11 reacted with multiple isoproteins containing sialic acid, whose pI and molecular weight correspond to those described for LCAT.

In highly sensitive immunoreactions, even our purest LCAT preparations were found to be heterogeneous, however we have been able to distinguish between the LCAT antigen and contaminating antigen. The contaminating protein reacted with antibodies 4E11 and 5G10 and had a molecular weight (about 35,000 daltons) that was different from that of LCAT (Fig. 1B). The contaminating protein was distinguished quantitatively from the 2H11 and 5D4 antigen by the radioimmunoassay shown in Table 1. Immunoreactivity between the contaminating protein and antibodies 4E11 and 5G10 was lost when the LCAT preparations were diluted to concentrations that conserved their immunoreactivity with 2H11 and 5D4. Antibodies 4E11 and 5G10 reacted in this radioimmunoassay with preparations of apo D supplied by Drs. McConathy and Fielding, and we have direct evidence that they are specific for apo D and other antigenically related proteins [36].

While antibodies 2H11 and 5D4 when preincubated with active LCAT preparations, failed to inhibit the enzyme activity, these antibodies linked to Sepharose could remove in a single passage about 80 and 50% respectively of the LCAT activity from partially purified preparations (Table 2). This demonstrates that antibodies 2H11 and 5D4 react with LCAT or with a protein specifically bound to LCAT. However the latter possibility appears highly unlikely since the analytical characteristics of the protein resemble very closely those described for LCAT.

In both solid phase radioimmunoassays and in immunoadsorber experiments, we have noted the critical importance of the carrier protein used in the respective dilution and elution buffers. While their presence is necessary to minimize non-specific adsorption and losses, the proteins appear to interfere with the antigen-antibody reaction. Therefore their concentrations were adjusted as a compromise between these two factors. Because we have not observed any similar effect of albumin on the interaction of antibodies against various apolipoproteins with their respective antigens [25,29] and because there is much evidence for a specific interaction between albumin and LCAT activity [2], we suggest that albumin may interact with LCAT in a specific manner which can mask the determinants recognized by antibodies 2H11 and 5D4.

In conclusion, monoclonal antibodies 2H11 and 5D4 have been shown to react with isoforms of a protein with all the analytical characteristics of human LCAT. We can also conclude that the antigenic determinants recognized by 2H11 and 5D4 either contain sialic acid residues or are dependent for their conformation upon the presence of sialic acid residues. However these determinants are different from the active site of the enzyme since the antibodies cannot inhibit its reaction in vitro in contrast with the polyclonal antibodies described previously[37,38]. The determinants also appear to be cryptic in plasma as the antigen could not be identified conclusively in plasma lipoproteins separated by gradient gel electroporesis. Therefore these antibodies have been satisfactory only in semi-purified systems and additional work will be required to find conditions that can expose LCAT antigenic determinants in the presence of plasma constituents. However LCAT enzymatic activity can be readily monitored in plasma and in assays with or without exogenous substrate, we were able to demonstrate that LCAT is associated not only with small HDL-like particles but also with a variety of other large lipoproteins. These large lipoproteins contain apo AI but they are inactive as substrates for LCAT. Thus they may represent either the precursors of LCAT substrates or the products of the exhausted substrates following LCAT reaction.

ACKNOWLEDGEMENTS

We are indebted to Dr. Alain Larouche and Thanh Dung N'guyen for their assistance in the purification of LCAT, to Louise Blanchette and Richard Théolis Jr. for their help in the screening of clones and preparation of ascitic fluids, to Mireille Hogue for her help with electrophoresis experiments, and to Louise Lalonde for typing the manuscript. This work was supported in part by grants from the Quebec Heart Foundation and from the Medical Research Council of Canada (PG-27). R.W. Milne is a scholar of the Fondation de la Recherche en Santé du Québec.

REFERENCES

1. Glomset J.A. and Norum K.R. (1973) Adv. Lipid Res. 11, 1-65.
2. Marcel Y.L. (1982) Adv. Lipid Res. 19, 85-135.
3. Rall S.C. Jr., Weisgraber K.H., Mahley R.W., Ogawa Y., Fielding C.J., Utermann G., Haas J., Steinmetz A., Menzel H.-J., and Assmann G. (1984) J. Biol. Chem. 259, 10063-10070.
4. Utermann G., Haas J., Steinmetz A., Paetzold R., Rall S.C. Jr., Weisgraber K.H. and Mahley R.W. (1984) Eur. J. Biochem. 144, 325-331.
5. Sparrow J.T. and Gotto A.M. Jr. (1982) C.R.C. Crit. Rev. Biochem. 13, 87-107.
6. Chen C.H. and Albers J.J. (1983) Biochim. Biophys. Acta 753, 40-46.
7. Soutar A.K., Garner C.W., Baker H.N., Sparrow J.T., Jackson R.L., Gotto A.M., and Smith L.C. (1975) Biochemistry 14, 3057-3064.
8. Albers J.J., Lin J.T. and Roberts G.P. (1979) Artery 5, 61-75.
9. Jonas A., Sweeny S.A. and Herbert P.N. (1984) J. Biol. Chem. 259, 6369-6375.
10. Steinmetz A. and Utermann G. (1984) J. Biol. Chem. 260, 2258-2264.
11. Furukawa Y. and Nishida T. (1979) J. Biol. Chem. 254, 7213-7219.
12. Boguski M.S., Elshourbagy N., Taylor J.M. and Gordon J.I. (1984) Proc. Natl. Acad. Sci. U.S.A. 81, 5021-5025.
13. Scanu A.M., Lagocki P. and Chung J. (1980) Ann. N.Y. Acad. Sci. 348, 160-173.
14. Jahani M. and Lacko A.G. (1981) J. Lipid Res. 22, 1102-1110.
15. Yamazaki, S., Mitsunaga T., Furukawa Y. and Nishida T. (1983) J. Biol. Chem. 258, 5847-5853.

16. Chen C.-H. and Albers J.J. (1982) Biochem. Biophys. Res. Commun. 107, 1091-1096.
17. Fielding P.E. and Fielding C.J. (1980) Proc. Natl. Acad. Sci. U.S.A. 77, 3327-3330.
18. Cheung M.C. and Albers J.J. (1984) J. Biol. Chem. 259, 12201-12209.
19. Havel R.J., Eder, H.A. and Bragdon J.H. (1955) J. Clin. Invest. 34, 1345-1353.
20. Morton, R.E. and Zilversmit D.B. (1981) Biochim. Biophys. Acta 663, 350-355.
21. Suzue, G., Vézina, C. and Marcel, Y.L. (1980) Can. J. Biochem. 58, 539-541.
22. Chen C.H. and Albers J.J. (1982) J. Lipid Res. 23, 680-691.
23. Stokke K.J., and Norum K.R. (1971) Scand. J. Clin. Lab. Invest. 27, 21-27.
24. Pattnaik N.M., Montes A, Hughes L.B. and Zilversmit D.B. (1978) Biochim. Biophys. Acta 530, 428-438.
25. Milne R.W., Douste-Blazy Ph., Retegui L., and Marcel Y.L. (1981) J. Clin. Invest. 68, 111-117.
26. Milne R.W., Weech P.K., Blanchette L., Davignon J., Alaupovic P. and Marcel Y.L. (1983) J. Clin. Invest. 73, 816-823.
27. Chalon M.P., Milne R.W. and Vaerman J.P. (1979) Europ. J. Immunol. 9, 747-751.
28. Mellman I.S. and Unkeless J.C. (1980) J. Exp. Med. 152, 1048-1069.
29. Marcel Y.L., Hogue M., Théolis R. Jr., and Milne R.W. (1982) J. Biol. Chem. 257, 13165-13168.
30. O'Farrell P.H. (1975) J. Biol. Chem. 250, 4007-4021.
31. Doi Y., and Nishida T. (1983) J. Biol. Chem. 258, 5840-5846.
32. Gambert P., Lallemant C., Athias A. and Padieu P. (1982) Biochim. Biophys. Acta 713. 1-9.
33. Weber K., and Osborne M. (1969) J. Biol. Chem. 244, 4406-4412.
34. Lowry O.H., Rosebrough N.J., Farr A.L. and Randall R.J. (1951) J. Biol. Chem. 193, 265-275.
35. Bensadoun A. and Weinstein D. (1976) Anal. Biochem. 70, 241-250.
36. Weech P.K., Camato R., Milne R.W. and Marcel Y.L. (1985) Manuscript submitted for publication.
37. Utermann G., Menzel, H.-J., Adler G., Dieker P. and Weber W. (1980) Eur. J. Biochem. 107, 225-241.
38. Albers J.J., Adolphson J.L. and Chen C.-H. (1981) J. Clin. Invest. 67, 141-148.
39. Chong K.S., Davidson L, Huttash R.G. and Lacko A.G. (1981) Arch. Biochem. Biophys. 211, 119-124.
40. McConathy W.J. and Alaupovic P. (1976) Biochemistry 15, 515-520.
41. Chajek T., and Fielding, C.J. (1978) Proc. Natl. Acad. Sci. USA 75, 3445-3449.

179

LECITHIN: CHOLESTEROL ACYLTRANSFERASE (LCAT) DEFICIENCY SYNDROMES

Jiri Frohlich and Roger McLeod

Shaughnessy Hospital
Lipoprotein Research Group
Department of Pathology, University of British
Columbia, Vancouver, B.C., Canada

There are two types of LCAT deficiency: the primary or familial disorder and that secondary to other diseases.

In man deficiency of the enzyme activity has been described in numerous conditions. Some of them are listed in table 1.

Table 1: Conditions in Which Decreased LCAT Activity
Has Been Reported

HYPERTENSION[1]	KIDNEY TRANSPLANTATION[1]
SURVIVORS OF M.I.[1,2]	HEREDITARY NEUROPATHY[12]
CORONARY ATHEROSCLEROSIS[3,4]	ABETALIPOPROTEINEMIA[13-15]
DIABETES MELLITUS[5,6]	TANGIER DISEASE[16-20]
MALABSORPTION[7]	A-I/C-III DEFICIENCY[21]
UREMIA[8-11]	

However, the data are difficult to interpret because in the majority of these reports no differentiation was made between deficiency in the enzyme mass, deficiency of the substrate, presence of inhibitors or lack of activators. In several patients with Tangier disease, and in a patient with A-I/C-III deficiency, both LCAT activity and mass (measured immunochemically) was low. In most cases the immunoassay values correlate well with those of the LCAT activity assay using exogenous, most commonly proteoliposome, substrate methods (J. Albers, personal communication).

<u>Familial LCAT deficiency</u>. Data on 37 patients diagnosed since 1967 are summarized in Table 2.

Additional six homozygotes from two Japanese (N. Murayama and J. Albers - personal communication) and one American family of Dutch-Mennonite origin (E. Janis, R. Rodeles and J. Frohlich: manuscript in preparation) are currently investigated. Patients with familial LCAT deficiency usually present with one or all of the following features:

1) Corneal changes. "Finely dotted cornea" with a peripheral arcus is one of the hallmarks of the disease. Chemical analysis of the corneal tissue in two patients revealed mostly extracellular

lipid deposits. Markedly increased free cholesterol and phospholipid content was observed in the tissue.[25,38]

2) Mild hemolytic anemia. Peripheral blood film from our patient S.F. (#17 in table II) shows numerous unusual cell shapes (fig. 1). Alterations of erythrocyte membrane composition and function have been described both in the homozygotes and heterozygotes[45,59-61,66,69]

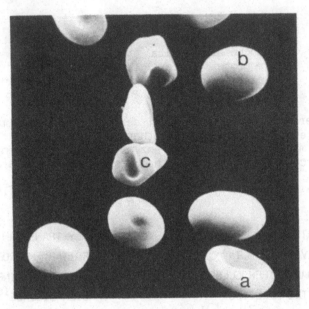

Fig. 1. Scanning electron micrograph of peripheral blood film from patient S.F. showing stomatocytes (a) spherostomatocytes (b), and a knizocyte (c).

Table 2: Patients with Familial LCAT Deficiency

FAMILY	PATIENT	SEX	DOB	ETHNIC ORIGIN	ANEMIA	RENAL INVOLVEMENT	REFERENCE
I	1	F	1936	Norwegian	+	+	
	2	F	1934	Norwegian	+	+	22
	3	F	1946	Norwegian	+	+	
II	4	F	1921	Swedish	+	+	22
	5	M	1935	Swedish	+	+	
III	6	F	1926	Norwegian	+	+	22
	7	M	1932	Norwegian	+	+	
IV	8	M	1918	Norwegian	+	+	22
	9	F	1914	Norwegian	+	–	
V	10	M	1942	Italian	+	+	23
	11	M	1944	Italian	+	+	
VI	12	M	1955	Indian	–	–	24,25
VII	13	M	1945	Eng-Cdn	+	?	26
	14	M	1940	Eng-Cdn	+	+	
VIII	15	F	1934	French	+	+	27
	16	F	1937	French	+	?	
IX	17	M	1959	Ital-Dutch	+	+	28
	18	F	1954	Ital-Dutch	+	–	
X	19	F	1946	Japanese	+	MIN	
	20	M	1948	Japanese	+	+	29
	21	M	1950	Japanese	+	+	
XI	22	M	1950	Eng-Irish	+	+	30
	23	F	1955	Eng-Irish	+	–	
XII	24	F	1932	Irish	+	+	
	25	F	1925	Irish	–	–	31
	26	M	1932	Irish	+	+	
	27	F	1931	Irish	+	–	
XIII	28	F	1954	Norwegian	+		32
XIV	29	F	1949	Germ-Irish	+	MIN	
	30	M	1958	Germ-Irish	–	+	33
	31	M	1955	Germ-Irish	–	+	
XV	32	M	1933	English	MIN	MIN	34
XVI	33	M	1965	Italian	–	+	35
XVII	34	F	1937	Japanese	+	–	36
	35	M	1933	Japanese	+	–	
XVIII	36	M	1934	Japanese	+	+	37
	37	F	1924	Japanese	+	–	

3)   Renal pathology. Most important clinically is the kidney
      lesion.[39] Figure 2 shows electron micrograph of the kidney
      lesion in LCAT deficiency. The typical EM findings include
      accummulation of electron dense material (corresponding to lipid
      e.g. free cholesterol and phospholipid) in the capillary lumen,
      in subendothelial or subepithelial spaces of renal arteries and
      in the glomerular basement membrane. Fusion of foot processes
      is also a prominent feature of both early and advanced lesions.
      Some of these changes were also observed in a biopsy of one
      patient without proteinuria[31] and in a normal kidney several
      months after its transplantation into a patient with familial
      LCAT deficiency.[40]

Fig. 2   Electron micrographs of a glomerulus from patient S.F. (patient
          17 in Table 2). Clear vacuoles (arrowheads) some of which
          contain very osmiophilic material are present in the glomerular
          basement membrane (arrows). There is widespread fusion of foot
          processes (CL = capillary lumen).

      In summary, the typical features of the clinical presentation are
corneal opacities, hemolytic anemia and various degrees of renal
involvement ranging from an increase in albumin and β_2 microglobulin
excretion[70] to proteinuria and uremia.
      Table 3 summarizes the frequency of these changes in the ten
patients from Scandinavia as compared to the 27 patients from other
ethnic backgrounds. The corneal opacities have been found in every
patient with the disease. Anemia, while present in all the Scandinavian
subjects, was absent in 5 of the other patients on initial examination.
Similarly, proteinuria was present in only 22 of the 37 patients listed
in table 2.

Table 3: Frequency of Clinical Findings in Two Groups of
Patients with Familial LCAT Deficiency

	Scandinavian (10 patients)	All other (27 patients)	Number Affected
Corneal opacities	100%	100%	37
Anemia	100%	82%	32
Proteinuria	80%	74%	22
Uremia	70%	14%	10
Death	60*%	4*%	7

* Retrospective diagnosis of LCAT deficiency made in one deceased sibling in each group.[22,31]

There is a marked difference in the incidence of uremia between the Scandinavian patients 70% of whom developed uremia and 6 of whom ultimately died of kidney failure, and the group of the other 27 patients only 14% of whom developed uremia.

These differences cannot be attributed to the age difference (table 4). The interval from the diagnosis of uremia to death was on average 3.5 years in these individuals. It is also of interest to point out that among the other patients two developed serious kidney disease at rather young age (patients 17 and 26 in table 2).

Table 4: Age at Diagnosis, Uremia and Death in the Two Groups of LCAT-Deficient Patient

	Diagnosis	Age at Uremia	Death
Scandinavians	39.0 + 6.2	44.5 + 9.5 (7)	47.8* + 6.9 (6)
Others	33.9 + 11.7	25, 34*, 47 (3)	34* (1)
Range	16 - 60	25 - 58	40 - 58

Mean + S.D. Number of patients in parentheses. *Retrospective diagnosis of LCAT deficiency in one deceased patient in each group.[22,31]

The following factors may play a role in the etiology of the kidney disease of familial LCAT deficiency:

1) Superimposed glomerulonephritis. Decreased clearance of immune complexes by the kidney was suggested by Borysiewicz et al.[31]. Apart from the suggestive clinical presentations and possible immune complex deposits found in the kidney biopsy of one patient[31] there is little else to support the concept that patients with LCAT deficiency are more susceptible to glomerulonephritis.

2) Possible relationships between hemolytic anemia and the kidney lesion as suggested in previous studies in the cholesterol fed guinea pig[41,42]. French et al[41,42] postulated that the severity of the kidney lesion is related to the degree of hemolytic anemia induced by cholesterol-feeding in guinea pigs. However, as shown in table 5 there is no significant difference in hemoglobin concentration between the group of patients with and without proteinuria. In addition at least three of the patients with anemia had no proteinuria and two of the patients without proteinuria were anemic. Thus, in man, it appears unlikely that the hemolytic anemia and the renal disease are causally related.

3)   The possibility that the development of the kidney lesion is
     accelerated by high protein (high fat) diet.[43]
     We examined the retrospective data on the level of BUN and
     creatinine at the time of the initial examination of all the
     Scandinavian patients and compared them with data available for
     the other patients (Table 6).

Table 5: Relationship Between Proteinuria and Anemia
In Familial LCAT Deficiency

	Proteinuria	
	Present	Absent
Initial hemoglobin (g/dl)	10.8	10.9
	± 1.5	± 2.1
Number of patients with anemia/total in category	19/22	11/13

Table 6: Initial BUN, Creatinine and BUN/Creatinine Ratios in Patients
with Familial LCAT Deficiency Related to the Clinical Outcome

PATIENT	BUN	CREATININE	BUN/CR	OUTCOME
1*	45	1.0	45	dead
2	44	1.1	40	dead
3	27	1.0	27	alive
4	51	1.0	46	dead
5	61	-	-	dead
6	58	1.4	41	dead
7	56	1.4	40	dead
8	254	6.0	42	transplanted
9	34	1.0	34	alive
28	15	.6	24	alive
All (5)	-	-	41.5 ± 2.4	dead
All (3)	-	-	28.3 ± 5.1	alive
All others	-	-	24.3 ± 9.4	alive

*numbers refer to Table 2

     It is apparent that patients who died of their kidney disease
presented with significantly higher (p 0.01) ratio of BUN to creatinine
in their plasma than the survivors. Other patients without kidney
disease had also BUN:creatinine ratios significantly (p 0.01) lower
than the affected Scandinavian patients.
     Since the level of BUN, in the absence of a significant clearance
defect (urea clearances were initially normal in these individuals), is
largely determined by the dietary protein intake these data suggest that
the patients with serious renal involvement were on substantially higher
protein diets than the others.
     Preliminary data from our laboratory show that addition of protein
to high cholesterol diet leads to more extensive structural and
functional renal impairment in guinea pigs. On the other hand
acetylphenylhydrazine-induced hemolytic anemia did not cause either the
lipid deposition or mesangial cell proliferation in the kidney
glomeruli.[44]
     The lipoprotein changes in LCAT deficiency have been thoroughly
reviewed by Glomset, Norum and Gjone[45]. Figures 3a and 3b show
electron micrographs of plasma LDL and HDL from LCAT deficient subject DH
(#18 in Table 2). Of note are the membraneous disc like structures in
LDL and the stacked discs and small particles in the HDL density region.

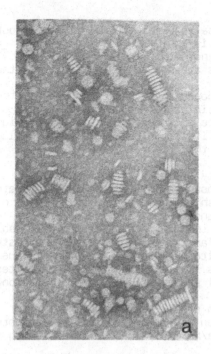 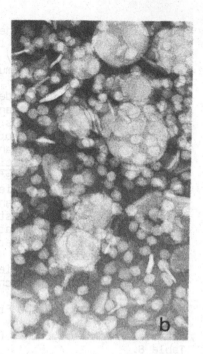

Fig. 3   Negative staining electron micrographs of HDL (a) and LDL (b)
           from subject DH (18 in table 2). Magnified x 60,000.

The plasma lipoprotein composition in patient SF (17 in Table 2)
who is hypertriglyceridemic and currently in renal failure and DH (whose
plasma lipids have always been low) are shown in Table 7. Regardless of
the plasma level of triglycerides the VLDL, LDL and HDL fractions are all
enriched in triglycerides.

Table 7:   Plasma and Lipoprotein Fraction Lipids in
           Two Patients with Familial LCAT Deficiency

	Plasma			VLDL			LDL			HDL		
	TC	TG	PL	TC	TG	PL	TC	TG	PL	TC	TG	PL
S.F.	218	738	402	92	708	146	82	25	145	7	–	24
D.H.	82	82	189	7	37	13	40	18	67	15	–	45
NORMAL	151	48	160	6	15	6	83	10	59	32	5	74

TC: total cholesterol, TG: triglycerides, PL: phospholipids.
VLDL, LDL, HDL: very low, intermediate, low and high density
lipoproteins, respectively. Data for IDL not shown. All values in
mg/dl.

Several hypotheses can be advanced why this is the case.
Norum[46] proposed that the enrichment of the triglyceride rich
particles with C-apoproteins is preventing their uptake by the liver
receptors. Another possibility is that the lower plasma concentration
of lysophosphatidylcholine may impair the dissociation of the C apo
proteins from these particles as shown in vitro by Windler et al.[47]
Finally, the lipoprotein lipase activity may be low in these
individuals. Data from Dr. J. Brunzell's laboratory show that both

the mass and activity of the lipase is within normal limits in our
patients. However, if the lipase is inhibited in vivo by free
cholesterol as suggested by Fielding[48] this inhibition may not be
apparent when artificial emulsions are used as a substrate for the in
vitro assay.

Finally, Walli & Seidel[71] showed that LP-X can compete with
the triglyceride rich remnant particles for uptake at the apo E
receptor in rat hepatocytes. Since LCAT-deficient plasma has high
concentration of this lipoprotein it is another possible mechanism of
delayed triglyceride clearance in this disorder.

Heterogeneity of LCAT Deficiency

Patients with LCAT deficiency are of diverse genetic
background. Of the 43 patients known to us at present approximately
46% (20) are of Scandinavian, Irish or English origin, 28% (12) of
Asian, 15% (7) have at least one ancestor of Italian or French
origin. Of interest is that 3 of the families (VII and IX in table 2
and the newly diagnosed American family) have at least one ancestor
from either Dutch or Swiss Mennonite background. Thus the frequency
of the gene for LCAT deficiency is likely quite high in Mennonites.

The clinical and biochemical presentations also differ among the
individual families.

Albers et al,[49] and Frohlich et al[50] established
immunoassays for LCAT. The data on LCAT mass and activity in both
artificial and endogenous, (Stokke/Norum[51]) substrates are shown in
Table 8.

Table 8: Range of LCAT Mass and Activity Values Reported
in the Homozygotes and Heterozygotes for Familial Deficiency
and in Normal Controls

	LCAT		
	MASS (mg/l)	ASA* (nmols/hr/ml)	SN**
Homozygotes	0 - 2.65	0 - 9.1	0 - 12
Heterozygotes	2.6 - 4.8	42 - 64	34 - 67
Controls	4.3 - 7.8	80 - 163	40 - 93

* ASA: artificial substrate assays
** SN: Stokke-Norum assay

Albers concluded that there are three types of families with
LCAT deficiency.[52]
1)   Those with the apparent absence of both LCAT activity and mass,
      e.g. family 9 in Table 2.[50,53]
2)   Families with functionally defective LCAT. In these families
      some LCAT protein is present but there is little or no activity,
      e.g. families 1 to 4, 13 and 17 in Table 2.[54]
3)   Those with small amount of apparently normal LCAT, e.g. family 5
      in Table 2.[55]

Table 9 summarizes the findings in the heterozygous, homozygous
and normal family members of family 9[28] and in a control
population. It should be pointed out that in this family both the
mass and the activity assay (using the artificial substrate)
distinguished between the normal and the heterozygous individuals.
However, the Stokke/Norum assay did not differentiate between
heterozygotes and normals. The most likely reason for this
discrepancy is the exposure of the substrate to the enzyme and,
possibly, suboptimal reaction rates in the Stokke/Norum assay.

188

Table 9: Summary of LCAT Assays in Italian-Dutch Family (#9, TABLE 2)

		MASS (mg/dl)	ASA (nmols/hr/ml)	SN
Homozygotes	(2)	0	0	0
Heterozygotes	(8)	3.3 + 0.1	69.8 + 13.7	88.4 + 21.8
Normal Relatives	(7)	5.5 + 0.5	117.9 + 10.4	67.5 + 12.1
Controls	(12-14)	5.4 + 0.5	101.0 + 10.9	70.7 + 13.7

Mean + SD, number of subjects in parentheses. Other abbreviations as in table 8.

Data in Table 10 show the other extreme of the spectrum of familial LCAT deficiency namely the findings in a Japanese family (XVIII in table 2) with partial LCAT deficiency. In these individuals a significant amount of relatively inactive enzyme is present in the plasma.[56] Also, the mass values in the heterozygotes are close to the normal range.

Table 10: Summary of LCAT Assays in Japanese Family (#18, TABLE 2)

		MASS	ASA	SN
Homozygotes	(2)	2.31; 2.65	8.3; 9.1	(15)
Heterozygotes	(6)	4.4 + .2	54 + 2.4	-
Controls	(19)	5.7 + .9	100 + 16	-

Symbols and units the same as in Tables 8 and 9.

Heterozygotes for LCAT Deficiency

No obvious lipoprotein abnormalities have been reported in these individuals so far. However, postprandial hypertriglyceridemia, lower HDL levels and increased apo B have now been reported in heterozygotes from family #9.[57] Very low plasma total and HDL-cholesterol levels were also observed in 6 obligatory heterozygotes in the currently investigated American family of Dutch-Mennonite background.

Other abnormalities found in the heterozygotes can be summarized as follows:
- Approximately half normal LCAT mass (between 2.6 and 4.6 mg/l) compared with normal 4.2 to 7.8 mg/l).
- Approximately half normal activity of LCAT in the artificial substrate assay but usually normal activity in Stokke-Norum assay.
- In many of these families hyperlipidemia especially hypercholesterolemia (in the Norwegian and Irish families) appears to be frequent.[58,31]
- Distinct abnormalities of erythrocyte membrane have been described in families 9 and 11.[59-61]
- Possible abnormalities of plasma lipoproteins in postprandial specimens as mentioned above.

Secondary LCAT Deficiency
1)   Liver Disease
In human liver disease, especially liver failure the levels of LCAT activity in plasma are low.[62,63] The situation is more controversial in cholestatic diseases. However, recent work from laboratory (J. Albers, personal communication) suggests a relationship between plasma LCAT mass, activity and the degree of biliary

obstruction. Their study on 12 patients with cholestatic jaundice should be supplemented with further data to substantiate the hypothesis that LCAT is a useful enzyme to measure for prognosis or even quantitation of the degree of cholestasis.

2) Experimental models of LCAT deficiency were established in
    1) galactosamine treated rat[64]
    2) praseodymium nitrate fed rat[65,66]
    3) cholesterol fed guinea pig[41,67]
    While useful for the study of lipoproteins, plasma erythrocyte and renal changes all of these models - in which liver damage leads to impaired LCAT secretion - are acute in nature. None of the typical progressive changes (e.g. corneal opacities and uremia) have been observed in these animals. Anatomical but no functional kidney lesions similar to those seen in LCAT deficiency were noted in cholesterol-fed guinea pig[41].
    Table 11 summarizes the current progress in our understanding of LCAT deficiency states.

Table 11: Summary of the Progress in the Study of LCAT Deficiency

Progress	Lack of Progress
- Genetics	- Control of LCAT synthesis and secretion
- Biochemistry	- Etiology and pathogenesis of the kidney disease
- Lipoprotein metabolism	- Study of the heterozygotes
	- Mechanism of the hypertriglyceridemia

With the availability of cDNA probe for LCAT[68] the genetic polymorphism in the deficient families can be investigated. Remarkable progress has also been made in our understanding of the changes in lipoprotein metabolism in this disorder.[45]
    However, little is known about the control of LCAT synthesis and secretion, about the etiology and pathogenesis of the kidney disease, or alterations of lipid metabolism in heterozygotes (i.e. what is the minimal amount of LCAT that is sufficient to maintain normal morphology and metabolism of lipoproteins?). Furthermore we also do not fully understand the mechanism of triglyceride enrichment of lipoproteins in patients with LCAT deficiency. By investigation of the LCAT deficiency states we should be able to further define the role of the LCAT reaction in plasma lipoprotein metabolism.

ACKNOWLEDGEMENTS
    This work has been supported by B.C. Heart and B.C. Health Care Research Foundations.

REFERENCES

1) Dobiasova, M. Lecithin-cholesterol acyltransferase and the regulation of endogenous cholesterol transport. Adv. Lip. Res. 20:107-194 (1983).
2) Soloff, L.H., Ruttenberg, H.L., Lacko, A.G. Plasma lecithin-cholesterol acyltransferase activities after myocardial infarction. Am. Heart J. 326:419-427 (1973).

3) Dobiasova, M., Vondra, K. Lecithin-cholesterol acyltransferase as a possible diagnostic tool in ischemic heart disease. Scand. J. Clin. Lab. Invest. 38 (Suppl 150):129-133 (1978).

4) Wallentin, L. and Moberg, B. Lecithin-cholesterol acyl transfer rate and high density lipoprotein in coronary artery disease. Atherosclerosis 41:155-165 (1982).

5) Misra, D.P., Staddon, G., Powell, N. and Crook, D. Lecithin-cholesterol acyltransferase activity in diabetes mellitus and effect of insulin on these cases. Clin. Chim. Acta 56:83-87 (1974).

6) Mattock, M.B., Fuller, J.H., Maude, P.S. and Keen, H. Lipoproteins and plasma cholesterol esterification in normal and diabetic subjects. Atherosclerosis 34:437-669 (1979).

7) Miller, J.P., Thompson, G.R. Lecithin-cholesterol acyltransferase activity in subjects with malabsorption. Eur. J. Clin. Invest. 3:401-406 (1973).

8) Guarnieri, G.F., Moracchiello, M., Companacci, L., Ursini, F., Ferri, L., Valente, M., Gregolin, C. Lecithin-cholesterol acyltransferase (LCAT) activity in chronic uremia. Kidney Int. 13 (Suppl 8):S26-S30 (1978).

9) Jung, K., Neumann, R., Precht, K., Nugel, E., Scholz, D. Lecithin-cholesterol acyltransferase activity, HDL-cholesterol and apolipoprotein A in serum of patients undergoing chronic hemodialysis. Enzyme 25:273-275 (1980).

10) Chan, M.K., Ramdial, L., Varghese, Z., Persand, J.W., Baillod, R.A., Moorhead, J.F. Plasma lecithin-cholesterol acyltransferase activities in uremic patients. Clin. Chim. Acta 119:65-72 (1982).

11) McLeod, R., Reeve, C.E., Frohlich, J. Plasma lipoproteins and lecithin-cholesterol acyltransferase distribution in patients on hemodialysis. Kidney Infl. 25:683-688 (1984).

12) Yao, J.K., Ellefson, R.D., Dyck, P.J. Lipid abnormalities in hereditary neuropathy. Part 1. Serum non-polar lipids. J. Neurol. Sci. 29:161-175 (1976).

13) Cooper, R.A., Gulbrandsen, C.L. The relationship between serum lipoproteins and red cell membranes in abetalipoproteinemia: Deficiency of lecithin-cholesterol acyltransferase. J. Lab. Clin. MEd. 78:323 (1971).

14) Kostner, G., Holasek, A., Bohlmann, H.G., Thiede, H. Investigation of serum lipoproteins and apoproteins in abetalipoproteinemia. Clin. Sci. Mol. Med. 46:457 (1974).

15) Scanu, A.M., Aggerbeck, L.P., Kruski, A.W., Lim, C.T., Kayden, H.J. A study of the abnormal lipoproteins in abetalipoproteinemia. J. Clin. Invest. 53:440 (1974).

16) Clifton-Bligh, P., Nestel, P.J., Whyte, H.M. Tangier disease: Report of a case and studies of lipid metabolism. N. Engl. J. Med. 286:567 (1972).

17) Scherer, R., Rukenstroth-Bauer, G. Untersuchung der lecithin-cholesterincyltransferase-aktivitat im Serum von drei Patienten mit Tangier-krankheit (Hyp-alpha-lipoproteinamie). Clin. Wochenschr. 51:1059 (1973).

18) Greten, H., Hannemann, T., Gusek, W., Vivell, O. Lipoproteins and lipolytic plasma enzymes in a case of Tangier disease. N. Engl. J. Med. 291:548 (1974).

19) Yao., J.K., Herbert, P.N., Fredrickson, D.S., Ellefson, R.D., Heinen, R.J., Forte, T., Dyck, P.J. Biochemical studies in a patient with a Tangier syndrome. J. Neuropathol. Exp. Neurol. 37:138 (1978).

20) Pritchard, H.P. Lipoprotein deficiency syndromes. Ed. Angel, A. and Frohlich, J., Plenum (1985).

21) Norum, R.A., Lakier, J.B., Goldstein, S. et al. Familial deficiency of apolipoproteins A-I and C-III and precocious coronary artery disease. New Engl. J. Med. 306:1513-1517 (1982).

22) Gjone, E., Norum, K.R., Glomset, J.A. Familial lecithin-cholesterol acyltransferase deficiency. In the Metabolic Basis of Inherited Disease. Eds. Stanbury, J.B., Wyngaarden, J.B., Fredrickson, D.S., 4th Edn. New York, McGraw-Hill, pp 589-603 (1978).

23) Utermann, G., Schoenborn, W., Langer, K.H. et al. Lipoproteins in LCAT deficiency. Humanagenetik 16:295-306 (1972).

24) Bron, A.J., Lloyd, J.K., Fosbrooke, A.S., Winder, A.F., Tripathi, R.C. Primary LCAT deficiency disease. Lancet 1:928-929 (1975)

25) Winder, A.G., Bron, A.J. Lecithin-cholesterol acyltransferase deficiency presenting as visual impairment, with hypocholesterolemia. and normal renal function. Scand. J. Clin. Lab. Invest. 38 (Suppl 150):151-155 (1978).

26) Bethell, W., McCulloch, C., Ghosh, M. Lecithin cholesterol acyltransferase deficiency. Light and electron microscopy findings from two corneas. Can. J. Opthalmol. 10:49-51 (1975).

27) Chevet, D., Ramee, M.P., LePogamp, P., Thomas, R., Garre, M., Alcindor, L.G. Hereditary lecithin cholesterol acyltransferase deficiency. Report of a new family with two afflicted sisters. Nephron. 29:212-219 (1978).

28) Frohlich, J., Godolphin, W.J., Reeve, C.E., Evelyn, K.A. Familial LCAT deficiency. Report of two patients from a Canadian family of Italian and Swedish descent. Scand. J. Clin. Lab. Invest. 38 (Suppl 150):156-161 (1978).

29) Iwamoto, A., Naito, C., Teramoto, T., Kato, H., Kako, M., Kariya, T., Shimizu, T., Oka, A., Oda, T. Familial lecithin-cholesterol acyltransferase deficiency complicated with unconjugated hyperbilirubinemia and peripheral neuropathy. Acta. Med. Scand. 204:219-227 (1978).

30) Shojania, M.A., Jain, S.K., Shohet, S.B. Hereditary lecithin-cholesterol acyltransferase deficiency. Report of 2 new cases and review of the literature. · Clin. Invest. Med. 6:49-45 (1983).

31) Borysiewicz, L.K., Soutar, A.K., Evans, D.J., Thompson, G.R., Rees, A.J. Renal failure in familial lecithin-cholesterol acyltransferase deficiency. Quarter J. Med. 204:411-26 (1982).

32) Gjone, E., Blomhoff, J.P., Holme, R., Torstein, H., Bjornar, O., Scarbovik, A.J., Teisbert, P. Familial lecithin-cholesterol acyltransferase deficiency. Report of a fourth family from north-western Norway. Acta. Med. Scand. 210:3-6 (1981).

33) Kane, J. Personal communication (1984).

34) Zech, L., Brewer, B. Personal communication (1984).

35) Vergani, C., Catapano, A.L., Roma, P., Guidici, G. A new case of familial LCAT deficiency. Acta. Med. Scand. 214:173-6 (1983).

36) Sakuma, M., Akanuma, Y., Kodama, T., Yamada, N., Murata, S., Murase, T., Itakura, H., Kosaka, K. Familial plasma lecithin-cholesterol acyltransferase deficiency. A family with partial deficiency of plasma LCAT activity. Acta. Med. Scand. 212:225-32 (1982).

37) Murayama, N., Asano, K., Kato, Y. et al. Effects of plasma infusion on plasma lipids, apoproteins, and plasma enzyme activities in familial lecithin-cholesterol acyltransferase deficiency. Eur. J. Clin. Invest. 14:122-129 (1984).

38) Winder, A., Garner, A., Sheraidah, G.A. Pathology of cornea in familial LCAT deficiency. Abstract, Intl. Conf. on Lipopr. Def. Syndromes, Vancouver, B.C., May 13-14 (1985).

192

39) Gjone, E. Familial lecithin-cholesterol acyltransferase deficiency - a new metabolic disease with renal involvement. Adv. Nephrol. 10:167-185 (1981).

40) Flatmark, A.L., Hovig, T., Mahre, E., Gjone, E. Renal transplantation in patients with familial lecithin-cholesterol acyltransferase deficiency. Transplant. Proc. 9:1665-1671 (1977).

41) French, S.W., Yamanaka, W., Ostwald, R. Dietary induced glomerulosclerosis in the guinea pig. Arch. Path. 83:204-210 (1967).

42) French, S.W., Wood, C., Benson, N., Ostwald, R. Mesangial proliferation in hemolytic anemia induced by a 1% cholesterol diet (unpublished).

43) Brenner, B.M., Meyer, T.W., Hostetter, T.H. Dietary protein intake and the progressive nature of kidney disease. N. Engl. J. Med. 307:652-658 (1982).

44) El Shebeb, T., Frohlich, J., Magil, A. Effect of dietary protein supplement on the renal lesion in cholesterol-fed guinea pigs. In preparation.

45) Glomset, J.A., Norum, K.R., Gjone, E. Familial lecithin-cholesterol acyltransferase deficiency. In Metabolic Basis of Inherited Disease. Eds. Stanbury, J., Wyngaarden, J., Fredrickson, D.S., Goldstein, J.L., Brown, M. 5th edition, McGraw Hill, 643-654 (1983).

46) Norum, K.R. Familial deficiency of lecithin-cholesterol acyltransferase for treatment of hyperlipoproteinemia. Eds. L.A. Carlson and A.G. Olsson. Raven Press, N.Y. (1984).

47) Windler, E., Preger, S., Greten, H. The role of lysophosphatidylcholine in the regulation of the catabolism of triglyceride-rich lipoproteins. In: Cologne Atherosclerosis Conference No. 2 Lipids. Ed. Parnham, J., Birkhauser Verlag, pp 101-105 (1984).

48) Fielding, C.J. Inhibition of lipoprotein lipase by free cholesterol. Bioch. Biophy. Acta 218:221-232 (1970).

49) Albers, J.J., Adolphson, J.C., Chen, C.H. Radioimmunoassay of human lecithin-cholesterol acyltransferase. J. Clin. Invest. 67:141-146 (1981).

50) Frohlich, J., Hon, K., McLeod, R. Detection of heterozygotes for familial lecithin-cholesterol acyltransferase deficiency. Am. J. Hum. Genet. 34:65-72 (1982).

51) Stokke, K.T., Norum, K.R. Determination of lecithin-cholesterol acyltransferase in human blood plasma. Scand. J. Clin. Lab. Invest. 27:21-27 (1971).

52) Albers, J.J., Bergelin, R.O., Adolphson, J.C. et al. Population-based reference values for lecithin-cholesterol acyltransferase (LCAT). Atherosclerosis 43:369-379 (1982).

53) Albers, J.J., Chen, C-H., Adolphson, J.C. Familial lecithin:cholesterol acyltransferase: Identification of heterozygotes with half-normal enzyme activity and mass. Clin. Genet. 58:306-309 (1981).

54) Albers, J.J., Gjone, E., Adolphson, J.C. et al. Familial lecithin:cholesterol acyltransferase deficiency in four Norwegian families. Acta Med. Scand. 210:455-459 (1981).

55) Albers, J.J., Uterman, G. Genetic control of lecithin-cholesterol acyltransferase: measurement of LCAT mass in a large kindred with LCAT deficiency. Am. J. Hum. Genet. 33:702-708 (1981).

56) Albers, J.J., Chen, C-H., Adolphson, J.C. et al. Familial lecithin:cholesterol acyltransferase deficiency in a Japanese family: Evidence for functionally defective enzyme in homo-zygotes and obligate heterozygotes. Hum. Genet. 62:82-85 (1982).

57) Frohlich, J., McConnathy, W., McLeod, R. Plasma lipoproteins and LCAT activity in heterozygotes for familial LCAT deficiency. Abstract, International Conference in Lipoprotein Deficiency Syndromes, Vancouver, B.C., May 13-14 (1985).

58) Torsvik, H., Gjone, E., Norum, K.R. Familial plasma cholesterol ester deficiency study in a family. Acta Med. Scand. 183:387-391 (1968).

59) Godin, D.V., Gray, G., Frohlich, J. Erythrocyte membrane alterations in lecithin:cholesterol acyltransferase deficiency. Scand. J. Clin. Lab. Invest. 38 (Suppl 150):162-167 (1978).

60) Maravajlia, B., Herring, F.G., Weeks, G., Godin, D.V. Detection of erythrocyte membrane abnormalities in lecithin:cholesterol acyltransferase deficiency using a spin label approach. J. Supramol. Struct. 11:1-7 (1979).

61) Jain, S.K., Mohandas, N., Sensabaugh, G.F. et al. Hereditary plasma lecithin:cholesterol acyltransferase deficiency: a heterozygous variant with erythrocyte membrane abnormalities. J. Lab. Clin. Med. 99:816-827 (1982).

62) Simon, J.B. Lecithin:cholesterol acyltransferase activity in liver disease. Scand. J. Clin. Lab. Invest. 33 (Suppl 137):107-111 (1974).

63) Harry, D.S., Day, R.C., Owen, J.S. et al. Plasma lecithin:cholesterol acyltransferase activity and the lipoprotein abnormalities of liver disease. Scand. J. Clin. Lab. Invest. 38 (Suppl 150):223-277 (1978).

64) Sabesin, S.M., Kuiken, L.B., Ragland, J.B. Lipoprotein and lecithin:cholesterol acyltransferase changes in galactosamine-induced rat liver injury. Science 190:1302-1304 (1975).

65) Von Lehmann, B., Grajewski, O. and Oberdisse, E. Correlation between serum lecithin:cholesterol acyltransferase activity and erythrocyte lipid content during experimental liver damage. Naunyn-Schmiedeberg's Arch. Pharmacol. 298:211-216 (1977).

66) Godin, D.V. and Frohlich, J. Erythrocyte attractions in praseodymium-induced lecithin: cholesterol acyltransferase (LCAT) deficiency in rat comparison with familial LCAT deficiency in man. Res. Commun. Chem. Pathol. Pharmacol. 31:555-556 (1981).

67) Drevon, C.A., Hovig, T. The effects of cholesterol fat feeding on lipid levels and morphological structures in liver, kidney and spleen in guinea pigs. Acta Path. Microbiol. Scand. Sect. A 85:1-18 (1977).

68) Humphries, S., McIntyre, N., et al. Manuscript in preparation.

69) Murayama, N., Asano, Y., Hosoda, S., et al. Decreased sodium influx and abnormal red cell membrane lipids in a patient with familial lecithin: cholesterol acyl transferase deficiency. Am. J. Hemat. 16:129-137 (1984).

70) Frohlich, J., Bergseth, M., Reeve, C.E. Plasma ß-2-microglobulin and urinary albumin: creatinine ratio in lecithin:cholesterol acyltransferase deficiency. Nephron 32:96 (1982).

71) Walli, A.K., Seidel, D. Role of CP-X in the pathogenesis of cholestatic hypercholesterolemia. J. Clin. Invest. 74:867-879 (1984).

72) Ritland, S., Gjone, E. Quantitative studies of lipoprotein-X in familial lecithin:cholesterol acyltransferase deficiency and during cholesterol esterification. Clin. Chim. Acta 59:109-119 (1975).

CHOLESTEROL EFFLUX: MECHANISM AND REGULATION

George H. Rothblat and Michael C. Phillips

Department of Physiology and Biochemistry
The Medical College of Pennsylvania
Philadelphia, PA

INTRODUCTION

The removal of cholesterol from peripheral cells by HDL is the first step in what has become known as reverse cholesterol transport. This process involves the efflux of free cholesterol from the membrane of cells and the transport of the cholesterol to the liver[1]. This report will present data obtained from our studies on cholesterol efflux and then, briefly, discuss some related data on the mechanism of cholesterol uptake by the liver.

Two general models can be presented for how cholesterol could leave a cell. The first model is termed the "collision complex model". In this model, the HDL would come in direct contact with the cell membrane and a collision complex would form. Within this complex the free cholesterol (FC) would partition from the plasma membrane into the HDL, after which the complex would dissociate and the HDL would leave carrying its load of cholesterol. The second model can be termed the "aqueous diffusion model". In this model, the cholesterol would desorb directly into the aqueous phase and the cholesterol molecules would then be picked up from the aqueous phase by the acceptor HDL.

The objective of the first series of studies was to determine which of the limiting models was responsible for cholesterol flux. Our first approach was to monitor the movement of cholesterol between two popula-tions of phospholipid vesicles, one population serving as cholesterol donors while the second population served as acceptors. It was felt that the interpretation of the kinetic data would be simplified by first studying cholesterol movement between vesicles or lipoproteins, and then extending these observations to the more complex cell systems. The data obtained from these studies have been published[2,3] and the results have been confirmed by other laboratories[4]. All of the data are consistent with the exchange or transfer of FC between vesicles or lipoproteins occurring by the aqueous diffusion mechanism. Among these observations are 1) the rate of exchange of cholesterol was first order with respect to cholesterol concentration in the donor particle. 2) the exchange of cholesterol was approximately five times faster than the movement of phosphatidylcholine (PC). This is consistent with the greater aqueous solubility of cholesterol (FC). 3) the rate of exchange of FC between populations of phospholipid vesicles or lipoproteins is zero order with respect to the concentration of acceptor particles. The interpretation of these and other results in the context of the aqueous diffusion model has previously been presented in detail[3].

CELLULAR FACTORS REGULATING EFFLUX

The objective of the next series of investigations was to determine if the aqueous solubility mechanism established for the movement of FC between vesicles or lipoproteins was applicable for the efflux of cholesterol from cells in culture. The general approach was to monitor the uni-directional movement of radiolabeled cholesterol from prelabeled monolayers of tissue culture cells to a variety of cholesterol acceptors[2,5]. The properties of some of these acceptors are presented in Table 1.

Table 1. Characteristics of Cholesterol Acceptor Particles

Composition	Shape	Diameter(A^O)
Egg PC vesicles	Unilamellar Spheres	250
Egg PC/apo HDL	Spheres	90
Egg PC/apo A-I or C	Discs*	170
Egg PC/Na taurocholate	Discs*	100

*Large dimension of discs measured by EM.

Although the various acceptor particles differed in size, shape and composition, none contained cholesterol and all contained phospholipid. Fig. 1 illustrates that upon the addition of any of these acceptors to the culture medium there was a rapid release of labeled cholesterol from the cells. Analysis of the kinetic data obtained from this type of study demonstrated that the release of cellular cholesterol was first order with respect to cholesterol concentration in the donor cells, thus a semi-exponential plot of the fraction of the original counts remaining in the cells against time is linear and can easily provide an estimate of the half-time ($t_{1/2}$) for the release of cell cholesterol[2,5].

Fig. 2 illustrates the relationship between increasing acceptor concentration and the rate of release of cellular cholesterol. The data in Fig. 2 were obtained by exposing J-774 mouse macrophage cells to increasing concentrations of an apo HDL/PC complex. As can be seen from Fig 2, the $t_{1/2}$ is a function of the acceptor concentration at low acceptor concentration and becomes independent of acceptor concentration at high acceptor concentrations. This pattern is different than that previously established in the studies on vesicles or lipoproteins where the transfer of cholesterol was independent of acceptor concentration[2,3]. The difference between the two experimental systems has been attributed to the presence of a large unstirred water layer around cells[2,5]. As represented by Fig. 3, when the concentration of acceptor particles is greater than A, desorption of cholesterol from the cells becomes rate limiting and the kinetics of cholesterol efflux become independent of acceptor concentration. When the acceptor concentration is less than A, efflux decreases with decreasing acceptor concentration because of limited mixing and

196

diffusion within the unstirred water layer which lowers the frequency of collisions between desorbed cholesterol molecules and acceptor particles. It is at such low acceptor concentrations that the presence of HDL binding sites on the surface of cells[6] could modulate the rate of release of cellular cholesterol by increasing the local concentration of acceptor particles near the cell surface.

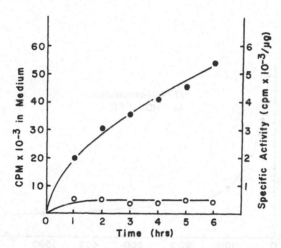

Figure 1. Release of cholesterol from prelabeled Fu5AH rat hepatoma cells in the absence (O) or presence (●) of 1 mg/ml egg phosphatidyl-choline added as unilamellar vesicles.

The objective of the next series of studies was to compare the rate of cholesterol release from different cell types. Although there was considerable differences between the various cell types, all cells produced essentially the same dose response pattern (see Fig. 2) demonstrating a large effect of acceptor dose at low concentrations, with efflux becoming independent of concentration as acceptors were increased. Fig. 4 indicates the $t_{1/2}$ of a number of different cell types exposed to apo HDL/PC complex at 1 mg PC/ml. Half-times range from 3 hr to greater than 25 hr. These differences in $t_{1/2}$ were shown to persist with both monolayers and suspension cultures and thus are characteristic of each cell line[5].

To determine if the rate of cholesterol release was a function of the cellular plasma membrane, a series of experiments were conducted comparing the $t_{1/2}$ for cholesterol efflux from whole cells and isolated plasma membranes[7]. Membrane vesicles were prepared by treatment with formalde-hyde[8]. The two cell lines selected for comparison were the Fu5AH rat hepatoma cells and the WIRL rat liver cells. Both cell lines are of rat liver origin, and both have a similar size and cholesterol content. As is illustrated in Fig. 4, the two cell lines differ considerably in $t_{1/2}$.

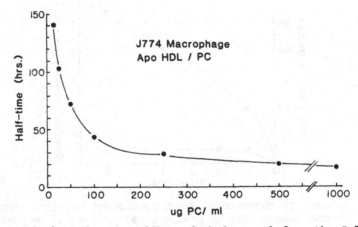

Figure 2. Half-times for the efflux of cholesterol from the J-774 mouse macrophage exposed to apo HDL/phosphatidylcholine complexes.

The difference in the rate of release of cholesterol observed with whole cells persisted when the $t_{1/2}$ for efflux from isolated membranes was compared[7]. These results demonstrate that the efflux of FC from cells is modulated at the level of the cellular plasma membrane and that differences in the plasma membrane structure account for differences in the kinetics of cholesterol release from different cell types. Among the structural features of membranes that may influence efflux are: 1) The cholesterol content. The rate of desorption is proportional to cholesterol concentration in the membrane. 2) Phospholipid unsaturation/saturation ratio. Studies with model membranes have shown increasing rates of flux

198

as unsaturation is increased. 3) PC/ sphingomyelin ratio. Preliminary studies have indicated a dramatic increase in $t_{1/2}$ of cholesterol exchange from sphingomyelin vesicles when compared to PC vesicles. 4) Transbilayer distribution of cholesterol. 5) Cholesterol interaction with membrane proteins and 6) receptor binding, particularly at low acceptor concentrations.

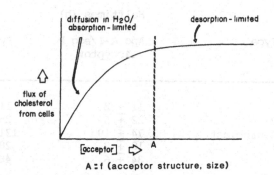

Figure 3. Schematic of the effect of acceptor concentrations on cellular cholesterol efflux. Details are discussed in text.

FACTORS DETERMINING ACCEPTOR EFFICIENCY

The studies described above were designed to establish those parameters of cells which influenced the rate of cholesterol release. Since the efflux of cholesterol is also a function of the specific acceptor particles to which the cell are exposed a series of studies were initiated to determine the characteristics of cholesterol acceptors which regulated their efficiency. This was accomplished by exposing cells to a variety of different acceptors[5]. Fig. 5 presents comparative data obtained upon exposure of Fu5AH hepatoma cells to increasing concentrations of PC liposomes, apo HDL/PC complexes and bile acid micelles (see table 1 for details). The data, which are normalized on the basis of phospholipid concentration in the incubation medium demonstrated, that at all concentrations, cholesterol efflux decreased in the order bile acid micelle > apo HDL/PC complex > egg PC vesicle. Maximum efflux was obtained with the micelles at a concentration of 250 ug PC/ml and the curves in Fig. 5 indicate that increasing the concentration of the HDL/PC complex and the PC vesicles caused the rate of cholesterol loss to approach the maximum value achieved with the bile acid/PC micelles. Analysis of comparative data obtained with acceptors which differ in composition and size revealed that the differences in $t_{1/2}$ between acceptors could be eliminated, to a large extent, by normalizing the data in terms of the product of acceptor particle number x particle radius[9].

Table 2. Comparison of Efflux from Different Cells Exposed to Apo A-I and Apo C Acceptors

| Cell Type | Halftimes (h)[*] | |
	Apo A-I/egg PC Acceptor	Apo C/egg PC Acceptor
Fu5AH	11 ± 2	11 ± 3
Hep G2	22 ± 5	25 ± 4
Skin fibroblast	74 ± 19	131 ± 37[**]
L-cell	18 ± 2	28 ± 1[**]
J-774	34 ± 6	48 ± 13[**]

[*]Acceptor concentration was 100 ug egg PC/ml.
 Incubation time was 9 h at 37°C.
 Mean ± standard deviation.

[**]$p < 0.01$

An important question related to acceptor efficiency is the role played by specific apoproteins in modulating the efflux of cellular cholesterol. We have compared the efflux from several cell types exposed to complexes of egg PC and either human apo A-I or apo C. These complexes were of defined size and stoichiometry. The data in Table 2 show that the different hepatoma cells give different $t_{1/2}$ values because the value for the human hepatoma Hep G2 is twice that of the rat hepatoma Fu5AH. In the case of these two cell lines, the nature of the apoprotein in the acceptor complex did not influence the rate of release of cellular cholesterol since particles containing either apo A-I or apo C gave the same $t_{1/2}$ values. This was not the case with the three non-liver cell lines shown in Table 2. With these cells the $t_{1/2}$ obtained in the presence of apo A-I containing acceptors is consistently shorter than that observed with apo C particles. One explanation for these differences could be that the three non-liver cells have receptors which recognize apo A-1 containing particles and thus increase the local concentration of acceptors near the surface of the cell. However, if the differences between apo A-I and apo C containing particles were due solely to an interaction with a specific binding site one would expect the differences between the two acceptors to

200

be most dramatic at low acceptor concentrations, below receptor saturation. This was not the case; the difference between the particles was
evident at all acceptor concentrations[9]. Alternatively, the differences
between the acceptors observed with some cell lines could be a consequence
of the apoprotein directly influencing the composition or structure of the
plasma membrane and thus modifying the rate of cholesterol desorption from
the membrane.

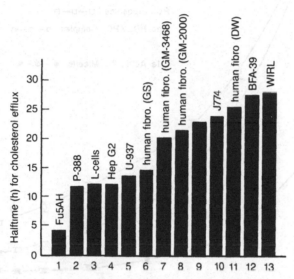

Figure 4.   Comparison of the half-times for cholesterol efflux for a variety
of cells incubated with apo HDL/egg PC acceptor particles at a
concentration of 1 mg PC/ml.

BIDIRECTIONAL FLUX

All of the experiments described above followed the unidirectional
movement of cholesterol from cells which served as cholesterol donors to
acceptor particles which contained no cholesterol. Under physiological
conditions where HDL or a specific sub-class of HDL would serve as an
acceptor the bidirectional flux of cholesterol would occur (for a review
see Bell[10]). The net movement of cholesterol between cells and lipoproteins would be a function of the relative rates of cellular uptake and
release. It has been proposed that changes in the composition of the
lipoprotein could effect the equilibrium between cells and lipoprotein
which would then result in either cellular cholesterol deposition or
depletion[10]. We have recently performed a detailed series of experiments
to test the hypothesis proposed by Jansen and Hulsmann[11] that hepatic
lipase, by depletng HDL of phospholipid, modulates the net transfer of HDL
FC into liver cells. The incubation of human HDL with either rat hepatic
lipase or snake venom phospholipase A-2 resulted in depletion of the HDL
phospholipids without changing either the cholesterol or apoprotein

content. Incubation of cultured rat hepatoma cells with this lipase modified HDL resulted in the net uptake of HDL FC relative to that observed with control HDL[12].

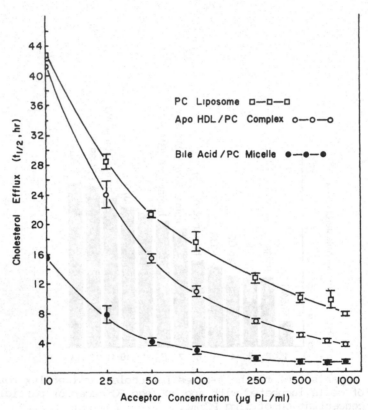

Figure 5. Effect of acceptor type and concentration on half-times for cholesterol efflux from Fu5AH hepatoma cells. ● , sodium taurocholate/egg PC; ○ , apo HDL/egg PC; □ , egg PC vesicles. See Table 1 for details on acceptor particles.

Investigations into the mechanism of this response indicated that the enhanced accumulation of cholesterol was not accompanied by increased apoprotein degradation or increased cellular uptake of HDL sphingomyelin or cholesteryl ester[13]. These observations suggested that the hepatic lipase was stimulating a physicochemical process and that the net accumulation of cellular cholesterol was being accomplished by either increasing cholesterol influx from the modified HDL or reducing cellular efflux. These two alternatives were tested in a recent series of experiments in which the initial rates of cholesterol movement between cells and HDL were quantitated[14]. Fig. 6 shows the results from a typical experiment in which Fu5AH rat hepatoma cells were exposed to control and phospholipid depleted HDL. Over the initial 90 min incubation period no significant differences were observed in the rate of cholesterol uptake. However, efflux of cholesterol was markedly reduced when the cells were incubated

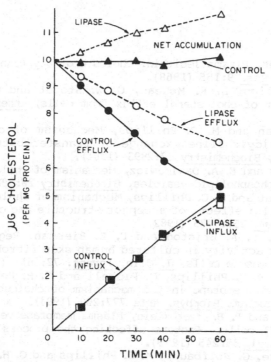

Figure 6. Bidirectional flux of cholesterol between HDL and rat hepatoma cells. Human HDL, labeled with $[H^3]$-free cholesterol, was treated with snake venom phospholipase A_2 and then postincubated with albumin. 62% of the HDL phospholipid was hydrolyzed and removed from the lipoprotein. Either lipase-modified HDL (open symbols) or control HDL (closed symbols) was incubated with Fu5AH rat hepatoma cells at a concentration of 25 ug HDL free cholesterol/ml. The ug of cholesterol transferred from the cells to the medium (efflux) or from the medium to the cell (influx) was calculated at each time point taking into account the dilution in specific activity. Influx and efflux values were determined from the slope of the lines over the initial 40 min. of the experiment. The net accumulation was determined as influx minus efflux. With the unmodified HDL, influx = 0.081 ug cholesterol/min/mg and efflux = 0.081 ug cholesterol/min/mg/efflux. With phospholipase A_2-modified HDL these values were 0.078 and 0.049 ug cholesterol/min/mg, respectively.

in the presence of lipase-modified HDL. This reduced efflux was responsible for the net accumulation of cellular cholesterol we had previously observed with lipase treated HDL. These results clearly demonstrate the importance of cholesterol efflux in maintaining cellular cholesterol homeostasis and illustrate that changes in the rate of release of free cholesterol can result in cellular cholesterol deposition as well as depletion.

References

1. J. A. Glomset, Plasma lecithin: cholesterol acyltransferase reaction. J. Lipid Res. 9:155 (1968).
2. M. C. Phillips, L. R. McLean, G. W. Stoudt and G. H. Rothblat, Mechanism of cholesterol efflux from cells, Atherosclerosis 36:409 (1980).
3. L. R. McLean and M. C. Phillips, Mechanism of cholesterol and phosphatidylcholine exchange or transfer between unilamellar vesicles, Biochemistry 20:2893 (1981).
4. J. M. Backer and E. A. Dawidowicz, Mechanism of cholesterol exchange between phospholipid vesicles, Biochemistry 20:3805 (1981).
5. G. H. Rothblat and M. C. Phillips, Mechanism of cholesterol efflux from cells: effects of acceptor structure and concentration, J. Biol. Chem. 257:4775 (1982).
6. J. F. Oram, E. A. Brinton and E. L. Bierman, Regulation of HDL receptor activity in cultured human skin fibroblasts and human arterial muscle cells, J. Clin. Invest. 72:1611 (1983).
7. F. Bellini, M. C. Phillips, C. Pickell and G. H. Rothblat, Role of the plasma membrane in the mechanism of cholesterol efflux from cells, Biochim. Biophys. Acta 777:209 (1984).
8. R. E. Scott and P. B. Maercklein, Plasma membrane vesiculation in 3T3 and SV3T3 cells. Factors affecting the process of vesiculation, J. Cell Sci. 35:245 (1979).
9. J. Delamatre, G. Wolfbauer, M. C. Phillips and G. H. Rothblat, Role of apolipoproteins in cellular cholesterol efflux, Submitted for publication (1985).
10. F. P. Bell, The dynamic state of membrane lipids: the significance of lipid exchange and transfer reactions to biomembrane composition, structure, function and cellular lipid metabolism, in "Biomembranes," L. Manson and M. Kates, eds., Plenum Press, New York City (1984).
11. H. Jansen and W. C. Hulsmann, Heparin-releasable (liver) lipase(s) may play a role in the uptake of cholesterol by steroid-secreting tissues, Trends Biochem. Sci. 5:265 (1980).
12. M. Bamberger, J. M. Glick and G. H. Rothblat, Hepatic lipase stimulates the uptake of high density lipoprotein cholesterol by hepatoma cells, J. Lipid Res. 24:869 (1983).
13. M. Bamberger, S. Lund-Katz, M. C. Phillips and G. H. Rothblat, Mechanism of the hepatic lipase induced accumulation of high density lipoprotein cholesterol by cells in culture, Biochemistry In press (1985).
14. W. Johnson, M. Bamberger, M. C. Phillips and G. H. Rothblat, Unpublished observation.

INHIBITION AND ACTIVATION OF CHOLESTERYL ESTER TRANSFER AND ITS

SIGNIFICANCE IN PLASMA CHOLESTEROL METABOLISM

Christopher J. Fielding

Cardiovascular Research Institute
University of California
San Francisco, CA

INTRODUCTION

Cholesteryl ester transfer in plasma can be defined as the catalyzed movement of preformed cholesteryl ester between lipoprotein particles. In mammalian plasma, this activity resides in one or more of a class of transfer proteins which show specificity for hydrophobic lipids such as cholesteryl and retinyl esters and triglycerides.[1-3] It has been suggested that the mechanism of catalysis is in the ability of such transfer proteins to increase the solubility of these lipids in plasma and in that way increase diffusion rates between lipid surfaces.[4] Alternatively, the transfer protein may facilitate complex formation between donor and acceptor lipoprotein particles.[5] Studies using labeled cholesteryl esters and triglycerides indicate that transfer is bidirectional.[6] However, the effective concentration of these lipids in different lipoproteins in usually not the same; as a result, the forward and back rates of transfer will not be identical and a net transfer of cholesteryl ester will result.

FUNCTION OF CHOLESTERYL ESTER TRANSFER IN NORMAL PLASMA

Present evidence suggests that the transfer of cholesteryl esters between lipoproteins in plasma is part of a directed sequence of reactions by means of which free cholesterol of peripheral origin is returned to the hepatocytes for catabolism. This series of reactions ("reverse cholesterol transport") appears to be important in maintaining whole body cholesterol homeostasis because inhibition of cholesteryl ester transfer is associated clinically with an increased risk of atherosclerosis.[7]

ORIGIN OF PLASMA CHOLESTERYL ESTERS

The major part of cholesteryl esters found in mammalian plasma is generated in the plasma itself by the action of lecithin:cholesterol acyltransferase (LCAT). In human plasma, essentially the whole of such esters is formed in this way.[8] While it was thought earlier that high density lipoproteins (HDL) in general might be substrates for LCAT, there is considerable recent evidence that this is not the case and that only a small part of HDL makes up the bulk of direct substrate for the enzyme:

205

1. The major part of LCAT is bound to a small fraction (about 10%) of total HDL, which is distinguished by containing only apo A-I but not A-II in its protein moiety.[9,10]

2. This fraction, which accumulates in congenital LCAT deficiency,[11] was found to be an excellent substrate for the enzyme in studies with isolated lipoproteins, while HDL containing apo A-II and other apolipoproteins, particularly the large HDL in the HDL fraction of plasma, are relatively poor substrates.[12]

3. Removal of apolipoproteins containing apo A-II from plasma by immunoaffinity chromatography has little effect on the rate of LCAT activity there.[9]

These results suggest that most of total HDL receives part or all of its cholesteryl ester component by transfer from the minor fraction which contains bound LCAT. At any moment in time, the following fluxes are operational (fluxes back to the fraction of HDL containing LCAT will be minimal because the effective concentration of cholesteryl ester will be highest there): from the LCAT-lipoprotein to VLDL and LDL, and from the LCAT-lipoprotein to HDL and on to VLDL and LDL.

These pathways are analogous to the "direct" and "indirect" routes of transfer suggested by Barter and colleagues,[13,14] except that LCAT is considered here, on the basis of the experimental data summarized above, to form a stable subfraction of HDL, and VLDL and LDL have been combined into a single population, in view of recent concepts of these fractions forming a structural and metabolic continuum.

Data obtained with recombined isolated lipoproteins indicated that while both pathways could be demonstrated, the major part of cholesteryl esters cycled through HDL.[13] It is as yet unclear in what proportions direct and indirect transfer take place in native plasma.

FATE OF LCAT-DERIVED CHOLESTERYL ESTERS

It seems clear that the major part of cholesteryl esters synthesized in plasma are transferred, directly or indirectly, to VLDL and LDL. As discussed above, the major part or all of such esters are generated by LCAT, and VLDL and LDL together contain about 75% of total plasma cholesteryl esters. Among the lipoproteins of normal plasma, VLDL are a more effective acceptor of esters than LDL,[15,16] while because of the much greater concentration of LDL, the greater mass of such esters is recovered in this fraction.[17]

Calculation of the number of cholesteryl ester molecules in VLDL and LDL had earlier led to the conclusion[18] that since there was a greater number of such molecules on average in the former, the uptake of cholesteryl ester into VLDL must be followed by a subsequent remodeling in which cholesteryl ester was disgorged again, as VLDL were converted to LDL. Such contradiction with the concept of continued cholesteryl ester transfer to LDL is likely to be more apparent than real, however. More recent evidence indicates that VLDL consist of two discrete subfractions in terms of cholesteryl ester transfer: one free of apo E, containing little cholesteryl ester, and one rich in both apo E and cholesteryl ester.[19] It seems most likely that it is cholesteryl ester poor, apo E-free VLDL that are converted to LDL with continuing accretion of cholesteryl ester via the LCAT/transfer mechanism, while apo E-containing VLDL, which contain the bulk of the cholesteryl ester, are removed as such from the plasma. In any case, in view of such heterogeneity, total VLDL cannot be used as the basis for calculation of lipid transfer during the VLDL-to-LDL conversion.

HYPERACTIVITY OF THE CHOLESTERYL ESTER TRANSFER SYSTEM

Cholesteryl ester transfer is increased in postprandial plasma.[20,21] Rates of transfer are approximately twofold higher, under conditions where the rate of LCAT activity is similarly increased. There is thus a considerable increase in the transport of cholesterol down the reverse cholesterol transport pathway under these conditions. Studies carried out in the presence of cultured cells indicate that a considerable proportion of this "extra" transport is of cholesterol derived from the cell membranes rather than from plasma lipoproteins. The explanation of this is most likely that in postprandial lipemia, because of the much greater influx of phospholipids (along with triglyceride) relative to cholesterol, the effective cholesterol concentration in the lipoproteins significantly decreases. Such lipoproteins are more likely to accept cholesterol themselves than donate it to LCAT for esterification.

Several explanations have been advanced for the greater rate of cholesteryl ester transfer observed. Firstly, the increased level of triglyceride-rich lipoproteins in VLDL (relative to LDL) itself is increased; since these are more effective acceptors of cholesteryl esters, transfer to plasma rich in VLDL should be more effective, other things being equal, than to plasma containing lower concentrations of triglyceride-rich lipoproteins. However, calculation[20] indicates that the difference in VLDL concentration observed could account for only a small proportion of the observed increase in cholesteryl ester transfer rates.

Secondly, just as the lower level of free cholesterol in VLDL and LDL may reduce their effectiveness as donors of free cholesterol for LCAT activity, so it might increase their ability to accept cholesteryl esters. Physical evidence[22] indicates that free cholesterol has important effects in determining the stability of lipid surfaces, and membranes rich in free cholesterol become condensed and inflexible to the transport of solutes. As discussed below, increased free cholesterol in VLDL and LDL evidently decreases their ability to accept cholesteryl esters. Extrapolation of the dependence of transfer rates on the free cholesterol content of VLDL and LDL acceptors, obtained with normal and pathological plasma samples, to postprandial lipoproteins shows an excellent data fit,[20] suggesting that this factor is an important one. Other potential factors include changes in the level of transfer protein itself in response to postprandial lipemia, changes in the composition or properties of HDL donor lipids, and the levels of free fatty acids.

The hyperactivity of cholesteryl ester transfer in postprandial lipemia, and the associated increase in total cholesteryl ester synthesis, may indicate an important role for dietary triglyceride in promoting whole body cholesterol homeostasis.

HYPOACTIVITY OF THE CHOLESTERYL ESTER TRANSFER SYSTEM

A decreased rate of cholesteryl ester transfer to VLDL and LDL has now been identified in the plasma of several patient groups, including noninsulin-dependent (type II) diabetics, patients with familial hypercholesterolemia or dysbetalipoproteinemia, certain hypertriglyceridemic subjects with documented vascular disease, and patients with end-stage renal disease.[7,19,23] Patients in these categories, who have disorders with quite different origins and with distinct plasma lipoprotein patterns, share a common significantly increased risk of atherosclerosis, and a common low rate of cholesteryl ester transfer, relative to the rate of total cholesteryl ester synthesis. This means that in such plasma, cholesteryl esters generated by

LCAT are more easily transferred to or retained in HDL than transferred to VLDL or LDL.

There has been extensive analysis of the composition of lipoproteins in these patient groups by many laboratories.[24] A consistent finding has been that these lipoproteins show an increased saturation of free cholesterol. While the origin of this abnormality is uncertain and may be different in the different patient groups, the rate of cholesteryl ester transfer is inversely correlated with free cholesterol saturation (or FC/PL mass ratio) in VLDL and LDL.[19]

Inhibition of transfer can also be mediated, in experiments in vitro, by incubation of plasma. VLDL and LDL increase their cholesteryl ester to a point that further net transfer does not take place. Such a content of VLDL cholesteryl ester increases about 300% under the same conditions.[17] While such "end-product" LDL have not been observed in native plasma, they are similar in cholesteryl ester content to the LDL of cholesterol-fed primates.[25] The cholesteryl ester content of "end-product" VLDL generated in vitro is similar to that of beta-VLDL (including an increased content of apo E, which passes to VLDL as cholesteryl ester accumulates). However, native lipoproteins are metabolized by a more complex set of conditions than pertain in vitro, including lipolysis by endothelial surface triglyceride lipases and the continuing entry of nascent lipoproteins. Such parallels may therefore be incomplete. Nevertheless, it is clear that in normal plasma cholesterol metabolism, circulating VLDL and LDL do not carry their maximum content of cholesteryl esters, while in pathological conditions, lipoproteins may accumulate which have an exaggerated content of transferred cholesteryl esters.

SIGNIFICANCE OF CHOLESTERYL ESTER TRANSFER

The transfer system permits the delivery of cholesteryl esters to different plasma lipoprotein classes to be regulated. In animals (such as the rat) in which the rate of cholesteryl ester transfer is low,[26] cholesteryl esters generated by LCAT must be retained at the site of synthesis, on HDL. In species that can transfer cholesteryl esters, a proportion, often large, of these esters can be transferred to VLDL and LDL. The evidence summarized above suggests that the proportion of cholesteryl esters so transferred largely depends on the concentration and composition of the different potential acceptors present, not on the rate of transfer as such.

In most animals that possess an active transfer system, the major part of cholesteryl esters is transferred to VLDL and LDL. These animals regulate their plasma cholesterol levels by the activity of specific high affinity lipoprotein receptors which bind VLDL and LDL. The removal of HDL appears to be much less well regulated. It can therefore be argued that the cholesteryl ester transfer system permits circulating levels of cholesterol (the major part of which is cholesteryl ester) to be tightly regulated by a receptor system adjustable by up- or down-regulation to buffer transient changes in concentration. Utilization of the VLDL-LDL system, whose initial functions lie in triglyceride transport, for reverse cholesterol transport also increases the capacity of the system overall to remove such cholesterol to the liver, site of the major part of lipoprotein receptors. On the other hand, the same species that have the ability to transfer cholesteryl esters to VLDL and LDL are atherosclerosis-prone, and LDL (whose cholesterol is largely in esterified form) are recognized as atherogenic lipoproteins, as are other products of possibly exaggerated transfer, such as beta-VLDL. Finally, there is some evidence that LDL whose content of cholesteryl ester is increased by incubation react less well with lipoprotein receptors than do LDL of normal composition. Cholesteryl ester transfer appears then to be at the center of those reactions in plasma which largely determine the composition and properties of the plasma lipoproteins. An inhibition of transfer is associated with the risk of atherosclerosis in many groups of human subjects.

ACKNOWLEDGMENTS

The personal research of the author described in this chapter was supported through the National Institutes of Health by Arteriosclerosis SCOR HL 14237 and by HL 23738.

REFERENCES

1. N. M. Pattnaik, A. Montes, L. B. Hughes, and D. B. Zilversmit, Cholesteryl ester exchange protein in human plasma. Isolation and characterization, Biochim. Biophys. Acta 530:428 (1978).

2. J. Ihm, J. L. Ellsworth, B. Chataing, and J. A. K. Harmony, Plasma protein-facilitated coupled exchange of phosphatidylcholine and cholesteryl ester in the absence of cholesterol esterification, J. Biol. Chem. 257:4818 (1982).

3. J. J. H. Albers, J. J. Tollefson, C.-H. Chen, A. Steinmetz, Isolation and characterization of human plasma lipid transfer protein, Arteriosclerosis 4:49 (1984).

4. L. R. McLean and M. C. Phillips, Mechanism of cholesterol and phosphatidylcholine exchange or transfer between unilamellar vesicles, Biochemistry 20:2893 (1981).

5. J. I. Ihm, D. M. Quinn, S. J. Busch, B. Chataing, and J. A. K. Harmony, Kinetics of plasma protein-catalyzed exchange of phosphatidyl choline and cholesteryl ester between plasma lipoproteins, J. Lipid Res. 23:1328 (1982).

6. R. E. Morton and D. B. Zilversmit, Inter-relationship of lipids transferred by the lipid-transfer protein isolated from human lipoprotein-deficient plasma, J. Biol. Chem. 258:11751 (1983).

7. P. E. Fielding, C. J. Fielding, R. J. Havel, J. P. Kane, and P. Tun, Cholesterol net transport, esterification and transfer in human hyperlipidemic plasma, J. Clin. Invest 71:449 (1983).

8. J. A. Glomset and K. R. Norum, The metabolic role of lecithin:cholesterol acyltransferase: perspectives from pathology, Adv. Lipid Res. 11:1 (1973).

9. P. E. Fielding and C. J. Fielding, A cholesteryl ester transfer complex in human plasma, Proc. Natl. Acad. Sci. USA 77:3327 (1980).

10. M. C. Cheung and J. J. Albers, Characterization of lipoprotein particles isolated by immunoaffinity chromatography. Particles containing A-I and A-II and particles containing A-I but no A-II, J. Biol. Chem. 259:12201 (1984).

11. C. Chen, E. Applegate, W. C. King, J. A. Glomset, K. R. Norum, and E. Gjone, A study of the small spherical high density lipoproteins of patients afflicted with familial lecithin:cholesterol acyltransferase deficiency, J. Lipid Res. 25:269 (1984).

12. C. J. Fielding and P. E. Fielding, Purification and substrate specificity of lecithin:cholesterol acyltransferase from human plasma, FEBS Lett 15:355 (1971).

13. P. J. Barter, Evidence that lecithin:cholesterol acyltransferase acts on both high density and low density lipoproteins, Biochim. Biophys. Acta 751:261 (1983).

14. P. J. Barter and G. J. Hopkins, Relative rates of incorporation of esterified cholesterol into human very low and low density lipoproteins. In vitro studies of two separate pathways, Biochim. Biophys. Acta 751:33 (1983).

15. T. Chajek and C. J. Fielding, Isolation and characterization of a human serum cholesteryl ester transfer protein, Proc. Natl. Acad. Sci. USA 75:3445 (1978).

16. Y. L. Marcel, C. Vezina, B. Teng, and A. Sniderman, Transfer of cholesteryl esters between human high density lipoproteins and triglyceride-rich lipoproteins controlled by a plasma protein factor, Atherosclerosis 35:127 (1980).

17. C. J. Fielding and P. E. Fielding, Regulation of human plasma lecithin: cholesterol acyltransferase activity by lipoprotein acceptor cholesteryl ester content, J. Biol. Chem. 256:2102 (1981).

18. R. J. Deckelbaum, S. Eisenberg, Y. Oschry, E. Butbul, I. Sharon, and T. Olivecrona, Reversible modification of human plasma low density lipoprotein towards triglyceride-rich precursors. A mechanism for losing excess cholesteryl esters. J. Biol. Chem. 257:6509 (1982).

19. C. J. Fielding, G. M. Reaven, G. Liu, and P. E. Fielding, Increased free cholesterol in plasma low and very low density lipoproteins in noninsulin-dependent diabetes mellitus: its role in the inhibition of cholesteryl ester transfer, Proc. Natl. Acad. Sci. USA 81:2512 (1984).

20. G. R. Castro and C. J. Fielding, Effects of postprandial lipemia on plasma cholesterol metabolism, J. Clin. Invest, 75:874 (1985).

21. A. Tall and D. Sammett, Mechanisms of enhanced cholesteryl ester (CE) transfer during alimentary lipemia, Circulation 72:III:287 (1985).

22. K. Bloch, Cholesterol: evolution of structure and function,, in "Biochemistry of Lipids and Membranes," D. E. Vance and J. E. Vance, eds., Benjamin/Cummings, California (1985).

23. H. Dieplinger, P. Y. Schoenfeld, and C. J. Fielding, Plasma cholesterol transport in end stage renal disease, Circulation 72:III:199 (1985).

24. C. J. Fielding, The origin and properties of free cholesterol potential gradients in plasma, and their relation to atherogenesis, J. Lipid Res. 25:1624 (1984).

25. A. R. Tall, D. M. Small, D. Atkinson, and L. L. Rudel, Studies on the structure of low density lipoprotein isolated from Macaca Fascicularis fed an atherogenic diet, J. Clin. Invest. 62:1354 (1978).

26. P. J. Barter, J. M. Gooden, and O. V. Rajaram, Species differences in the activity of a serum triglyceride transferring factor, Atherosclerosis 33:165 (1979).

APOLIPOPROTEIN C-II DEFICIENCY

W. Carl Breckenridge

Department of Biochemistry, Dalhousie University

Halifax, Nova Scotia B3H 4H7

INTRODUCTION

The lipolysis of chylomicrons and very low density lipoproteins is responsible for the delivery of energy in the form of fatty acids to adipose or muscle tissue. The fatty acids are transported in the form of triacylglycerols in these lipoproteins following their synthesis in the intestine or liver. Defects in this process result in massive accumulations of triacylglycerols in plasma. A deficiency of lipoprotein lipase, the major enzyme responsible for the hydrolysis of lipoprotein triacylglycerols in peripheral tissues, was recognized [1] over 20 years ago as one cause of hyperchylomicronemia. Subsequently [2,3] the specific plasma component necessary for normal lipoprotein lipase activity was isolated from very low density lipoproteins (VLDL) and identified as apolipoprotein C-II (apo C-II). The apolipoprotein, which contains 79 amino acids, influences lipoprotein lipase activity by reducing the Km of the enzyme for triacylglycerol emulsions [4] or apo C-II deficient VLDL [5,6]. In one instance apo C-II also increased the Vmax of the reaction [6].

The importance of apo C-II for normal rates of lipolysis, in vivo, was established with our discovery [7] of a subject with severe hypertriglyceridemia who lacked apo C-II as defined by lack of immunoprecipitation with polyclonal anti-apo C-II, by lack of activation of lipoprotein lipase by the patient's plasma and by the absence of apo C-II on polyacrylamide gel electrophoresis. Post-heparin lipolytic activity was latent but could be stimulated by the addition of normal plasma or apo C-II to the incubation. Finally infusion of plasma [7] or apo VLDL preparations [8] into the patient resulted in a marked lowering of plasma triacyglycerols. Since the description of this kindred [7-12] eight other kindreds (Table I) have been reported [13,14,15,16,17,18,19,20,21,22] and further subjects have been found in France (Rebourcet et al., personal communication). Apo C-II deficiency should be considered in all instances where there is fasting hyperchylomicronemia and low post-heparin lipolytic activity in the absence of exogenously added apo C-II. Although the defect may be considerably less frequent in the population than lipoprotein lipase deficiency recent studies suggest that a number of variants of apo C-II may exist which may affect triglyceride clearance.

Table 1. Kindreds for Apo C-II Deficiency.

| Kindred | Reference | Country | Obligate | | Nature of Defect[a] | | |
			Homozygotes	Heterozygotes	Apolipoprotein C-II Absent	Apolipoprotein C-II Variant	Gene Present
1	Breckenridge et al. (7-12)	Canada, USA, Caribbean	15	28	N.D.	Apo CII X, Y	?
2	Yamamura et al. (13)	Japan	2	2	N.D.	?	?
3	Crepaldi et al. (16,17)	Italy	2	2	N.D.	Apo CII Padua	+
4	Miller et al. (14)	England	1	–	N.D.	?	?
5	Stalenhof et al. (18)	Holland	4	6	N.D.	?	+
6	Catapano et al. (15,19)	Italy	1	2	N.D.	?	?
7	Saku et al. (20)	USA	1	–	+		?
8	Fojo et al. (21)	USA	1	–		Apo CII Bethesda	+
9	Hayden et al. (22)	Canada, Italy	2	2		?	?, +[b]

a N.D. = not detectable by immunoassay, ? = unknown.

b This subject yields a faint reaction on immunoblots with anti apo C-II to a component with an electrophoretic mobility that is between apo C-II and apo C-III on isoelectric focussing gels. No apo C-II is detectable by immunoassay or isoelectric focussing of apo VLDL.

Table 2. Clinical Complications of Homozygotes for Apo C-II Deficiency.

Kindred[a]	Pancreatitis	Episodic Abdominal Pain	Hepato-Splenomegaly	Xanothomata	Diabetes Mellitus	Coronary Heart Disease	Other
1	9/14	9/14	3/14	-	1/14	1/14b	Anemia[c]
2	-	-	-	-	-	-	-
3	-	1/2	2/2	1/2	-	-	-
4	-	-	-	1/4	-	-	Thyroxine binding globulin deficiency, apoprotein E deficiency, lipoprotein lipase deficiency
5	1/4	3/4	4/4	-	-	-	-
6	-	-	+	+	-	-	Thalessemia minor
7	-	++	++	++	-	-	-
8	+	++	++	++	-	-	-

a Source of data is from references given in Table 1.
b Male, age 59 with angina
c Eight of fourteen homozygotes had anemia as defined by a hemoglobin concentrations of less than 14 g % for men and 12 g % for women.

CLINICAL COMPLICATIONS

The clinical complications reported for kindreds with apo C-II deficiency show many similarities but some differences (Table 2) when compared with familial lipoprotein lipase deficiency. In the original kindred, which has 15 homozygotes, pancreatitis is very common in those subjects with very severe elevations of triacylglycerols. In some kindreds the occurrence of pancreatitis is low. Hepatosplenomegaly has been observed in 5 kindreds. Xanthomata have not been observed in most kindreds. In the case of kindred 4 one instance of eruptive xanthomata may be related to the fact that these patients also appear to have very low amounts of lipoprotein lipase (18) even after the addition of apo C-II to post-heparin plasma. Diabetes mellitus is not common in the kindreds. It is interesting to note that apo C-II deficiency resembles lipoprotein lipase deficiency by the low incidence of cardiovascular disease in adult subjects. Only 1 of 14 subjects in kindred 1 demonstrated some angina at age 59. Thus despite the very high levels of cholesterol in chylomicrons but very low levels of HDL there appears to be very little prevelance of premature atherosclerosis. Obligate heterozygotes for apo C-II deficiency have no serious complications (8).

LIPOPROTEIN ABNORMALITIES

Apo C-II deficiency is characterized by fasting hypertriglyceridemia with normal or elevated cholesterol concentrations (Table 3). Plasma triacyglycerols vary widely from 550 to 5000 mg/100 ml. In instances where very low fat diets were instituted for the patient the plasma triacylglycerols decreased slowly over several days (7). The major portion of lipid is present as chylomicrons and very low density lipoproteins. The chylomicron fraction is usually elevated to a much greater degree than VLDL. The low concentrations of LDL cholesterol (10-66 mg/100 ml) and HDL (4-26 mg/100 ml) are very consistent throughout all the kindreds.

Within each lipoprotein class there are a number of abnormalities as analyzed by analytical ultracentrifugation. The Sf 20-400 fraction may be elevated or normal in the homozygote. In kindred 1 the Sf 0-12 and 12-20 lipoproteins show two peaks of equal size in contrast to 1 large peak in the Sf 0-12 fraction in normal LDL (11). Other kindreds have proportionally more mass in the Sf 0-12 than Sf 12-20 (14,19), but both frac-

Table 3. Plasma Lipid and Lipoprotein Concentrations in Homozygotes for Apo C-II Deficiency

Kindred[a]	N	Triacyl-glycerols	Cholesterol				
			Total	Chylo	VLDL	LDL	HDL
			mg/100 ml				
1	14	555-5270	108-530	94-310	30-188	12-44	4-26
2	1	1100	160-200	100-120	35-45	23	17
3	1	2900	320	190	42	66	19
5	1	700	150	71	34	21	12
6	2	2000	183-235	136-181		10-20	8-11
7	1	2500	200			38	11

[a] Source of data from references quoted in Table 1.

214

tions are much lower than normal values. In all kindreds that have been studied for the distribution of HDL mass both HDL2 and HDL3 are reduced substantially (11,14,19).

In addition to major changes in the amount of LDL and HDL the composition of these lipoproteins is abnormal (11,14,19). In the neutral lipid core a major portion of the cholesteryl ester is replaced by triacylglycerols. This alteration is also found in lipoprotein lipase deficiency and is probably due to the neutral lipid exchange protein and the vast excess of triacylglycerols found in these subjects, since there is a high degree of correlation of the percent weight of triacylglycerol in the lipoprotein with the absolute mass of plasma triacylglycerol (Table 4).

Normally the ratio of unesterified cholesterol to phospholipid is low in chylomicrons, high in chylomicron remnants, intermediate in VLDL and LDL and low in HDL (24). The characteristic differences in this ratio are possibly due to the size as well as the metabolism of the lipoprotein. Inherent phospholipase activity of lipoprotein lipase or hepatic lipase may also influence this ratio (25,26,27). However the ratio is quite distinct in subjects with apo C-II deficiency where it is extremely high for chylomicrons and lower than normal for LDL (Table 5). Perhaps the inadequate lipolytic system in apo C-II deficiency plays a role in influencing this ratio.

Table 4. Correlation of Triacylglycerol Content of Lipoproteins with Plasma Triacylglycerol Concentrations in Homozygotes for Apo C-II Deficiency[a].

	Lipoprotein Fraction		
	1.006 Inf.	LDL	HDL
r	0.84	0.83	0.90

[a] Plasma triacylglycerols ranged from 875 to 3300 mg/100 ml. The proportion of triacylglycerols as a weight percent of total lipid was 30–54% in LDL and 19–40% in HDL (11).

Table 5. Cholesterol/Phospholipid Molar Ratios in Lipoproteins[a].

Subjects	N	Chylo	VLDL	LDL	HDL
Normal	6	–	0.61	0.75	0.25
			±0.07	±0.06	±0.02
Apo CII Deficiency	6	0.94	0.71	0.51	0.23
		±0.10	±0.10	±0.07	±0.4

[a] Lipids were analyzed by gas liquid chromatography as described previously (11).

Lipids and lipoprotein concentrations are usually within the normal range for obligate heterzygotes in most kindreds. However, a comparison by percentile ranking (8) of 23 obligate heterozygotes in kindred 1 with a North American population demonstrated that plasma triacylglycerols and VLDL were more frequently above the 50% percentile ranking while HDL was more frequently below the 50% percentile when compared with the normal population. One heterozygote had transient hypertriglyceridemia. These observations suggest that the rate of lipolysis may be limiting VLDL catabolism slightly in some heterozygotes due to the reduced apo C-II/apo C-III ratio.

APOLIPOPROTEIN ABNORMALITIES

Kindred 1 lacked detectable apo C-II by several criteria. However an unusual apolipoprotein with an apparent pI very similar to that of apo A-I was present in VLDL (10). Analysis of the apoproteins of VLDL by sodium dodecyl sulphate-glycerol polyacrylamide gel electrophoresis indicated no difference in the amount of apo C-II. Two dimensional polyacrylamide gel electrophoresis of apo VLDL from normal and apo C-II deficient subjects revealed (23) that additional apolipoproteins were present in the VLDL of homozygotes and obligate heterozygotes which had a molecular weight similar to apo C-II but a much higher pI than normal apo C-II. In addition to these observations a number of antibodies raised to apo C-II revealed, in some instances, a weak reactivity to some of the homozygotes, while some obligate heterozygotes failed to yield properly formed rockets in the electroimmunoassay (Breckenridge, unpublished results). These observations suggested that a variant of apo C-II might be present in this kindred. This idea was confirmed (23) by isoelectric focussing of apo VLDL from the subjects followed by transfer and immobilization of the apolipoprotein on nitrocellulose and detection of apo C-II by immunoblotting with goat polyclonal anti-apo C-II. With normal VLDL a single component is observed by protein staining corresponding to normal apo C-II (Figure 1, 2). In aged preparations a second or third component (Figure 1, 2) is observed by immunoblotting and additional components may be seen in purified preparations of apo C-II. Isomorphs of apo C-III can be clearly distinguished from apo C-II isomorphs by a second immunoblot with anti-apo C-III (Figure 1). In some subjects apo C-II$_2$ is also observed with the protein stain (Figure 2).

In the homozygotes for apo C-II deficiency two other components are detected by the antibody (Figure 2) and correspond to the two components identified by two dimensional polyacrylamide gel electrophoresis (23). While one form (apo C-IIY) is present in considerable quantities the second form (apo C-IIX) is present in minor amounts. Thus it is apparent that kindred 1 has a variant which has a molecular weight very similar to normal apo C-II but a net charge and antigenic reactivity quite distinct from apo C-II. The apoprotein shows other similarities to normal apo C-II. It tends to bind to triacylglycerol-rich lipoproteins and is solubilized by acetone extraction of VLDL (Figure 2). It is possible that the amino terminal portion of the protein, which is responsible for lipid binding is unaltered while the carboxyl terminal region of the protein, which possesses the sequences responsible for activating the enzyme may be altered. It would also appear that a sufficient number of the immunologic determinants for the polyclonal antibodies are lost thereby preventing precipitin reactions but a sufficient number of sites are retained to allow localization of protein if the protein is immobilized on nitrocellulose.

Apo C-II variants have also been reported in subjects from Italy and United States (21). In these cases very small amounts of distinct apo C-II variants have been observed. Kindred 6 (20) has been reported as C-II anapolipoproteinemia while other kindreds have not been carefully assessed for the possibility of variant forms of apo C-II.

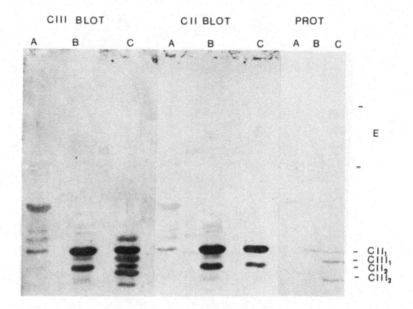

Fig. 1.   Isoelectric focussing of apolipoproteins. Apolipo-
           proteins were subjected to isoelectric focussing
           followed by protein staining or transfer to nitro-
           cellulose and detection with immunoblotting as
           described previously (10,23): A, Purified apo C-II
           subjected to cleavage with cyanogen bromide; B, puri-
           fied apo C-II; C, apo VLDL solubilized in 8 M urea.
           The three lanes on the right represent gels stained
           with Coomassie blue. The centre 3 lanes represent an
           immunoblot of the gel with anti apo C-II. The three
           lanes on the left represent the anti apo C-II blot
           which was subsequently subjected to an immunoblot to
           anti-apo C-III. The isomorphs of apo C-II and apo
           C-III are readily distinguished. Lane A demonstrates
           that the products of cyanogen bromide cleavage have
           some immunoreactivity with the antibody.

The absence of functional apo C-II is associated with marked alterations in apolipoprotein levels (Table 6). Apo A-I, A-II, and B are all
reduced compared to normal plasma values while apo C-III and E are
elevated 2-5 fold over normal plasma concentrations. These changes agree
with the observed changes in lipid and lipoprotein concentrations where
chylomicrons and VLDL are elevated but LDL and HDL are reduced compared

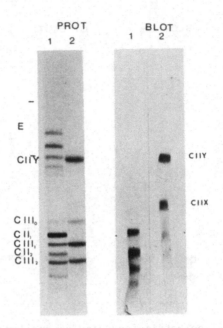

Fig. 2. Isoelectric focussing of apolipoproteins from apo C-II
deficiency. Conditions as described in Figure 1.
Lane 1 represents urea solubilized apo VLDL from
pooled hyperlipidemic subjects. Lane 2 represents the
acetone soluble fraction of triacylglycerol rich lipoproteins from kindred 1 for apo C-II deficiency. Protein staining of gels is on the left with immunoblots
for apo C-II on the right. It can be seen that the apo
C-II deficient subjects possess two components, apo
C-IIY and apo C-IIX which are selectively solubilized
with the other C apolipoproteins by acetone extraction.
While there is considerable mass of apo C-IIY, apo
C-IIX is a very minor component. The multiple bands
for the apo C-II immunoblot of hypertriglyceridemic
VLDL represent a second isomorph (apo C-II$_2$) plus
additional components that appear when apo VLDL is
stored.

with normal subjects. Thus the accumulation of chylomicons due to defective lipolysis acts as a sink for the accumulation of apo C-III and E whereas the extent of apo B accumulation in these lipoproteins is not as high as that observed normally for LDL. The reason for low apo A-I values remains unexplained. Very little apo A-I is associated with the chylomicron fraction (23). Perhaps the lack of active lipolysis with the concomitant delivery of cholesterol and phospholipid to HDL promotes a more rapid clearance of apo A-I from the circulation.

Apo C-II concentrations in obligate heterozygotes are reduced to about 50% of the normal mean value (10). The values for the ratio of apo C-II/apo C-III in VLDL and the ability of plasma to activate skim milk lipoprotein lipase are also reduced. However the absolute level of apo C-II is highly dependent on plasma triacylglycerol concentrations. There is considerable overlap in the values for absolute amounts of apo C-II for obligate heterozygotes when compared with the normal population (10, Breckenridge unpublished observations) In the case of kindred 1 the best criteria for identification of the heterozygote is the appearance of normal apo C-II as well as the variant form apo C-IIY (23). Apo C-III, B, E and A-II are essentially normal while apo A-I may be slightly reduced in concentration in obligate heterozygotes. In summary there are relatively few abnormalities in the heterozygote. However the data on lipid, lipoprotein, and apolipoprotein levels and the accumulation of apo C-III in VLDL (10) suggest that there may be defective lipolysis. Until fat tolerance tests are completed on heterozygotes the possibility of some impairment in lipolysis can not be excluded.

LIPOPROTEIN METABOLISM IN APO C-II DEFICIENCY

Lipoprotein metabolism in normal subjects appears to be driven by the production and catabolism of chylomicrons and VLDL (28). While chylomicrons loose about 90% of their triacylglycerol content and are removed as chylomicron remants, VLDL are converted in part to LDL. Excess surface material consisting of apo C, cholesterol and phospholipid have been demonstrated to be transferred to HDL where the lipids may serve as substrates for lecithin cholesterol acyl transferase for the production of cholesteryl ester. Cholesteryl ester transfer to VLDL is facilitated by the neutral lipid exchange protein (29). Apo E displacement from VLDL to

Table 6. Apolipoprotein Concentrations in Homozygotes for Apo C-II Deficiency.

Kindred[a]	N	Apolipoprotein				
		A-I	A-II	B	C-III	E
		mg/100 ml				
1	11[b]	70 ± 12	24 ± 7	59 ± 26	22 ± 8	23 ± 8
3	1	58	26	53	39	58
5	1	94	27	49	15	–
6	2	90–115	–	66–88	–	–
7	1	50	–	92	–	17

[a] Source of data given in references quoted in Table 1.
[b] N = 11 for apo A-I, B and E; N = 7 for apo A-II.

HDL is also promoted by lipolysis (27,30,31). On the basis of in vitro studies it is possible that this process also may involve loss of neutral lipid from the VLDL (27,28). Thus the levels of the product lipoproteins LDL and HDL, should be influenced in part by the rate of lipolysis. The apo C-II deficient subjects represent an in vivo model for studying clearance of chylomicrons. Under present concepts of lipoprotein metabolism LDL and HDL mass should be augmented by the infusion of apo C-II.

Post-Heparin Lipolytic Activities

The inital proband in kindred 1 demonstrated very high levels of lipoprotein lipase activity following addition of normal plasma to the assay (7). This observation suggested that the amount of tissue lipoprotein lipase might be stimulated by the lack of supply of fatty acid to the tissue as has been observed in rat heart preparations after fasting. This observation was not consistent in other members of the kindred or in other kindreds (Table 7). However, adipose tissue lipoprotein lipase was high in the one apo C-II deficient subject of kindred 4 (14).

Hepatic lipase activity was quite low in a number of homozygotes in kindred 1. It is interesting to note that subjects in this kindred with a low level of hepatic lipase have relatively higher concentrations of IDL and HDL2 than those individuals with normal values for the enzyme activity (11). These effects are similar to those noted in subjects with hepatic lipase deficiency where there is an increase in beta VLDL and HDL2 mass (32). The reason for the high degreee of variablity in hepatic lipase concentration may be explained by the observation of Klose et al. (33) that low hepatic lipase activity is associated with pancreatitis.

Influence of Plasma or Apo C-II Infusions on Lipoprotein Metabolism

The original proband in kindred 1 received a number of infusions of plasma (7) and apo VLDL (8). There was a rapid clearance of the triacylglycerol-rich lipoproteins with little change in LDL or HDL. Similar observations were made in kindred 5. However studies by Miller et al. (14) indicated a small increase in LDL and HDL2 while Catapano et al. (19) noticed a slightly larger increase in LDL and HDL. A number of factors may account for the lack of large increases in LDL and HDL in some sub-

Table 7. Post Heparin Lipolytic Activities in Homozygotes for Apo C-II Deficiency.

Kindred[a]	N	Lipoprotein Lipase	Hepatic Lipase
		% control mean	
1	5	56-300	12-125
2	2	120	43
3	1	*b	137
4	4	14-40	20-48
5	1	66	54
6	2	80-100	100-200
7	1	45	100

[a] Source of data from references given in Table 1.
[b] Adipose tissue lipoprotein lipase was elevated 20-fold over normal values.

jects. In kindred 1 the chylomicrons are extremely rich in apo E and
C-III and possess a very high cholesterol/phospholipid ratio (Table 5).
There is evidence from animal models to indicate that the clearance of
chylomicron remnants is influenced by the ratio of apo C-III/apo E (34).
With the infusion of apo C-II it is possible that following a limited
amount of lipolysis and loss of apo C-III the majority of the lipid is
rapidly removed by the liver and little augmentation of LDL and HDL mass
results. On the basis of our studies in perfused rat heart (27), any surface material, generated during lipolysis, would have a very high
cholesterol/phospholipid ratio and would not serve as an ideal substrate
for incorporation into HDL.

GENETICS OF APO C-II DEFICIENCY

    For a detailed description of the isolation and cloning of the gene
for apo C-II the reader is referred to the article by Hayden et al. (22)
in this symposium. A number of investigators have achieved the isolation
and cloning of the gene and located it on chromosome 19 along with apo E
and the LDL receptor (35-40). There is good agreement of the predicted
amino acid sequence from the cDNA sequence with the observed amino acid
sequence. In all kindreds for apo C-II deficiency studied at present
(Table 1) the apo C-II gene is present (21,22,41) and no major deletions
or insertions are found.

    In the case of kindred 1 it is interesting to speculate on the nature
of the molecular defect in these subjects (Figure 3). In contrast to
other apolipoprotein variants (21), there are fairly large amounts of the
apo C-IIY variant. Thus it does not appear to be poorly synthesized or
rapidly catabolized. Since the molecular weight is very similar to normal

```
                        50                                    60
Val Asp Glu Lys Leu Arg Asp Leu Tyr Ser Lys Ser Thr Ala Ala Met
GTA GAT GAG AAA CTC AGG GAC TTG TAC AGC AAA AGC ACA GCA GCC ATG
    190         200         210         220         230

                                    70
Ser Thr Tyr Thr Gly Ile Phe Thr Asp Gln Val Leu Ser Val Leu Lys
AGC ACT TAC ACA GGC ATT TTT ACT GAC CAA GTT CTT TCT GTG CTG AAG
        240         250         260             270         280

        79
Gly Glu Glu Stop
GGA GAG GAG TAA CAG CCA GAC CCC CCA TCA GTG GAC AAG GGG AGA GTC
        290         300         310         320
```

```
Ala Leu Thr Gln Ala Phe Leu Leu Thr Lys Phe Phe Leu Cys Stop
GCA CTT ACA CAG GCA TTT TTA CTG ACC AAG TTC TTT CTG TGC TGA
```

Fig. 3. Amino acid and cDNA nucleotide sequence for the carboxyl
         terminal portion of apo C-II from reference 36. The hypo-
         thetical sequence shown at the bottom of the figure
         assumes that deletion of adenine 234 has occurred. The
         new sequence results in a stop codon at nucleotide 276-279.
         Other deletions or insertions in this region could yield the
         same result.

apo C-II there does not appear to be a major deletion of a portion of the sequence. However, there is a very significant charge shift; too large to be accounted for by a single substitution of a nucleotide base and an alteration in one amino acid. An examination of the nucleotide sequence for the carboxyl terminal portion of the cDNA for apo C-II may indicate a potential explanation for the variant (Figure 3). If an appropriate base deletion or insertion occurs after nucleotide residue 233 a frame shift results with a new nucleotide sequence (for one example of many see Figure 3) and a stop codon occurs at residue 279. In this example a new protein is formed which is smaller by 5 amino acids. Since the last two amino acid residues at the carboxyl terminal of normal apo C-II are glutamic acid there is a substantial shift in charge. If a similar frame shift occurs earlier in the sequence a stop codon could be encountered at residues 231-234 which would yield a much smaller protein. Isolation and sequencing of the apolipoprotein or the cDNA from these subjects will allow the testing of this possibility for kindred 1.

With the description of apo C-II variants in at least three of the kindreds it would seem pertinent to consider whether other apo C-II variants might account for some forms of hypertriglyceridemia commonly referred to as Type V. A study of the sequence of apo C-II (Figure 4) and

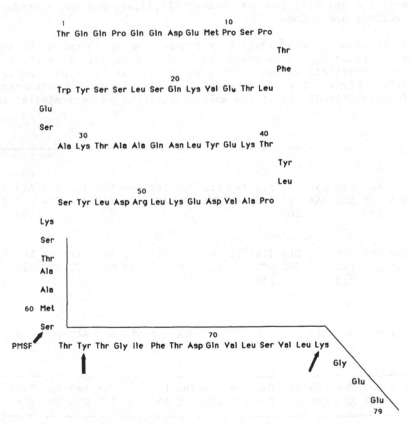

Fig. 4. Amino acid sequence of apolipoprotein C-II. The region of apo C-II responsible for activation of lipoprotein lipase is contained in the sequence from Ser 56 to Glu 79. Derivatization of Ser 61, or Lys 76 or replacement of Tyr 63 eliminates or greatly reduces the ability of native apo C-II or the carboxyl terminal peptides to activate lipoprotein lipase.

the regions important in the activation process (42,43) indicates that many variants might exist. For example derivatization of serine 61 of intact apo C-II or its peptides with phenyl methane sulfonyl fluoride results in loss of activation of lipoprotein lipase (44). Other residues such as lysine 76 and tyrosine 63 seem important for full activity (43,45) while alterations in the region of apo C-II responsible for activating lipoprotein lipase might alter the folding of the protein and reduce activation (43). Thus unrecognized variants of apo C-II may show normal immunologic activity, relatively normal electrophoretic characteristics in single dimension gels but completely or partially reduced ability to activate lipoprotein lipase. In view of the characteristics of the sequence of apo C-II responsible for activation of lipoprotein lipase (43) it is also possible to envisage that there may be variants of apo C-II which may act as competitive inhibitors of normal apo C-II (in a heterozygote) for activation of lipoprotein lipase. Thus it would seem important in the future to review thoroughly the characteristics of apo C-II in hypertriglyceridemic subjects commonly characterized as Type V to determine whether further variants of apo C-II may occur. The techniques of molecular biology should facilitate greatly this type of investigation.

ACKNOWLEDGEMENTS

The author is indebted to J.A. Little, D.W. Cox, P. Alaupovic, A. Kuksis, and G. Kakis for extensive collaboration during the studies of apo C-II deficiency. The assistance of G.F. Maguire, A. Chow, P. Lee, T. Daniels and R. Abraham in lipoprotein and lipid analyses and Lisa Laskey in the preparation of the manuscript is gratefully acknowledged. Research on lipolysis of lipoproteins and apo C-II deficiency in the author's laboratory is supported by the Medical Research Council of Canada and the Nova Scotia Heart Foundation.

REFERENCES

1. R.J. Havel, and R.S. Gordon, Idiopathic hyperlipemia: Metabolic studies in an affected family, J. Clin Invest. 39:1777 (1960).
2. J.C. LaRose, R.I. Levy, P. Herbert, S.E. Law, and D.S. Fredrickson, A specific activator for lipoprotein lipase, Biochem. Biophys. Res. Commun 41:57 (1970).
3. R.J. Havel, V.G. Shore, B. Shore, and D.M. Dier, Role of specific glycopeptides of human serum lipoproteins in the activation of lipoprotein lipase, Circ. Res. 27:595 (1970).
4. C.J. Fielding, Kinetics of lipoprotein lipase activity: Effects of the substrate apoprotein on reaction velocity, Biochim. Biophys. Acta 316:66 (1973).
5. N. Matsuoka, K. Shirai, J.D. Johnson, M.L. Kashyap, L.S. Srivastava, T. Yamamura, N. Yamamoto, and Y. Saito, A. Kumagai and R.L. Jackson, Effects of apolipoprotein C-II on the lipolysis of very low density lipoprotein from apo C-II deficient patients, Metabolism 30:818 (1981).
6. W. Haberbosch, A. Poli, G. Baggio, R. Fellin, A. Gnasso, and J. Augustin, Apolipoprotein C-II deficiency: The role of apolipoprotein CII in the hydrolysis of triacylglycerol-rich lipoproteins, Biochim. Biophys. Acta 793:49 (1984).
7. W.C. Breckenridge, J.A. Little, G. Steiner, A. Chow and M. Poapst, Hypertriglyceridemia associated with deficiency of apolipoprotein C-II, N. Engl. J. Med. 298:1265 (1978).
8. J.A. Little, D.W. Cox, W.C. Breckenridge, V.M. McGuire, Introduction to deficiencies of apolipoproteins C-II and E-III with some clinical findings, in: "Atherosclerosis V", A.M. Gotto, Jr., L.C. Smith and

223

B. Allen, eds., Springer-Verlag, New York (1980).

9. D.W. Cox, W.C. Breckenridge, and J.A. Little, Inheritance of apolipoprotein C-II deficiency with hypertriglyceridemia and pancreatitis, N. Engl. J. Med. 299:1421 (1978).

10. W.C. Breckenridge, P. Alaupovic, D.W. Cox and J.A. Little, Apolipoprotein and lipoprotein concentrations in familial apolipoprotein C-II deficiency, Atherosclerosis 44:223 (1982).

11. W.C. Breckenridge, J.A. Little, P. Alaupovic, D.W. Cox, F.T. Lindgren and A. Kuksis, Abnormal triglyceride clearance secondary to familial apolipoprotein C-II deficiency, in: "The Adipocyte and Obesity: Cellular and Molecular Mechanisms", A. Angel, C.H. Hollenberg and D.A.K. Roncari, eds., Raven Press, New York (1983).

12. W.C. Breckenridge, D.W. Cox, and J.A. Little, Apolipoprotein CII deficiency, in: "Atherosclerosis V", A.M. Gotto, Jr., L.C. Smith, and B. Allen, eds., Springer-Verlag, New York (1980).

13. T. Yamamura, H. Sudo, K. Ishikawa, and A. Yamamoto, Familial type I hyperlipoproteinemia caused by apolipoprotein C-II deficiency, Atherosclerosis 34:53 (1979).

14. N.E. Miller, S.N. Rao, P. Alaupovic, N. Noble, J. Slack, J.D. Brunzell, and B. Lewis. Familial apolipoprotein C-II deficiency: Plasma lipoproteins and apolipoproteins in heterozygous and homozygous subjects and the effects of plasma infusion, Europ. J. Clin. Invest. 11:69 (1981).

15. A. Capurso, L. Pace, L. Bonomo, A. Catapano, G., Schiliro, M. La Rosa and G. Assman, New case of apolipoprotein C-II deficiency, Lancet 1:268 (1980).

16. G. Crepaldi, R. Fellin, G. Baggio, J. Augustin and H. Greten, Lipoprotein and apoprotein, adipose tissue and hepatic lipoprotein lipase levels in patients with familial hyperchylomicronemia and their immediate family members, in: "Atherosclerosis V", A.M. Gotto, Jr., L.C. Smith and B. Allen, eds., Springer-Verlag, New York (1980).

17. R. Fellin, G. Baggio, A. Poli, J. Augustin, M.R. Baiocchi, G. Baldo, M. Singaglia, M. Greten, and G. Crepaldi, Familial lipoprotein lipase and apo C-II deficiency, Atherosclerosis 49:59 (1983).

18. A.F.H. Stalenhof, A.F. Casparie, P.N.M. Demacker, J.T.J. Stouten, J.A. Lutterman and A. van't Laar. Combined deficiency of apolipoprotein C-II, and lipoprotein lipase in familial hyperchylomicronemia, Metabolism 30:919 (1981).

19. A.L. Catapano, G.L. Mills, P. Roma, M. LaRosa, and A. Capurso, Plasma lipids, lipoproteins and apoproteins in a case of apo C-II deficiency, Clin Chim. Acta 130:317 (1983).

20. K. Saku, C. Cedres, B. McDonald, B.A. Hynd, B.W. Liu, L.S. Srivastava and M.L. Kashyap, C-II Apolipoproteinemia and severe hyperglyceridemia, Am. J. Med. 77:457 (1984).

21. S.S. Fojo, S.W. Law, G. Baggio, D.L. Sprecher, L. Taam and H.B. Brewer Jr. Apolipoprotein C-II deficiency. Identification of two kindreds with an abnormal apolipoprotein C-II. Clin. Res. 33:569A (1985).

22. M.R. Hayden, C. Vergani, S.E. Humphries, L. Kirby, R. Shukin and R. McLeod, The genetics and molecular biology of apolipoprotein C-II, in: "Lipoprotein Deficiency Syndromes", A. Angel, ed., Plenum Press, New York (1986).

23. G.F. Maguire, J.A. Little, G. Kakis and W.C. Breckenridge, Apolipoprotein C-II deficiency associated with non functional mutant forms of apolipoprotein C-II, Can. J. Biochem. Cell Biol. 62:847 (1984)

24. V. Skipski, Lipid composition of lipoproteins in normal and diseased states, in: "Blood Lipids and Lipoproteins: Quantitation, Composition and Metabolism", G.J. Nelson, eds., Wiley-Interscience, New York (1972).

25. P.H.E. Groot, and A. van Tol, Metabolic fate of the phosphatidyl choline component of very low density lipoproteins during catabolism

224

by perfused rat heart, Biochim. Biophys. Acta. 530:188 (1978).

26. A. van Tol, T. Van Gent, and H. Jansen, Degradation of high density lipoprotein by heparin releasable liver lipase. Biochim. Biophys. Res. Commun. 94:101 (1980).

27. S.P. Tam, and W.C. Breckenridge, Apolipoprotein and lipid distribution between vesicles and HDL-like particles formed during lipolysis of human very low density lipoproteins by perfused rat heart, J. Lipid Res. 24:1343 (1983).

28. W.C. Breckenridge, The catabolism of very low density lipoproteins, Can. J. Biochem. Cell. Biol. 63:890 (1985).

29. Y.L. Marcel, C. Vezina, B. Teng, and A. Sniderman, Transfer of cholesteryl esters between human high density lipoproteins and tri-glyceride-rich lipoproteins controlled by a plasma protein factor, Atherosclerosis 35:127 (1980).

30. C. Blum, Dynamics of apolipoprotein E metabolism in humans, J. Lipid Res. 23:1308 (1982).

31. A. Rubinstein, J.C. Gibson, J.R. Paterniti, G. Kakis, J.A. Little, H.N. Ginsberg and W.V. Brown, Effect of heparin-induced lipolysis on the distribution of apolipoprotein E among lipoprotein subclasses, J. Clin. Invest. 75:710 (1982).

32. W.C. Breckenridge, J.A. Little, P. Alaupovic, C.S. Wang, A. Kuksis, G. Kakis, F. Lindgren, and G. Gardiner, Lipoprotein abnormalities associated with a familial deficiency of hepatic lipase, Atherosclerosis 45:161 (1982).

33. G. Klose, J. Augustin, and H. Greten, Low hepatic triglyceride lipase in pancreatitis, N. Engl. J. Med. 299:553 (1978).

34. E.T.T. Windler, Y. Chao, and R.J. Havel, Determinants of hepatic uptake of triglyceride with lipoproteins and their remnants in the rat, J. Biol. Chem. 255:5475 (1980).

35. C.L. Jackson, G.A.P. Bruns, and J.L. Breslow, Isolation and sequence of human apolipoproteins C-II cDNA clone and its use to isolate and map to chromosome 19 the gene for apolipoprotein C-II, Proc. Natl. Acad. Sci. 81:2945 (1984).

36. O. Myklebost, B. Williamson, A.F. Markham, S.R. Myklebost, J. Rogers, D.E. Woods and S.E. Humphries, The isolation and charac-terization of cDNA clones for human apolipoprotein C-II, J. Biol. Chem. 259:4401 (1984).

37. S.S. Fojo, S.W. Law, and H.B. Brewer, Jr., Human apolipoprotein C-II: Complete nucleic acid sequence of preapolipoprotein C-II, Proc. Natl. Acad. Sci. USA 81:6354 (1984).

38. S.E. Humphries, K. Berg, L. Gill, A.M. Cumming, F.W. Robertson, A.F.H. Stalenhoef, R. Williamson, and A.L. Borneser, The gene for apolipoprotein C-II is closely linked to the gene for apolipoprotein E on chromosome 19, Clin. Genet. 26:389 (1984).

39. S.J. Fojo, S.W. Law, H.B. Brewer, Jr., A.Y. Sakaguchi, and S.L. Naylor, The localization of the gene for apolipoprotein C-II to chromosome 19, Biochim. Biophys. Res. Commun. 122:687 (1984).

40. U. Francke, M.S. Brown, and J.L. Goldstein, Assignment of the human gene for the low density lipoprotein receptor to chromosome 19: Syntery of a receptor, a ligand and a genetic disease, Proc. Natl. Acad. Sci. USA 81:2826 (1984).

41. S.E. Humphries, L. Williams, O. Myklebost, A.F.H. Stalenhoef, P.N.M. Demacker, G. Baggio, G. Crepaldi, D.J. Galton, and R. Williamson, Familial apolipoprotein C-II deficiency: A preliminary analyses of the gene defect in two independent families, Hum. Genet. 67:151 (1984).

42. P.K.J. Kinnunen, R.L. Jackson, L.C. Smith, A.M. Gotto, Jr., and J.T. Sparrow, Activation of lipoprotein lipase by native and synthetic fragments of human apolipoprotein C-II, Proc. Natl. Acad. Sci. 74:4848 (1977).

43. L.C. Smith, J.C. Yoyta, A.L. Catapano, P.K.J. Kinnunen, A.M. Gotto,

Jr., and J.T. Sparrow, Activation of lipoprotein lipase by synthetic fragments of apolipoprotein C-II, in: "Lipoprotein Structure", Ann. N.Y. Acad. Sci. 348:213 (1980).

44. P. Vainio, J.A. Virtanen, P.K.J. Kinnunen, A.M. Gotto, Jr., J.T. Sparrow, F. Pattus, P. Bougis, and R. Verger, Action of lipoprotein lipase on mixed triacylglycerol/phosphatidyl choline monolayers: Activation by apolipoprotein C-II, J. Biol. Chem. 258:5477 (1983).

45. T.A. Musliner, P.N. Herbert, and E.C. Church, Activation of lipoprotein lipase by native and acylated peptides of apolipoprotein C-II, Biochim. Biophys. Acta. 573:501 (1979).

PRIMARY LIPOPROTEIN LIPASE DEFICIENCY

John D. Brunzell, Per-Henrik Iverius, Mark S. Scheibel,
and Wilfred Y. Fujimoto

University of Washington
Seattle

Michael R. Hayden, Roger McLeod, and Jiri Frolich

University of British Columbia
Vancouver

INTRODUCTION

The enzyme lipoprotein lipase (LPL) is a glycoprotein located on the luminal surface of capillary endothelial cells (see reviews: 1-4). It is bound to a glycosaminoglycan on the endothelium, can be displaced into plasma by intravenous heparin or other polyanions, and binds to heparin-Sepharose gels. The enzyme has an apparent monomeric molecular weight on SDS gel of over 60,000 and 48,300 by sedimentation-equilibrium ultracentrifugation, and probably functions as a dimer in vivo. It has binding sites for heparin, for the cofactor apolipoprotein CII, and for lipid, and has a separate catalytic site for triglyceride hydrolysis. It is inhibited by serine proteases inhibitors, by protamine, and by high ionic strength. The enzyme appears to be synthesized in a number of different parenchymal cells including monocyte-derived macrophages, Kupfer cells, adipocytes, and cells in cardiac and skeletal muscle. The enzyme is secreted from the adipocyte and transported in an unknown fashion to the plasma surface of the capillary endothelial cell, where it has several functions in humans.

Bound to the endothelial cell, in conjunction with apolipoprotein CII contained in chylomicrons and very low density lipoproteins, LPL hydrolyses the triglyceride in the core of these lipoproteins to free fatty acids and glycerol. The removal rate of triglyceride-rich lipoproteins from plasma correlates with the level of LPL in adipose tissue (Fig. 1). The free fatty acids are utilized for energy by muscle or other tissue or is stored in adipose tissue. The enzyme seems to play a "gatekeeper" role for energy storage; fat cell size is proportional to enzyme activity levels in adipose tissue (6).

The core remnants remaining after triglyceride hydrolysis are then further processed in the liver. The chylomicron remnant is completely degraded, while some of the very low density lipoprotein remnant is processed to form low density lipoproteins (LDL).

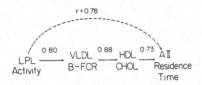

Fig. 1. Interrelations among adipose tissue LPL activity,
VLDL catabolism (apo-b fractional catabolic rate)
and plasma HDL apoprotein catabolism in humans
[from reference (5)].

A small portion of core triglyceride from chylomicrons and very low
density lipoproteins (VLDL) can be transferred to the high density lipoproteins (HDL). However, more important contributors to HDL are the surface
remnants of the triglyceride-rich lipoproteins that accumulate as a result
of the hydrolysis of the core triglycerides. A relationship between LPL
activity and HDL cholesterol has been noted in many clinical situations by
Nikkila and his coworkers (7). The transfer of the triglyceride-rich lipoprotein surface containing free cholesterol and phospholipid to HDL prolongs
the residence time of these apolipoprotein AI- and AII-containing particles
(5) (Fig. 1).

Thus LPL functions at an important junction in lipoprotein metabolism,
regulating the distribution of energy in the form of free fatty acids
and the distribution of cholesterol to LDL and HDL. In humans and other
mammals, LPL has also been shown to be present in the breast, the placenta,
the brain, and the lung. Specialized emphasis of the various functions of
LPL occurs in each of these tissues.

CLASSICAL LIPOPROTEIN LIPASE DEFICIENCY

The clinical syndrome associated with lactescent plasma in childhood
was first described in 1932 by Bürger and Grütz in a young male with eruptive xanthomas and hepatosplenomegaly (see review: 8). In 1960 Havel and
Gordon (9) demonstrated defective clearance of triglyceride-rich lipoproteins in this disorder related to diminished lipolytic activity in plasma
after intravenous heparin. Harlan et al. (10) found decreased LPL activity
in adipose tissue to be associated with the decrease in postheparin plasma
lipolytic activity in two patients. In 1974 Krauss et al. (11) separated
the post-heparin plasma activities into LPL and hepatic lipase (46).

Over fifty patients have been described who have a primary abnormality
in LPL activity (see reviews: 8,12-14). Most of these patients can be
characterized by a familial syndrome of chylomicronemia which will here
be termed "classical lipoprotein lipase deficiency." Variations of this
syndrome will be considered below. The disease may be detected in early
infancy and present with colicky pain, hepatosplenomegaly, or failure to
thrive. Often it is not recognized until later in childhood with the occurrence of abdominal pain, the development of eruptive xanthomas, the presence
of lactescent plasma, or the development of other symptoms and signs of the
chylomicronemia syndrome (15).

The lactescent plasma is due to the accumulation of dietary fat in plasma as triglyceride-rich chylomicrons. The severity of the hypertriglyceridemia is related to the amount of ingested long chain fatty acids; thus the term fat-induced or exogenous hypertriglyceridemia (16). Although it is difficult to separate endogenous, triglyceride-rich VLDL from chylomicrons (particularly when the latter are markedly elevated), VLDL are usually not elevated in children and young adults with LPL deficiency. However, compelling data exist which indicate that VLDL triglyceride is nonetheless catabolized abnormally in LPL deficiency (see review: 17), and therefore factors in addition to LPL deficiency must exist for the presence (phenotype V) or absence (phenotype I) of VLDL with the chylomicronemia. Low density and high density lipoproteins have altered composition and are decreased in amount (18).

LPL activity in plasma is absent or low in these patients following a bolus (11,14) of heparin (Fig. 2) and is not responsive to higher doses of heparin or more prolonged infusions of heparin (19). In contrast, postheparin plasma HL activity is present in normal or low normal amounts (11, 14). Adipose tissue LPL activity is very low (10) or undetectable (19). LPL activity was also missing in cultured monocyte-derived macrophages from a subject with the classical form of LPL deficiency (20).

Consanguinity is exceptionally common in classic LPL deficiency (12,14) suggesting that the abnormal allelle(s) for the defective LPL activity are very rare (21). Multiple affected siblings, the lack of parent-to-child transmission, and equal involvement of both sexes are observations consistent with an autosomal recessive pattern of inheritance. Mental retardation has been seen in affected individuals in several families with LPL deficiency (22) (personal observation-MRH); whether this is related to an increased prevalence of mental retardation in consanguinous matings (23) or to the LPL deficiency per se is unknown.

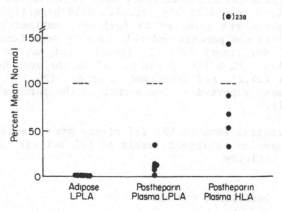

Fig. 2. Various lipase activities plotted as percent of mean normal values (---) in subjects with classical LPL deficiency.

229

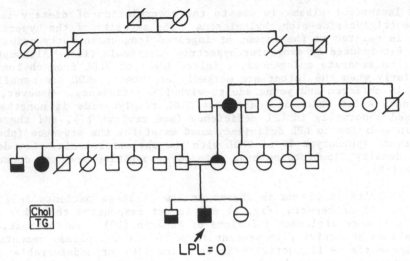

Fig. 3. Pedigree of patient with LPL deficiency with moderate hyperlipidemia present in relatives of both
parents.

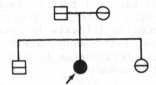

Fig. 4. Pedigree of patient with LPL deficiency; all first degree relatives
had normal lipid levels.   The
families of both parents emigrated
from a small town in East Germany.

The obligate heterozygote parents and other family members of affected
individuals with classical LPL deficiency have demonstrated mild lipoprotein
abnormalities or have been entirely normal. Mild hyperlipidemia and variable lipoprotein phenotype (similar to familial combined hyperlipidemia)
have been observed in the parents and relatives in some families of probands
with classical LPL deficiency (Fig. 3) (24-27), but not in other families
(Fig. 4) (24,27,28). LPL activity was normal in the post-heparin plasma of
some heterozygotes (24,27) and decreased in others (26,28). Adipose tissue
LPL activity has been reported as low normal in the parents of two affected
individuals (10,27).

Thus, the classical form of LPL deficiency presents in childhood with
chylomicronemia, has low to absent levels of LPL activity in all tissues,
and has normal HL activity.

LIPOPROTEIN LIPASE DEFICIENCY VARIANTS

In addition to the heterogeneity among individuals with classical forms
of LPL deficiency, there are unusual patients with uniquely different variations in the syndrome.

Site specific lipoprotein lipase deficiency

While most patients have decreased LPL activity in all synthetic sites, two subjects have been noted who may have a tissue site-specific abnormality in the enzyme (19). One of these patients was diagnosed as LPL deficiency following pancreatitis during pregnancy at age 28 (29). She was found to have less than 10% of normal LPL activity in plasma shortly after the injection of heparin (60 units/kg). She had, however, normal levels of LPL activity in adipose tissue and in plasma during the last two hours of a 6-hour, high-dose heparin infusion (19). The fatty acid composition of this patient's breast milk following a successful pregnancy suggests that her LPL defect also included the mammary gland (29). From observations made in this patient, it has been postulated that the LPL appearing in plasma during the latter phase of the heparin infusion arises from adipose tissue and represents the LPL present in this patient, while the LPL released into plasma during the early phase of the heparin infusion comes from other tissue sources, including enzyme bound to capillary endothelium (Fig. 5).

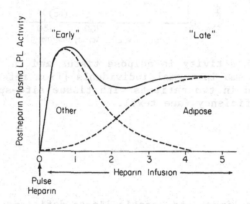

Fig. 5. Model of tissue site contribution of lipoprotein lipase to postheparin plasma LPL activity during a heparin infusion.

In support of the above hypothesis, a second patient has been described with very low adipose tissue LPL activity and low activity during the late phase of the heparin infusion; however, the LPL activity released early during the infusion was normal (19). This patient presented at the age of 20 years with mild abdominal discomfort and lipemic plasma. He had been investigated 10-15 years earlier at the University of California at San Francisco, but had remained almost symptom free without overt dietary modification. A second patient, seemingly similar to this, has been described by Burton and Nadler (30).

These patients can be easily detected by simultaneous measurement of adipose tissue and 10-minute post-heparin plasma LPL activities (Fig. 6). Both of these patients appear to have less severe problems than most patients with classical LPL deficiency since they were able to eat essentially normal diets and did not become symptomatic until young adult life.

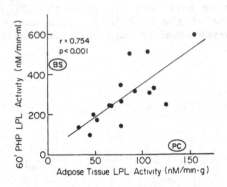

Fig. 6.   LPL activity in adipose tissue and in postheparin plasma in normal individuals [from reference (31)] and in two patients with tissue site-specific LPL deficiency (see text).

Combined lipoprotein lipase and hepatic lipase deficiency

A family has been described with HL deficiency in whom LPL activity was found to be normal (32). Affected individuals accumulate excess levels of cholesterol-rich remnants of VLDL and chylomicrons and demonstrate triglyceride enrichment of low density and high density lipoproteins. A model for combined deficiency of both HL and LPL activity has been discovered in mice (33). LPL mass in these mice is present, but the enzyme has defective functional activity (34).

A 41-year-old male has been noted to have decreased levels of both LPL and HL activities (19). He differs from patients with classical LPL deficiency in that he first presented with symptoms at age 32; he has a marked increase in VLDL as well as chylomicrons (lipoprotein phenotype V); and he developed the clinical symptoms of a myocardial infarction at age 41, confirmed by typical electrocardiogram and plasma enzyme changes.

Combined lipoprotein lipase and apoprotein CII deficiency

Patients with LPL deficiency have elevated levels of apoprotein CII, a cofactor necessary for hydyrolysis of triglyceride by LPL, while patients with apo CII deficiency often have elevated adipose tissue LPL activity (35) (personal observation: JF and JDB). Four of nine siblings in one family have been demonstrated to have decreased levels of both LPL activity in post-heparin plasma and plasma apoprotein CII (36). These affected individuals appear similar to those with moderately symptomatic classical LPL deficiency, except that they may have had greater elevations in VLDL levels (37). Incidentally, all four affected siblings were apoprotein E_2 homozygotes, as was their father; and thyroid binding globulin deficiency was also noted in two of the four, the father, and two normal siblings (36,37).

Transient lipoprotein lipase deficiency

Very low levels of LPL activity in post-heparin plasma were found in a nine-year-old boy several weeks after recovering from an episode of pancreatitis associated with lactescent plasma (38). His triglyceride level was 7280 mg/dl and his post-heparin plasma hepatic lipase activity was normal. Following institution of moderately severe dietary fat restriction, his LPL activity returned to normal and triglyceride fell to 57 mg/dl. Both parents had borderline hyperlipidemia.

A previously unreported patient may have had a similar problem. A 61-year-old male presented with moderate abdominal pain and lactescent plasma. He had no family history of coronary artery disease or pancreatitis, but was found to have monogenic familial hypertriglyceridemia with moderate hypertriglyceridemia present in many relatives. On presentation he was markedly hypertriglyceridemic (plasma triglyceride 4890 mg/dl, Table 1) with a very low level of plasma post-heparin lipolytic activity (PHLA) compatible with LPL deficiency. The PHLA returned to normal with a marked improvement in

Table 1.
Possible Transient LPL Deficiency

Date	TG mg/dl	PHLA*	Comments
5/19/71	4890	0.13	Start Clofibrate
9/2/71	245	0.39	
	(200–860)		
1/16/73			Stop Clofibrate
1/22/73	2960		Start one gram fat eucaloric diet
2/9/73	1543	0.01	Add 600 kcal diet
2/21/73	486	0.32	
2/23/73	433	0.34	Restart Clofibrate and normal diet
	(103–239)		
12/18/73	231	0.49	

*PHLA: Total plasma post-heparin lipolytic activity (µEq FFA/ml/min) ten minutes after heparin; 1971:10 units/kg weight, 1973: 60 units/kg weight. N = 25 controls: low dose 0.47±.16 (x±SD, range 0.22–0.85), high dose 0.58±0.18 (range 0.24–1.03).

plasma triglyceride levels (Table 1) while on clofibrate therapy and on a separate occasion while on a calorically restricted diet. Although it might be argued that clofibrate normalized the LPL activity, it is unlikely that severe caloric restriction would do the same. An alternative explanation would be an improvement in LPL activity secondary to the marked decrease in plasma triglyceride levels. Perhaps triglyceride-rich lipoproteins were competing for the enzyme loosely bound to the endothelial surface. Infusion of lipid emulsions into rodents (39) and into humans (40) has been associated with decreases in lipolytic activity.

Other unusual patients

One patient has been noted with very low levels of post-heparin lipolytic activity after moderate doses of intravenous heparin, but normal lipolytic activity after high doses of heparin (41). Another unusual patient was not diagnosed to have LPL deficiency until the age of 75 years (42).

Implications of genetic variants

The simultaneous deficiencies of HL activity and LPL activity in one individual and of LPL activity and apoprotein CII in individuals of another family suggests that the genes for HL, LPL, and apo CII may be linked. A focal defect might cause these abnormalities, such as seen with the deletion of part of chromosome 11 associated with apo AI and apo CIII deficiency (43). However, as low levels of HL and LPL activity and of apo CII mass have been reported for these patients, these genes are present and are either expressed at markedly reduced levels, or an altered protein is produced with diminished activity and/or immunoreactivity. HL and LPL may be homologous proteins arising from a common ancestral gene; such a defect probably would be a much older mutation than one involving only one of the proteins. And finally, it is possible that a defect is present in some other common features such as the interaction with heparin or sensitivity to serine protease inhibitors.

The individuals with tissue site-specific LPL deficiency may have mutations at sites involved in the regulation of expression of LPL rather than in regions encoding structural domains of the enzyme protein. It is known that the enzymes in adipose tissue and in muscle are regulated in opposite ways during fasting and feeding, although the mechanisms to account for this regulation are not understood.

MEASUREMENT OF LIPOPROTEIN LIPASE MASS

The studies of LPL deficiency have previously been solely dependent on measurement of enzyme activity. Recently, however, it has been possible to identify enzyme mass (44-50). Utilizing an enzyme-linked immunosorbent assay with murine monoclonal antibodies to LPL, it was possible to demonstrate a relationship between enzyme mass and activity (48) in normal individuals (Fig. 7). Most patients with lipoprotein lipase deficiency apparently have enzyme mass by this immunological technique (48) ranging from low normal levels to above normal levels (Fig. 8). The low specific activity of the enzyme protein in these patients is compatible with a defect due to a point mutation(s) in the gene for LPL.

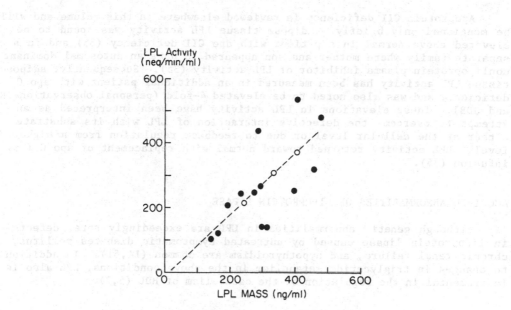

Fig. 7. LPL activity mass in postheparin
plasma in 13 normal females and
3 normal males [from reference
(47)].

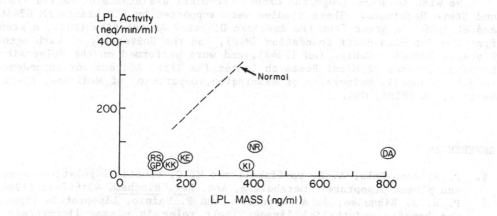

Fig. 8. LPL activity and apparent mass in 7 patients with
deficiency of LPL activity.

LIPOPROTEIN LIPASE ACTIVITY IN APOPROTEIN CII DEFICIENCY

Apoprotein CII deficiency is reviewed elsewhere in this volume and will be mentioned only briefly. Adipose tissue LPL activity was found to be elevated above normal in a patient with apo CII deficiency (35) and in a separate family where mother and son appeared to have an autosomal dominant nonlipoprotein plasma inhibitor of LPL activity (35). Subsequently, adipose tissue LPL activity has been measured in an additional patient with apo CII deficiency and was also noted to be elevated 4-fold (personal observation-JR and JDB). These elevations in LPL activity have been interpreted as an attempt to overcome the defective interaction of LPL with its substrate either at the cellular level or due to feedback regulation from a higher level. LPL activity returned toward normal with replacement of apo CII by infusion (35).

ACQUIRED ABNORMALITIES OF LIPOPROTEIN LIPASE

Although genetic abnormalities in LPL are exceedingly rare, defects in lipoprotein lipase caused by untreated symptomatic diabetes mellitus, chronic renal failure, and hypothyroidism are common (14,51). In addition to changes in triglyceride metabolism in the above conditions, LPL also is instrumental in the regulation of the catabolism of HDL (5,7).

SUMMARY

The enzyme lipoprotein lipase plays a central role in the processing of energy in the form of calorically dense triglyceride. Classical LPL deficiency usually presents in childhood with the multiple manifestations related to chylomicronemia. Many patients with genetic variations have been noted who differ in one of many ways from the classical patients. With the development of techniques to measure enzyme mass and to study gene expression, the molecular defects in each of these families should become evident.

ACKNOWLEDGEMENTS

We wish to acknowledge the expert technical assistance of Martha Kimura, and Steve Hashimoto. These studies were supported by NIH grants AM 02456 and HL 30086, a grant from the American Diabetes Association (PHI), a grant from the Canadian Heart Foundation (MRH), and the University of Washington Diabetes Research Center (AM 17047), and were performed on the University Hospital General Clinical Research Center (RR-37). Address correspondence to J.D. Brunzell, University of Washington, Department of Medicine, RG-26, Seattle, WA 98195, USA.

REFERENCES

1. P. Nilsson-Ehle, A. S. Garfinkel, and M. C. Schotz, Lipolytic enzymes and plasma lipoprotein metabolism, Ann. Rev. Biochem. 49:667-93 (1980).
2. P. K. J. Kinnunen, J. A. Virtanen, and P. Vainio, Lipoprotein lipase and hepatic endothelial lipase: Their roles in plasma lipoprotein metabolism, in: "Atherosclerosis Reviews," vol. 11, A. M. Gotto, Jr., and R. Paoletti, eds., Raven Press, New York (1983), pp. 65-105.
3. R. L. Jackson, Lipoprotein lipase and hepatic lipase, in: "The Enzymes," vol. 16, P.D. Boyer, ed., Academic Press (1983), pp. 141-181.
4. A. Cryer, Tissue lipoprotein lipase activity and its action in lipoprotein metabolism, Int. J. Biochem. 13:525-541 (1981).
5. P. Magill, S. N. Rao, N. E. Miller, A. Nicoll, J. Brunzell, J. St.

Hilaire, and B. Lewis, Relationships between the metabolism of high density and very-low-density lipoproteins in man: Studies of apolipoprotein kinetics and adipose tissue lipoprotein lipase activity, Eur. J. Clin. Invest. 12:113-120 (1982).

6. R. S. Schwartz, and J. D. Brunzell, Adipose tissue lipoprotein lipase and obesity, in: "Recent Advances in Obesity Research: III," P. Bjorntorp, M. Cairella, and A. N. Howard, eds., John Libbey and Co., London (1981), pp. 94-98.

7. E. A. Nikkila, M. R. Taskinen, and M. Kekki, Relation of plasma high density lipoprotein cholesterol to lipoprotein-lipase activity in adipose tissue and skeletal muscle of man, Atherosclerosis 29:497-501, (1978).

8. R. S. Lees, D. E. Wilson, G. Schonfeld, and S. Fleet, The familial dyslipoproteinemias, in: "Progress in Medical Genetics," vol. 9, A. G. Steinberg, and A. G. Bearn, eds., Grune and Stratton, New York (1973), pp. 237-259.

9. R. J. Gordon and R. S. Havel, Jr., Idiopathic hyperlipemia: Metabolic studies in an affected family, J. Clin. Invest. 39:1777-1790 (1960).

10. W. R. Harlan, Jr., P. S. Winesett, and A. J. Wasserman, Tissue lipoprotein lipase in normal individuals and in individuals with exogenous hypertriglyceridemia and the relationship of this enzyme to assimilation of fat, J. Clin. Invest. 46:239-247 (1967).

11. R. M. Krauss, R. I. Levy, and D. S. Fredrickson, Selective measurement of two lipase activities in postheparin plasma from normal subjects and patients with hyperlipoproteinemia, J. Clin. Invest. 54:1107-1124 (1974).

12. D. S. Levy, and R. I. Fredrickson, Familial hyperlipoproteinemia, in: "The Metabolic Basis of Inherited Disease," 3rd ed., J. B. Stanbury, J. B. Wyngaarden, and D. S. Fredrickson, eds., McGraw-Hill, New York (1972), pp. 545-611.

13. W. V. Brown, M. L. Baginsky, and C. Ehnholm, Primary type I and type V hyperlipoproteinemia, in: "Hyperlipidemia: Diagnosis and Therapy," B. M. Rifkind, and R. I. Levy, eds., Grune and Stratton, New York/ San Francisco/London, (1977), pp. 93-112.

14. E. A. Nikkilä, Familial lipoprotein lipase deficiency and related disorders of chylomicron metabolism, in: "Disorders of Lipoprotein and Lipid Metabolism," vol. 5, J. B. Stanbury, J. B. Wyngaarden, D. S. Fredrickson, J. L. Goldstein, and M. S. Brown, eds., McGraw-Hill Book Co., New York (1983), pp. 622-642.

15. J. D. Brunzell, and E. L. Bierman, Chylomicronemia syndrome: Interaction of genetic and acquired hypertriglyceridemia, in: "Medical Clinics on Lipid Disorders," R. J. Havel, ed., W. B. Saunders Co., Philadelphia; Med. Clin. N. Am. 66:455-468 (1982).

16. J. L. Knittle, and E. H. Ahrens, Jr., Carbohydrate metabolism in two forms of hyperglyceridemia, J. Clin. Invest. 43:485-491 (1964).

17. A. Chait, and J. Brunzell, Very low density lipoprotein kinetics in familial forms of hypertriglyceridemia, in: "Lipoprotein Kinetics and Modeling," M. Berman, S. Grundy, and B. Howard, eds., Academic Press, New York (1982), pp. 69-76.

18. E. Manzato, R. Marin, A. Gasparotto, G. Baggio, R. Fellin, and G. Crepaldi, The plasma lipoproteins in familial chylomicronemia: Analysis by zonal ultracentrifugation, J. Lab. Clin. Med. 104:778-788 (1984).

19. J. D. Brunzell, A. Chait, E. A. Nikkilä, C. Ehnholm, J. K. Huttunen, and G. Steiner, Heterogeneity of primary lipoprotein lipase deficiency, Metabolism 29:624-629 (1980).

20. A. Chait, P.-H. Iverius, and J. Brunzell, Lipoprotein lipase secretion by human monocyte-derived macrophages, J. Clin. Invest. 69:490-493 (1982).

21. F. Vogel, and A.G. Motulsky, "Human Genetics," Springer-Verlag, New York (1982), pp. 416-422.

22. H. J. Sternowsky, U. Gaertner, N. Stahnke, and E. Kaukel, Juvenile familial hypertriglyceridemia and growth retardation: Clinical and bio-

chemical observations in three siblings, Eur. J. Pediat. 125:59-70 (1977).

23. E. Seemanover, A study of children of incestuous mating, Human Genetics 21:108-128 (1971).

24. C. Gagne, D. Brun, S. Moorjani, and P.-J. Lupien, Hyperchylomicronémie familiale: Etude de l'activité lipolytique dans une famille, Un. Med. Canada 106:333-338 (1977).

25. J. M. Potter, and W. B. Macdonald, Primary type I hyperlipoproteinaemia: A metabolic and family study, Aust. N.Z. J. Med. 9:688-693 (1979).

26. R. Fellin, G. Baggio, A. Poli, J. Augustin, M. R. Baiocchi, G. Baldo, M. Sinigaglia, H. Greten, and G. Crepaldi, Familial lipoprotein lipase and apolipoprotein C-II deficiency: Lipoprotein and apoprotein analysis, adipose tissue and hepatic lipoprotein lipase levels in seven patients and their first degree relatives, Atherosclerosis 49:55-68 (1983).

27. D. E. Wilson, C. Q. Edwards, and I.-F. Chan, Phenotypic heterogeneity in the extended pedigree of a proband with lipoprotein lipase deficiency, Metabolism 32:1107-1114 (1983).

28. Y. Kondo, I. Kurobane, K. Omura, R. Sano, R. Abe, N. Chida, and K. Tada, Postheparin plasma lipoprotein lipase activity in heterozygotes of familial lipoprotein lipase deficiency, Tohoku J. Exp. Med. 145:1-6 (1985).

29. G. Steiner, J. J. Myher, and A. Kuksis, Milk and plasma lipid composition in a lactating patient with type I hyperlipoproteinemia, Am. J. Clin. Nutr. 41:121-128 (1985).

30. B. K. Burton, and H. L. Nadler, Primary type I hyperlipoproteinemia with normal lipoprotein lipase activity, J. Pediat. 90:777-779 (1977).

31. P.-H. Iverius, and J. D. Brunzell, Human adipose tissue lipoprotein lipase: Changes with feeding and relation to postheparin plasma, Am. J. Physiol. 249:E107-E114 (1985).

32. W. C. Breckenridge, J. A. Little, G. Steiner, A. Chow, and M. Poapst, Hypertriglyceridemia associated with deficiency of apolipoprotein C-II, N. Eng. J. Med. 298:1265-1273 (1978).

33. J. R. Paterniti, W. V. Brown, H. N. Ginsberg, and K. Artzt, Combined lipase deficiency (cld): A lethal mutation on chromosome 17 of the mouse, Science (Washington, D.C.) 221:167-169 (1983).

34. T. Olivecrona, S. S. Chernick, G. Bengtsson-Olivecrona, J. R. Paterniti, Jr., W. V. Brown, and R. O. Scow, Combined lipase deficiency (cld/cld) in mice: Demonstration that an inactive form of lipoprotein lipase is synthesized, J. Biol. Chem. 260:2552-2557 (1985).

35. J. D. Brunzell, N. E. Miller, P. Alaupovic, R. J. St. Hilaire, C. S. Wang, D. L. Sarson, S. R. Bloom, and B. Lewis, Familial chylomicronemia due to a circulating inhibitor of lipoprotein lipase activity, J. Lipid Res. 24:12-19 (1983).

36. A. F. H. Stalenhoef, A. F. Casparie, P. N. M. Demacker, J. T. J. Stouten, J. A. Lutterman, and A. van't Laar, Combined deficiency of apolipoprotein C-II and lipoprotein lipase in familial hyperchylomicronemia, Metabolism 30:919-926 (1981).

37. S. W. J. Lamberts, A. F. Casparie, K. Miedema, G. Hennemann, and H. A. M. Hulsmans, Thyroxine binding globulin deficiency in a family with type I hyperlipoproteinaemia, Clin. Endocrinol. 6:197-206 (1977).

38. I. J. Goldsberg, J. R. Paterniti, Jr., B. H. Franklin, H. N. Ginsberg, F. Ginsberg-Fellner, and W. V. Brown, Case report: Transient lipoprotein lipase deficiency with hyperchylomicronemia, Am. J. Med. Sci. 286:28-31 (1983).

39. E. Shadfir, and Y. Biale, Effect of experimental hypertriglyceridemia on tissue and serum lipoprotein lipase activity, Eur. J. Clin. Invest. 1:19-24 (1970).

40. A. H. Kissebah, P. W. Adams, and V. Wynn, Plasma free fatty acid and triglyceride transport kinetics in man, Clin. Sci. Mol. Med. 47:259-278 (1974).

41. A. Horst, J. Paluszak, K. Zawilska, and S. Sobisz, Three variants of postheparin lipoprotein lipase activity in idiopathic hyperlipoprotein-emia, Bull. Acad. Pol. Sci. 21:199-202 (1973).
42. J. M. Hoeg, J. C. Osborne, Jr., R. E. Gregg, and H. B. Brewer, Jr., Initial diagnosis of lipoprotein lipase deficiency in a 75-year-old man, Am. J. Med. 75:889-892 (1983).
43. S. K. Karathanasis, R. A. Norum, V. I. Zannis, and J. L. Breslow, A mutation in the human apo A-I gene locus related to the development of atherosclerosis, Nature 301:718-720 (1983).
44. L. Jonasson, G. K. Hansson, G. Bondjers, G. Bengtsson, and T. Olive-crona, Immunohistochemical localization of lipoprotein lipase in human adipose tissue, Atherosclerosis 51:313-326 (1984).
45. J. C. Voyta, D. P. Via, P. K. J. Kinnunen, J. T. Sparrow, A. M. Gotto, Jr., and L. C. Smith, Monoclonal antibodies against bovine milk lipo-protein lipase: Characterization of an antibody specific for the apo-lipoprotein C-II binding site, J. Biol. Chem. 260:893-898 (1985).
46. L. Socorro, and R. L. Jackson, Monoclonal antibodies to bovine milk lipoprotein lipase: Evidence for proteolytic degradation of the native enzyme, J. Biol. Chem. 260:6324-6328 (1985).
47. M. Scheibel, P. Iverius, J. Brunzell, and W. Fujimoto, Measurement of human lipoprotein lipase by enzyme-linked immunosorbent assay (ELISA) using a single monoclonal antibody, Fed. Proc. 44:1156 (1985).
48. M. S. Scheibel, P.-H. Iverius, J. D. Brunzell, and W. Y. Fujimoto, An enzyme-linked immunosorbent assay for human postheparin plasma lipo-protein lipase, Submitted.
49. M. E. Pedersen, M. Cohen, and M. C. Schotz, Immunocytochemical locali-zation of the functional fraction of lipoprotein lipase in the perfused heart, J. Lipid Res. 24:512-521 (1983).
50. J. Etienne, L. Noé, M. Rossignol, C. Arnaud, N. Vydelingum, and A. H. Kissebah, Antibody against rat adipose tissue lipoprotein lipase, Biochim. Biophys. Acta 834:95-102 (1985).
51. J. D. Brunzell, Endocrine disorders and adipose tissue lipoprotein lipase, in: "Lipoprotein Metabolism and Endocrine Regulation," L. W. Hessel, and H. M. J. Krans, eds., Elsevier/North Holland Biomedical Press, New York (1979), pp. 27-34.

THE GENETICS AND MOLECULAR BIOLOGY OF APOLIPOPROTEIN CII

M.R. Hayden[1], C. Vergani[2], S.E. Humphries[3], L. Kirby[4],
R. Shukin[4], R. McLeod[4]

1 Department of Medical Genetics, Vancouver, B.C.
2 University of Milan, Italy
3 Charing Cross Research Medical Centre, London
4 Department of Pathology, Vancouver, B.C.

INTRODUCTION

Medium and long chain fatty acids which are released during digestion
of dietary fat, circulate in the form of chylomicrons. Triglycerides are
also transported as very low density lipoproteins (VLDL) which are produced
in the liver. Lipoprotein lipase (LPL) catalyses the hydrolysis of
triglycerides within the vascular space and this process is pivotal in the
removal of triglycerides from the circulation.

LPL is activated by apolipoprotein CII (apo CII) which converts it from
a less to a more active form (1). Since lipoprotein lipase does hydrolyze
triglyceride in the absence of apo CII (2, 3), a true coenzyme function can
be excluded.

Apo CII is a 79 aminoacid polypeptide (4, 5) which is synthesized in
the liver and resides in chylomicrons, VLDL and high density lipoprotein
(HDL) fractions. Studies using tryptic fragments of apo CII have revealed
that the carboxy terminal aminoacids 55 - 79 are necessary for maximal lipoprotein activation (5). It is, therefore, likely that persons with markedly
elevated fasting triglycerides due to apo CII deficiency have defects within
the portion of the gene which codes for these aminoacid residues.

Recent advances in molecular biology have utilized the knowledge
obtained from protein biochemistry and facilitated isolation of the cDNA and
genomic clones for this apoprotein.

STRATEGIES FOR THE CLONING OF APO CII

Knowledge of the primary aminoacid structure of apo CII has allowed
identification of aminoacid sequences specified by relatively unambiguous
codons (4,6,7). cDNA oligomers which correspond to the range of mRNA's that
could code for these aminoacids have been synthesized. Human liver cDNA
libraries were then screened by these radioactively labelled synthetic
oligonucleotides and positive clones showing hybridization were then
characterized by restriction endonuclease mapping and DNA sequence analysis.
Finally, these positive clones were then used to screen a human genomic
library (Fig.1). The clones identified from the genomic library have been
digested with multiple restriction endonucleases and subjected to further
DNA sequencing.

In a similar analysis, DNA from a normal person was prepared from peripheral blood leukocytes, digested with multiple restriction enzymes and subjected to southern blot analysis with a radiolabelled cDNA for CII. This revealed a similar restriction map for the CII gene in the human genome and the cloned gene (4) confirming that the identified clone was indeed the cDNA for CII.

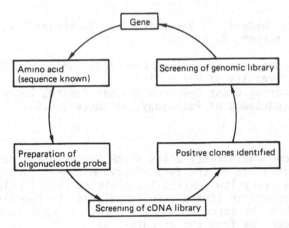

Fig 1: Synthesis of oligonucleotides has allowed identification of cDNA's for different lipoproteins.

From the DNA sequence of the cDNA, it was possible to deduce the aminoacid sequence of the entire mature form of apo CII. The aminoacid sequence predicted differed slightly from that previously determined (8). The DNA sequence specified a polypeptide of 79 aminoacid residues rather than 78 previously reported. In addition, other small differences in aminoacid sequence were identified which could either reflect apo CII polymorphisms or may be true corrections to the aminoacid sequence (4).

Studies at the DNA level have indicated common features between the genetic organization of different apolipoproteins including apo E, apo A-1 and apo CII (9). All genes have three introns in similar locations and the fourth exon consists largely of repeated DNA sequences that may have arisen by intragenic duplication (9). This is evidence supporting the concept that these apolipoproteins were derived from a common evolutionary precursor (10) and would explain the clustering of apolipoprotein genes such as A-I and CIII on chromosome 11, CI, CII and apo E on chromosome 19 (see below). The dispersal of genes for other apolipoproteins to different regions of the genome could have arisen via recombination (11).

EXPRESSION OF THE GENE:

Isolation of the cDNA and genomic clones has allowed determination of tissue levels, size and structure of CII mRNA. Human apo CII mRNA is approximately 500 base pairs in size. The CII cDNA, when used as probe, shows intense hybridization to apo CII mRNA in liver with significantly less hybridization to intestinal mRNA. No apo CII mRNA was detected in the kidney (13).

This confirms that the liver is the major site for production of CII with significantly less synthesis in the intestine. This had previously only been demonstrated by immunological techniques (13).

Isolation of the human cDNA clones for CII has facilitated comparison of the degree of homology for CII between mammalian species. Use of the human CII clone in hybridization experiments with CII mRNA from the liver and intestine of cynamologus macaques has revealed marked complimentarity between monkey CII and that of man. However, the concentration of CII mRNA in the liver of the monkey is more than twofold higher than that seen in human liver samples. This increase is also reflected in a twofold higher serum level in the monkey than seen in humans (14). Apo CII mRNA concentration in monkey liver is 60-70 fold greater than in the intesting (14).

It has recently been demonstrated that hormonal factors can markedly affect the expression of CII in the hep G2 cell line derived from a human hepatocarcinoma. Treatment with low levels of estrogen results in a doubling of the apo CII mRNA concentration. The hep G2 cell line has been proposed as a good model of normal hepatocyte function as this cell line synthesizes a similar spectrum of serum proteins, including the major apolipoproteins (15). If there are persons with CII deficiency due to underproduction of CII, it is theoretically possible that one could increase CII production with low dose oestrogen therapy. However, a doubling of synthetic rates is probably unlikely to alter the clinical course of the disease caused by CII deficiency.

MAPPING OF THE GENE FOR CII:

Isolation of the cDNA for CII has allowed chromosomal localization and assessment of linkage to other genes. Southern blot analysis of DNA from human rodent somatic cell hybrids revealed that CII was on chromosome 19 in the region 19pter-->q13. Other genes previously localized in the region include that for the LDL receptor (16, 17), complement component 3 (16), blood group markers lutheran (Lu), secretor (Se) (18) and lewis (Le) (19), another apolipoprotein, Apo E (17) and the loci for peptidase D (20) and myotonic dystrophy (21).

The gene for apo CII and Apo E are extremely closely linked. Humphries et al have used a common Taq1 restriction fragment polymorphisms (Fig 3.) to study linkage with Apo E protein variants. The genes may be separated by only a few thousand base pairs similar to the distance between apo AI and apo CIII on chromosome 11 (22). Close proximity of the genes for these lipoproteins provides further evidence for a common ancestral origin and may imply some relationship in their metabolic functions.

Further examination of the linkage relationships of the gene for apo CII with loci on chromosome 19 has revealed that apo CII is not linked to the LDL receptor gene and shows only loose linkage to the gene for complement component 3. Apo CII is, however, closely linked to the blood group loci, Lu and Se (23) and also shows linkage, to the gene for myotonic dystrophy (24) and Le (21).

The probable order for these loci on chromosome 19 is shown below:

```
          LDL --- C3 --- DM -Le-(Lu, Se, CII, CI, Apo E)
telomere _____short arm chromosome 19_____centromere
```

APO CII DEFICIENCY:

This rare lipoprotein disorder is characterized by elevated fasting triglyceride levels with absolute or functional absence of CII in the plasma. The clinical features show some differences from those of patients with chylomicronemia due to familial LPL deficiency. Later onset of symptoms, absence or infrequent occurence of xanthomata, together with an infrequent occurence of hepatomegaly or splenomegaly are some differentiating features. This probably reflects the fact that LPL has some intrinsic activity even in the absence of apo CII (2,3). The primary defects responsible for CII are unknown. However, the availability of a cDNA probe has allowed preliminary investigation at the molecular level.

APO CII DEFICIENCY: DOES THE DEFECT RESIDE IN THE APO CII GENE?

Apo CII deficiency could result from a defect within the CII gene or may be due to a post translational problem resulting in altered processing, rapid degradation and/or an increased uptake with consequent functional CII deficiency. This differentiation is most important as sequencing of the gene for CII would be fruitless if the primary problem was due to a metabolic defect outside of the gene.

Sequencing of the apo A-I gene in Tangier disease and comparison with normal apo A-I has revealed no differences showing that the A-I deficiency in this disorder is due to post translational modification and not due to a primary defect within the apo A-I gene itself (25).

A common DNA polymorphism using the enzyme Taq1 and the cDNA for CII as probe (Fig. 2) has been identified (26). In the general population, about 60% of CIII genes are found on a 3.8 Kb Taq1 fragment and 40% are found on a 3.5 Kb Taq1 fragment. Analysis of the segregation of the DNA polymorphism and the mutant CII allele in two CII deficient families has shown that in these instances, the lesion is inherited with one allele of the CII gene (27). In other words, the presence or absence of the endonuclease site for

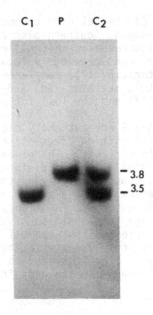

C₁   P   C₂

$-$ 3.8
$_-$ 3.5

Fig 2:   The Taq1 polymorphism gives fragments of 3.5 and 3.8 Kb in size.
The Vancouver patient (P) has two 3.8 fragments.

Taq1 (the DNA polymorphism) is linked to the mutation causing CII with odds in favour of linkage greater than 1000:1 (Lod score 3.31). This is most consistent with the mutation causing the defect being in or near the apo CII gene.

MOLECULAR ANALYSIS OF CII DEFICIENCY:

Four families with CII deficiency have been studied using the cDNA for CII (27,28,29). One family originated from Holland, two are of Italian descent whilst the other proband resides in the U.S. Two families have previously been fully reported (27) whilst results from the second Italian family will be presented here. The proband in this family is currently living in Vancouver, Canada. The results after isoelectric focussing of this patients' chylomicron/ VLDL apoproteins and a control are shown in Figure 3. Absence of CII in this patient (column A) and his brother (column B) with intermediate levels in his mother and father (column C and D) are apparent.

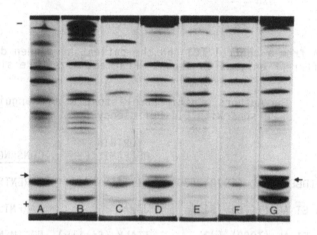

Fig 3: Isoelectric focusing of hypertriglycerdemic control (G) and different family members of our family with CII deficiency. A - Homozygote (proband Fig 5), B - Homozygote, C - Mother - Obligate heterozygote, D - Father - Obligate heterozygote, E - Heterozygotes, F - Heterozygote. Arrows indicate location of CII band.

Digestion of control and this patients' DNA with multiple restriction enzymes and hybridization with the CII probe has revealed gene fragments of equal sizes (Fig.4). The restriction map for CII in the normal and the CII deficient patient were equivalent. This demonstrates that the defect in this patient is not due to a major deletion or insertion in or around the apo CII gene. Similar results have been obtained from analysis of the other probands from 4 different families.

Analysis of the molecular defect in patients with CII deficiency will await sequencing of identified genomic clones and comparison with the previously described normal DNA sequence for CII.

THE GENETICS OF CII DEFICIENCY:

Seven different families with CII deficiency have been published (Table 1).

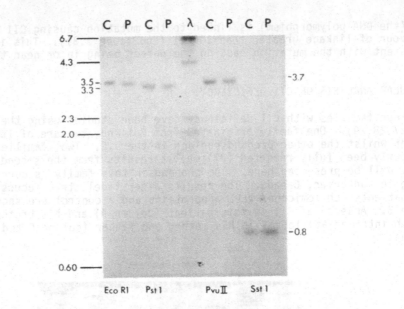

Fig 4: DNA from a control (C) and the patient have been digested with different enzymes as shown. Fragment sizes are similar

Table 1. The Authors, Origin and History of Consanguinity in Persons With CII Deficiency

AUTHORS	ORIGIN OF FAMILIES	CONSANGUINITY
1. BRECKENRIDGE ET AL (1978) (30)	BRITAIN	PARENTS 2ND COUSINS
2. YAMAMURA ET AL (1979) (31)	JAPAN	PARENTS 2ND COUSINS
3. CREPALDI ET AL (1980) (32)	ITALY (Sicily)	NOT MENTIONED
4. STALENHOEF ET AL (1981) (33)	HOLLAND	NONE
5. MILLER ET AL (1981) (34)	ENGLAND	NONE
6. CATAPANO ET AL (1981) (35)	ITALY (Sicily)	YES
7. SAKU ET AL (1984) (36)	PUERTO RICO	NONE
8&9 REBOURCEK ET AL (unpublished) (37)	FRANCE FRANCE	PARENTS 2ND COUSINS PARENTS 1ST COUSINS
10 OUR PATIENT	ITALY (Sicily)	PARENTS 1ST COUSINS
TOTAL NO. OF FAMILIES:	10	
FREQUENCY OF CONSANGUINITY:	50%	

Our patient (Fig.5) was the offspring of a first cousin consanguineous mating. Five of the 10 known families with CII deficiency have a history of consanguinity. These findings together with the biochemical and DNA results conclusively show that CII deficiency is inherited as an autosomal recessive trait.

The lower the gene frequency for an autosomal recessive genetic disease, the higher the frequency of consanguinity. For example, alkaptonuria is a rare genetic disease with the homozygote frequency in the order of 1 per million. The history of consanguinity is present in approximately 25% of affected persons. Cystic fibrosis, on the other hand, occurs in 1 of 2000 persons and the frequency of consanguinity is only between 1 - 2%.

The occurrence of consanguinity in 50% of subjects with CII deficiency suggests that this is a very rare disease with the frequency of homozygotes certainly less than 1 per million.

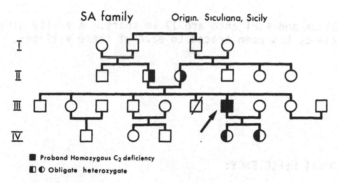

Fig 5: Family tree revealing that the proband is an offspring of a first cousin consanguinous marriage

It is of interest that 3 families are of Italian descent with all of them originating from Sicily. Analysis of our patient has revealed that his CII gene is found on the 3.8 Kb Taq1 fragment. The CII gene is also found on the 3.8 Taq1 fragment in another Sicilian family (27,32). This is different to the family from Holland where the mutant CII gene is found on the 3.5 Kb Taq1 fragments.

Our family originates from a small Sicilian village called Siculiana. The kindred reported by Catapano (35) also originates from Sicily from a village called Agrigento which is approximately 17 km from Siculiana (Fig.6). There is no known genetic relationship between these three families. Nevertheless, the close proximity of Siculiana and Agrigento and the fact that the proband from Siculiana has the mutant CII gene on the same Taq1 fragments as the third family from Italy suggests that these families may, in fact, be related. This could be explained by a single mutational event in the families of Sicilian origin with an independent mutation in the Dutch family. This hypothesis will be tested when the DNA sequence causing the mutation in different families is available.

The report of 3 of 10 families with CII deficiency from Sicily raised the possibility that the gene frequency for this disorder is increased in that area.

Fig 6: Siculiana and Agrigento are 17 km apart. A family with CII deficiency has been traced to each of these villages.

HETEROGENEITY IN CII DEFICIENCY:

An isomorph of apo CII with a pI different from that of the major isoform yet indistinguishable in amnioacid composition and ability to activate lipoportein lipase has been reported (38). This represents heterogeneity of the normal CII protein.

Functional CII deficiency could, in general principle, be due to decreased synthesis of normal CII or due to the production of mutant non-functional CII.

Maguire et al have recently described mutant apo CII which pI markedly different to that seen with normal CII. The decrease or absence of normal CII is accompanied by the appearance of two protein bands not seen in control subjects. These bands have pI's different to that seen normally with apo CII (39). These proteins have partial immulological reactivity with antibodies raised against apo CII but do not form normal precipitation reactions with the antibodies and, thus, can be regarded as mutant CII's. Although they are present in significant amounts in the plasma of homozygotes for CII deficiency, they do not activate lipoprotein lipase (Fig.7).

It is of interest that no CII was detected in homozygotes based on immunodiffusion and electroimmunoassay and activation assays (30). These studies were performed on the first family described with CII deficiency, which is of British descent. The Italian proband living in Vancouver, Canada has also been investigated using similar immunological and isoelectric focusing techniques. In contrast to the findings in homozygotes in the family of British descent, this patient had markedly decreased amounts of CII close to the pI of normal CII. No extra protein bands were detected (40) (Fig.7).

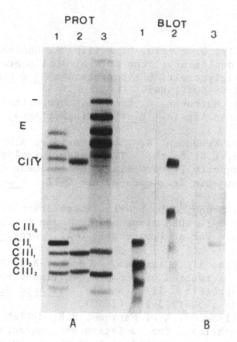

Fig 7: Isoelectric focusing gels with comassie blue staining on the left (A) and an immunoblot with polyclonal anti CII on the right (B).
    Lane 1: Urea soluble normal VLDL
    Lane 2: Acetone soluble chylomicron and VLDL from affected patient from Toronto.
    Lane 3: Urea soluble VLDL from Vancouver patient.

The defect in the Caribbean family of British extraction has resulted in the production of a mutant CII with no ability to activate LPL, whilst plasm in the Italian proband cannot stimulate LPL activity due to an absolute deficiency in CII.

This represents obvious biochemical heterogeneity which will be reflected in different molecular defects. Biological activity for activation of the LPL resides in the peptide sequence between positions 55-79 (5). Since the mutant forms of CII in the family of British descent bind to lipoproteins, it is most likely that the defect occurs in the aminoacid fragments between 55-79 (39). The CII deficiency in the Italian family is not due to a major DNA deletion or insertion but could be due to production of an unstable message, a transcriptional defect or a base mutation which has resulted in impaired synthesis.

Two probands with functional CII deficiency from two other families have also been shown to have apo CII with abnormal mobility after immunoblots of plasma separated by two dimensional gel electrophoresis were performed (29).

Similar to many other genetic disorders, it is obviously apparent that biochemical and molecular heterogeneity underlie CII deficiency.

ACKNOWLEDGEMENTS:

I would like to thank Dr. C. Breckenridge who did the immuloblots and supplied Figure 7.

REFERENCES

1. W. Haberbosch, A. Poli, G. Baggio, R. Fellin, A. Gnasso, J. Augustin. Apolipoprotein C-II deficiency: the role of apolipoprotein C-II in the hydrolysis of triacylglycerolrich lipoproteins. <u>Biochimica et Biophysica Acta.</u> 793:49-60 (1984).

2. P.K.J. Kinnunen, J.K. Huttunen, C. Ehnholm. Properties of purified bovine milk lipoprotein lipase. <u>Biochem Biophys Acta.</u> 450:342-351 (1976).

3. A.L. Catapano, P.K.J. Kinnunen, W.C. Breckenridge, A.M. Gotto, R.L. Jackson, J.A. Little, L.C. Smith, J.T. Sparrow. Lipolysis of Apo C-II deficient very low density lipoproteins: enhancement of lipoprotein lipase action by synthetic fragments of Apo C-II. <u>Biochem Biophys Res.</u> Commun 89:951-957.

4. C.L. Jackson, G.A.P. Bruns, J.L. Breslow. Isolation and sequence of a human apolipoprotein C-II cDNA clone and its´ use to isolate and map to chromosome 19 the gene for apolipoprotein C-II. <u>Proct Nat´l Acad Sci.</u> 81:2945-2949 (1984).

5. L.C. Smith, J.C. Voyta, A.L. Catapano, K. Paavo, K.J. Kinnunen, A.M. Gotto, J.T. Sparow. Actuation of lipoprotein lipase by synthetic fragments of apolipoprotein C-II in Annals. <u>NY Acad Sci.</u> A.M. Scanu & F.R. Landshesger, Eds., 348:213-221 (1980).

6. O. Myklebost, B. Williamson, A.F. Markham, S.R. Myklebost, J. Rogers, P.E. Woods, S.E. Humphries. The isolation and characterization of cDNA clones for human apolipoprotein CII. <u>J Biol Chem.</u> 259:4401-4404 (1984).

7. S.F. Fojo, S. Law, G. Baggio, R.E. Gregg, H.B. Brewer. Complete nucleotide and sequence of preapolipoprotein CII and an analysis of the apolipoprotein CII gene in Apo CII deficient patients. <u>Arteriosclerosis.</u> 4:562A (1984).

8. R.L. Jackson, H.N. Baker, E.B. Gilliam, A.M. Gotto. Primary structure of very low density apo CII of human plasma. <u>Proc Nat´l Acad Sci:</u> USA 74:1942-1945 (1977).

9. J.L. Breslow. Molecular genetics of lipoprotein disorders. <u>Circulation.</u> 69:(6)1190-1195 (1984).

10. W.C. Barker, M.O. Dayhoff. <u>Comp Biochem Physicol.</u> S7B:309-315 (1977).

11. G.A.P. Bruns, S.K. Karathanasis, J.L. Breslow. The human apolipoprotein A-I - CII gene complex is located on chromosome 11. <u>Arteriosclerosis.</u> 4:97-102 (1984).

12. S.K. Karathanasis, J. McPherson, V.I. Zannis, J.L. Breslow. Linkage of human apolipoproteins A-I and CII genes. <u>Nature.</u> 304:371-373 (1983).

13. G. Schonfeld, N. Grimme, D. Alpers. Detection of apolipoprotein C in human and rat enterocytes. <u>J Cell Biol.</u> 86:562-567 (1980).

14. T.K. Archer, S.P. Tam, K.V. Deugau, R.G. Deeley. Apolipoprotein CII mRNA levels in primate liver. <u>J Biol Chem.</u> 260:1676-1681 (1985).

15. B.B. Knowles, C.C. Howe, D.P. Aden. Human hepatocellular carcinoma lines secrete the major plasma proteins and hepatitis B surface antigen. <u>Science.</u> 209:497-499 (1980).

16. K. Berg, A. Heiberg. Linkage between familial hypercholesteremia and the C3 polymorphism confirmed. <u>Cytogenet Cell Genet.</u> 22:621 (1978).

17. U. Francke, M.S. Brown, J.L. Goldstein. Assignment of the human gene for the low density lipoprotein receptor to chromosome 19: Syntery of a receptor, a ligand and a genetic disease. <u>Proc Nat´l Acad Sci:</u> USA 81:2826-2830 (1984).

18. J. Mohr. Estimation of linkage between the Lutheran and Lewis blood groups. <u>Acta Pathol Microbiol Scand.</u> 29:339-344 (1951).

19. L.R. Weitkamp, E. Johnston, S.A. Guttormsen. Probable genetic linkage between the loci for the Lewis blood group and complement C3. <u>Cytogenet Cell Genet.</u> 13:183-184 (1974).

20. P.J. McAlpine, T. Mohandas, M. Ray, H. Wang, J.L. Hamerton. Assignment of the peptidase D gene locus to chromosome 19 in man. <u>Cytogenet Cell Genet.</u> 16:204-205 (1975).

21. K. Simola, A. de la Chapelle, A. Pirkola, P. Karli, P.I.L. Cook, P.A. Tippett. Data on DM-Le Linkage. Cytogenet Cell Genet. 32:317 (1982).
22. S.E. Humphries, K. Berg, L. Gill, A.M. Cumming, F.W. Robertson, A.F.H. Stalenhoef, R. Williamson, A.L. Borneser. The gene for apolipoprotein CII is closely linked to the gene for apolipoprotein E on chromosome 19. Clin Genet. 26:389-396 (1984).
23. J.A. Donald, S.C. Wallis, A. Kessling, P. Tippett, E.B. Robson, S. Ball, K.E. Davies, P. Scambler, K. Berg, A. Heiberg, R. Williamson, S.E. Humphries. Linkage relationships of the gene for apolipoprotein CII with loci on chromosome 19. Hum Genet. In press (1985).
24. A. Roses. Personal Communication.
25. V.L. Zannis, J.L. Breslow, J. Ordovas, S.K. Karathanasis. Isolation and sequence of Tangier apo A-1 gene. Arteriosclerosis. 4:562A (1984).
26. S.E. Humphries, N.I. Jowett, L. Williams, A. Rees, M. Vella, A. Kessling, O. Myklebost, A. Lydon, M. Seed, D.J. Galton, R. Williamson. A DNA polymorphism adjacent to the human apolipoprotein CII gene. Mol Biol Med. 1:463-467 (1983).
27. S.E. Humphries, L. Williams, O. Myklebost, A.F.H. Stalenhoef, D.N.M. Demarker, G. Baggio, G. Crepaldi, D.J. Galton, R. Williamson. Familial apolipoprotein CII deficiency: A preliminary analysis of the gene defect in two independent families. Hum Genet. 67:151-155 (1984).
28. M.R. Hayden. Unpublished results.
29. S.S. Fojo, S.W. Law, G. Baggio, D.L. Sprecher, L. Taam, H.B. Brewer. Apolipoprotein CII deficiency: Identification of two kindred with an abnormal apolipoprotein CII. Clin Res. 33:569A (1985).
30. W.C. Breckenridge, J.A. Little, G. Steiner, A. Chow, M. Poapst. Hypertriglyceridemia associated with deficiency of apolipoprotein CII. N Engl J Med. 298:1265-1273 (1978).
31. T. Yamamura, H. Sudo, K. Ishikawa, A. Yamamoto. Familial type I hyperlipoproteinemia caused by apolipoprotein CII deficiency. Atherosclerosis. 34:53-65 (1979).
32. G. Crepaldi, R. Fellen, G. Baggio, J. Augustin, H. Greten. Lipoprotein and apoprotein, adipose tissue and hepatic lipoprotein lipase levels in patients with familial hyperchylomicronemia and other immediate family members. In: Atherosclerosis, Springer-Verlag 250-254 (1980).
33. A.F.H. Stalenhoef, A.F. Casparie, P.N.M. Demacker, J.T.J. Stouten, J.A. Lutterman, A. van't Laar. Combined deficiency of apolipoprotein CII and lipoprotein lipase in familial chylomicronemia. Metabolism. 30:919-926 (1981).
34. N.E. Miller, S.N. Rao, P. Alaupovic, N. Noble, J. Slack, J.D. Brunzell, B. Lewis. Familial apolipoprotein CII deficiency: plasma lipoproteins and apolipoproteins in heterozygous and homozygous subjects and the effects of plasma infusion. Eur J Clin Invest. 11:69-76 (1981).
35. A.L. Catapano, G.L. Mills, P. Roma, M. LaRosa, A. Capurso. Plasma lipids, lipoproteins and apoproteins in a case of apo CII deficiency. Clin Chem Acta. 130:317-327 (1983).
36. K. Saku, C. Ledres, B. McDonald, B.A. Hynd, B.W. Liu, C.S. Srivastaia, M.L. Kashyap. CII, an apolipoproteinemia and severe hypertriglyceridemia. Am J Med. 77:457-462 (1984).
37. R. Rebouriet, J.L. Bresson, F. Rey, R. Benbrahem, J. Rey. Apolipoprotein CII deficiency and type V hyperlipoproteinemia. (submitted).
38. J.R. Havel, L. Kotite, J.P. Kane. Biochem Med. 21:121-128 (1979).
39. G.F. Maguire, A. Little, G. Katis, W.C. Breckenridge. Apolipoprotein CII deficiency associated with nonfunctional mutant forms of apolipoprotein CII. Can J Bioch Cell Biol. 62:847-851 (1984).
40. W.C. Breckenridge, R. McLeod, M.R. Hayden. Unpublished results.

FAMILIAL HEPATIC LIPASE DEFICIENCY

J. Alick Little and Philip W. Connelly

Lipid Research Clinic, University of Toronto
1 Spadina Crescent
Toronto, Canada M5S 2J5

INTRODUCTION

We have described a kindred with familial hepatic lipase
(HL) deficiency [1]. The proband presented with marked hyperlipoproteinemia and β-VLDL and was originally thought to have
Type III hyperlipoproteinemia. However, he was investigated
further because of an unusually high proportion of triglyceride
in low density and high density lipoproteins and a slightly
reduced post-heparin lipolytic activity. Subsequently he, his
brother and other family members were discovered to have a
marked reduction in hepatic lipase, with gross changes in
plasma lipoprotein composition and premature vascular disease.
Characterization of the lipoprotein abnormalities has allowed
us to evaluate the role of HL in human lipoprotein metabolism
and to make inferences about the relationships between the
composition of their lipoproteins and atherogenicity.

CLINICAL PRESENTATION

The two middle aged brothers had obesity and elevated
lipids for several years, with no causes for secondary
hyperlipoproteinemia and no evidence of liver disease. The
proband had premature corneal arcus, palmar crease xanthomata
and eruptive xanthomata and both had coronary heart disease.

Examination of their lipoproteins by agarose electrophoresis revealed the presence of β-VLDL particles in the
d<1.006 g/ml fraction. Their plasma lipid levels were variable.
The proband's total cholesterol (C) ranged between 260-1495
mg/dl and triglycerides (TG) between 395-820 mg/dl. Despite
the presence of β-VLDL, the ratio of VLDL-C to total TG was
always normal (0.16-0.22). This inconsistency was explained by
the presence of unusually large amounts of TG in the d>1.006
g/ml fraction. Isoelectric focusing of the urea soluble
apolipoproteins of VLDL revealed that both brothers had an
E3:E2 phenotype. Furthermore their E apoproteins were found to
bind normally in a fibroblast system (K. Weisgraber, personal
comm.). Thus although initial clinical and laboratory findings
suggested Type III hyperlipoproteinemia, further investigation
showed that they did not have this syndrome.

Examination of post-heparin lipolytic activities [2] 30
minutes after 60 units of intravenous heparin revealed normal
or slightly increased lipoprotein lipase activity (17 and 38
μmole FFA/ml/hour) and a virtual absence of hepatic lipase (HL)
activity, (0.0 and 0.02 μmole FFA/ml plasma/hour). The absence
or near absence of HL activity was consistent on replicate
sampling and using other methods.

Since our preliminary results indicated that there were
significant changes in the composition of the lipoprotein
fractions greater than and less than density 1.006 g/ml, a more
detailed study of all lipoprotein subfractions was undertaken.
The lipoproteins of the proband and brother were evaluated by
analytical ultracentrifugation in the laboratory of Dr. Frank
Lindgren. There was an increased concentration of S_f 20-400
particles (VLDL) to 897mg/dl and S_f0-12 particles (LDL_2) to
504mg/dl with a lesser increase in S_f12-20 particles (LDL_1 or
IDL) to 160mg/dl. The HDL fractions were unusual, considering
that the patients had hypertriglyceridemia. $S_{f1.20}3-9$ (HDL_2)
was increased while $S_{f1.20}0-3$ (HDL_3) was slightly increased.
The quantities of the plasma apolipoproteins, determined by Dr.
P. Alaupovic's laboratory were consistent with these
lipoprotein levels, i.e., B, C and E, major components of VLDL
and LDL, were elevated 2-4 fold while AI, AII and D, major
components of HDL, were normal or slightly elevated (Table I).

254

Table I

Concentration (mg/dl) of Apolipoproteins in Proband and Brother

Plasma Apolipoprotein	Proband	Brother	Normal Male
AI	110.1	151.0	127 $\pm$ 18
AII	69.4	78.7	62 $\pm$ 11
B	248.9	220.2	117 $\pm$ 25
CI	13.8	20.5	6 $\pm$ 1
CII	14.4	11.9	3.4 $\pm$ 2
CIII	69.1	35.0	8.6 $\pm$ 1
D	7.3	7.2	8.9 $\pm$ 2
E	33.5	32.7	12.3$\pm$ 6

Normal values given are for male subjects 40-50 years of age.

The composition, molar ratios and calculated size [3] of the lipoprotein fractions separated by preparative ultracentrifugation are shown in Tables II,III and IV. For VLDL there is an enrichment with CE similar to that found in Type III hyperlipoproteinemia. The particle size is larger than normal and the number of CE molecules is increased relative to TG. IDL (d1.006-1.019 g/ml) was present in such relatively small amounts that accurate compositional analysis of this fraction could not be carried out.

LDL$_2$ has a four fold enrichment with TG apparently at the expense of CE. The diameter is significantly larger than normal and there are a greater number of molecules of each lipid class per particle. The most striking increases are in PL and TG. The ratio of TG/CE is ten times normal whereas UC to PL is only slightly less than normal. The absolute amounts of CE are not reduced.

HDL$_2$ also appears to be enriched in TG at the expense of CE (Table II and III) and the molar ratios indicate that PC is increased relative to sphingomyelin (SPM). The particle diameter is greater than normal. There is a large increase in the number of TG molecules per particle with lesser increases in the other lipid classes. HDL$_3$ in contrast to HDL$_2$ appeared to be normal.

Table II

Weight % Composition of Lipoproteins from a HL Deficient Subject

	VLDL		LDL$_2$		HDL$_2$		HDL$_3$	
	Proband	Normal	Proband	Normal	Proband	Normal	Proband	Normal
UC	5.1	5 ± 1	5.6	9 ± 1	2.3	4 ± 1	1.3	2 ± 1
CE	19.3	12 ± 1	22.0	40 ± 2	15.3	19 ± 1	8.6	14 ± 1
TG	51.8	55 ± 3	29.8	6 ± 1	15.5	2 ± 1	1.6	2 ± 1
PL	16.0	18 ± 1	18.7	25 ± 1	20.8	28 ± 1	19.0	21 ± 1
Protein	7.8	9 ± 1	23.8	20 ± 1	46.1	47 ± 3	69.5	61 ± 2

* Normal values taken from Kuksis et al. J. Chromatography, 224 (1981) 1-23
UC, unesterified cholesterol; CE, cholesteryl ester; TG, triglyceride;
PL, phospholipid.

Table III

Lipid Molar Ratios of Lipoproteins from a HL Deficient Subject

	VLDL		LDL$_2$		HDL$_2$		HDL$_3$	
	Proband	Normal[a]	Proband	Normal	Proband	Normal	Proband	Normal
UC/TC	0.39	0.44	0.30	0.27	0.20	0.24	0.20	0.19
UC/PL	0.65	0.62	0.60	0.72	0.23	0.26	0.12	0.19
UC/PC	0.86	0.70	0.67	0.95	0.26	5.44	0.15	
UC/SPM	5.16	5.5	3.36	2.9	3.9	2.39	1.90	1.84
PC/SPM	5.93	7.7	5.0	2.77	7.5	5.44	12.9	7.11

[a]Kuksis et al. J. Chromatogr. 224 (1981) 1-23. UC, unesterified cholesterol; TC,
total cholesterol; PL, phospholipid; PC, phosphatidylcholine; SPM, sphingomyelin.

It is significant that TG and PC, which are known to be
substrates for HL [4], accumulated in plasma LDL$_2$ and HDL$_2$ but
not in the VLDL of either of the two brothers. The composition
of the VLDL fraction suggests that HL is not required for the
formation of β-VLDL. The enrichment of HDL$_2$ with both TG and
PC in HL deficiency is consistent with the concept that HL
plays a role in the conversion of HDL$_2$ to HDL$_3$. Through its
action it is possible to reduce both the core and surface of
HDL$_2$. The enrichment of LDL$_2$ with TG and PC in HL deficiency
suggests that LDL$_2$ is a primary substrate for HL.

Table IV

Calculated Size and Molecules/Particle of the Lipoproteins of a
HL Deficient Subject

	VLDL		LDL_2		HDL_2	
	Proband	Normal	Proband	Normal	Proband	Normal
DIAM*	43.1	36.1	28.2	19.0	19.6	11.5
UC	4026	2562	1210	527	220	63
PL	6159	4509	1973	717	974	215
CE	10 008	3881	2793	1396	872	180
TG	19 915	13 155	2793	153	654	14
TG/CE	1.99	3.39	1.0	0.11	0.75	0.08
UC/PL	0.65	0.57	0.61	0.73	0.23	0.29

*Diameter in nanometers.

INVESTIGATION OF OTHER FAMILY MEMBERS

We studied 16 family members plus 5 control subjects in an attempt to determine the heritability of hepatic lipase deficiency and the associated clinical and lipoprotein abnormalities. Eleven of the 16, the mother, the proband, 4 sibs, one child and 4 nephews and nieces, have HL deficiency based on plasma post- heparin HL activity <5 µmole FFA/ml/hour. Only three, the proband and 2 sibs, had β-VLDL and this was associated with plasma post-heparin HL activity <1 µmole FFA/ml/hour. The proband, three sibs and both their parents had ischemic vascular disease.

Table 5 shows the composition of the lipoproteins of 16 family members plus five control subjects. They are divided into a group with low HL activity ≤ 5.0 µmole FFA/ml/hour and a group with HL activity > 5. By definition there are no mis-classed subjects using HL activity. Note that LDL C/TG has zero misclassification whereas other values for LDL, VLDL and HDL misclassify 40 - 95% of subjects. The high misclassification by total HDL C/TG is consistent with our observation in the proband that TG enrichment occurs primarily in HDL_2 and not in HDL_3. Thus for total HDL the absence of compositional changes in HDL_3 may obscure the compositional changes in HDL_2. The standard preparative ultracentrifuge top and bottom

Table V
Lipoprotein Composition (range and mg%) in Relation to Hepatic Lipase Activity

Hepatic Lipase[*]	0.3-4.3	5.5-15.8	%Misclassification
LDL-C	82-166	97-199	81
LDL-TG	28-237	14-49	38
LDL-C/TG	0.70-3.75	4.06-9.64	0
HDL-C	27-79	33-70	95
HDL-TG	10-63	9-39	86
HDL-C/TG	0.71-5.90	0.85-5.56	85
VLDL-C/Plasma TG	0.18-0.30	0.12-0.40	76

*μmoleFFA/ml plasma/hour. C, total cholesterol; TG, triglyceride;

fractions from the 1.006 g/ml spin can be used to determine the LDL C/TG ratio as a screening test for hepatic lipase deficiency. From the data in our HL deficient family, LDL-C/TG ratios less than 4 suggest HL deficiency.

DISCUSSION

Originally the proband was mistaken for Type III hyperlipo-proteinemia because of β-VLDL, but the normal apo E3 and the ample amounts of LDL$_2$ did not fit. Despite the enrichment of VLDL with cholesterol, the ratio of VLDL-cholesterol over total plasma triglyceride was less than 0.25 because of the enrichment of LDL and HDL with triglyceride.

Administration to animals of antibodies to hepatic lipase over short periods have produced increased phospholipid content in HDL$_2$ and increased TG in VLDL in rats [5,6], increased half-life of VLDL and delayed conversion to IDL and LDL in monkey [7] and an accumulation of IDL in non-fasting rats [8]. Possibly the duration of the antibody induced hepatic lipase deficiency in these animals was not sufficient to produce triglyceride enrichment of LDL and HDL, as in our patients. This variation may also reflect species differences in the role of hepatic lipase [9]. In general terms, these animal experiments simulate our findings in this family.

The net result of hepatic lipase deficiency in our family has been the accumulation of enlarged LDL_2 and HDL_2 particles, enriched in triglyceride and to a lesser extent in phospholipids.

Figure 1 depicts the role of HL in lipoprotein catabolism. As suggested by Patsch et al [10] we hypothesize that LPL mediated lipolysis of TG rich lipoproteins produces an intermediate density remnant and helps to enrich HDL with lipids and apoproteins. The remnant lipoprotein is acted on further by LPL to produce a TG-rich LDL_2. And it is this particle that is the substrate for HL which converts it to LDL_2. HL also hydrolyzes HDL_2-TG to produce an HDL_3 or and HDL_2 depending upon its TG and CE content. HL deficiency would result in the accumulation of triglyceride and phospholipid-rich LDL_2 and HDL_2.

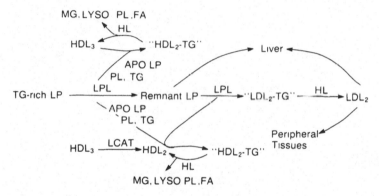

Figure 1. Schematic of the role of HL in normal lipoprotein catabolism. MG, monoglyceride; Lyso PL, lysophospholipid; FA, free fatty acid; LP, lipoprotein; LCAT, lecithin:cholesterol acyltransferase.

REFERENCES

1. Breckenridge, W.C., J.A. Little, P. Alaupovic, C.S. Wang, A. Kuksis, G. Kakis, F. Lindgren and G. Gardiner (1982) Atherosclerosis 45:161.

2. Krauss, R.M., R.I Levy and D.S. Fredrickson (1974) J. Clin. Invest. 54:1107.

3. Shen, B.W., A.M. Scanu and F.J. Kezdy (1977) Proc. Natl. Acad. Sci.(USA) 74:837.

4. Jackson, R.L. (1983) in "The Enzymes" Vol. XVI (Academic Press) pp.141-181.

5. Kuusi, T., P.K.J. Kinnunen and E.A. Nikkila (1979) FEBS Lett. 104:384.

6. Jansen, H., A. Van Tol and W.C. Hulsmann (1980) Biochem. Biophys. Res. Comm. 92:53.

7. Goldberg, I.J., N.A. Le, J.R. Paterniti,Jr., H.N. Ginsberg, F.T. Lindgren and W.V. Brown (1982) J. Clin. Invest. 50:1184.

8. Murase, T. and H. Itakura (1981) Atherosclerosis 39:293.

9. Jansen, H. and W.C. Hulsmann (1985) Biochem. Soc. Trans. 13:24

10. Patsch, J.R., S. Prasad, A.M. Gotto,Jr. and G. Bengtsson-Olivecrona (1984) J. Clin. Invest. 74:2017.

THE APO E-SYSTEM: GENETIC CONTROL OF PLASMA LIPOPROTEIN CONCENTRATION

Gerd Utermann

Institut für Medizinische Biologie und Genetik der
Universität Innsbruck
Schöpfstrasse 41, A-6020 Innsbruck (Austria)

FUNCTION OF APOLIPOPROTEIN E

Apolipoprotein E from human plasma is a glycoprotein composed of 299 amino acids of known sequence[1] that is synthezised in the liver and in several peripheral tissues of the body[2-4]. In plasma apo E is present on different lipoprotein particles. It is a constituent of plasma chylomicrons that carry exogenous lipids and of their catabolic products called remnants that are generated from chylomicrons by hydrolysis of core triglycerides through the action of lipoprotein lipase (LPL). Such remnants are not present in the plasma of healthy fasting humans since they are rapidly taken up by the liver through receptor mediated endocytosis (Fig. 1). The ligand on the surface of chylomicrons that is recognized by the remnant-receptor on liver cells is apo E[5].

A second class of lipoproteins that contains apo E are the very low density lipoproteins (VLDL) that are the transport vehicles for endogenous cholesterol and triglycerides and that are assembled in the liver. VLDL initially have the same fate in plasma as chylomicrons. Their core triglycerides are hydrolyzed by LPL resulting in intermediate density lipoproteins (IDL=VLDL-remnants; see Fig. 1). The further processing of VLDL-remnants is different from that of chylomicron-remnants. A fraction of IDL may be taken up directly by the liver. However at least in healthy humans most IDL-particles are converted to LDL[6]. Both pathways seem to depend on the presence of functional apo E molecules on IDL but the precise role of apo E in these steps has not yet been defined with certainty. The factors that regulate the metabolic channeling of IDL are presently unknown but may include the activities of lipases as well as those of specific liver cell surface receptors. Liver cell plasma membranes exhibit two specific receptors that are able to bind IDL in vitro, the classical LDL-receptor that recognizes apo B-100 and apo E - both of which are constituents of IDL - and the apo E or remnant receptor [5,6]. The relative contribution in vivo of both types of receptors for IDL uptake has yet to be defined.

Third apo E is present on a subpopulation of high density lipoproteins called HDL-1 or HDL_c when induced by cholesterol feeding[5]. This HDL fraction is believed to deliver cholesterol from extrahepatic tissues to the liver where it is taken up through the specific hepatic receptors. In summary apo E is involved in three important pathways of cholesterol e.g. 1. the uptake of dietary cholesterol by the liver, 2. the transport of endogenous cholesterol back to the liver and 3. in the transport of cholesterol from peripheral cells to the liver, a pathway called re-

Cholesterol Transport In Men

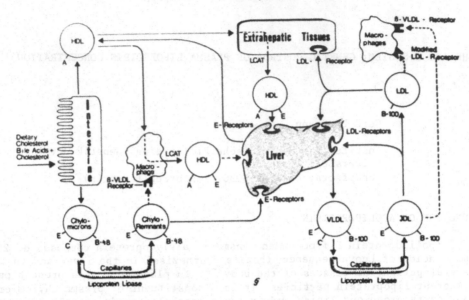

Cholesterol Transport In Apo E2 Homozygotes

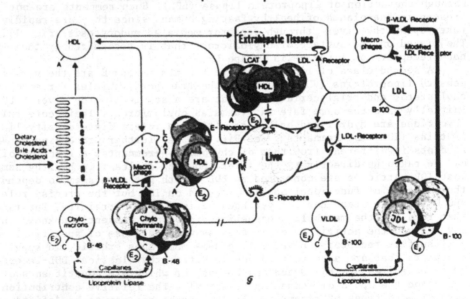

Fig. 1. Receptor mediated transport of cholesterol in human plasma. For simplicity only those apoproteins that play a known role in the pathways indicated are shown.

reversed-cholesterol transport. Evidently mutations in the apo E gene that impair any of these functions will have profound effects on cholesterol homeostasis.

APO E EXHIBITS GENETIC POLYMORPHISM

Extensive family and population studies have shown that apo E exhibits genetic polymorphism that is controlled by the three common allelic genes $\mathcal{E}2$, $\mathcal{E}3$ and $\mathcal{E}4$ at the apo E gene locus[7-9]. These genes specify for the three homozygous phenotypes E 4/4, E 3/3 and E 2/2 and for the three heterozygous phenotypes 4/3, 4/2 and 3/2 (Fig. 2). The frequencies of the common apo E alleles vary considerably between certain populations but $\mathcal{E}3$ is by far the most frequent allele in all populations studied (Table 1). Compared to Germans that have been studied most extensively the allele $\mathcal{E}4$ is significantly more frequent in the Finnish population (Ehnholm, Lukka, Kuusi, Nikkilä and Utermann, submitted) whereas the gene $\mathcal{E}3$ is more common in the population of Singapore that includes Chinese, Indians and Malays (Saha and Utermann, in preparation). This observation seems significant in view of the observed associations of apo E alleles with plasma lipid and apolipoprotein concentrations and with dyslipoproteinemic states (see below). The differences between the common apo E

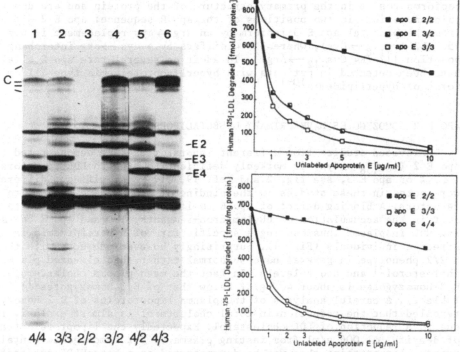

Fig. 2. left: Identification of the six common apo E phenotypes by
isoelectric focussing. Major isoforms are labelled.
right: Competition of apo E/phospholipid complexes for specific degradation of ^{125}I-LDL by Hela cells (from G.Utermann, W.Weber, S.Motzny, U.Beisiegel, H.J.Menzel, in preparation)

Phenotype	Population (%)		
	Finns (N=408)	Germans (N=1031)	Chinese, Indians, Malays (N=188)
4/4	6.6	2.8	1.1
4/3	30.4	22.9	16.0
4/2	0.5	1.5	1.1
3/3	57.4	59.8	60.1
3/2	4.9	12.0	20.2
2/2	0.2	1.0	1.6
Gene			
ε 4	0.221	0.150	0.096
ε 3	0.750	0.773	0.782
ε 2	0.029	0.077	0.122

[a]Data from Utermann et al, 1982; Ehnholm, Lukka, Nikkilä and Utermann, submitted; Saha and Utermann, in
preparation

isoforms reside in the primary structure of the protein and are due to
point mutations in two positions of the apo E sequence: apo E 2 differs
from the parental apo E 3 isoform by an arg $\rightarrow$ cys replacement in position
158 (E2 $Arg_{158} \rightarrow$ Cys), whereas E 4 differs by a cys $\rightarrow$ arg interchange in
position 112 (E4 $Cys_{112} \rightarrow$ Arg)[1]. In addition several rare apo E alleles
have been detected in patients with hyperlipoproteinemia type III or other
forms of hyperlipidemia[10-18].

APO E 2 HOMOZYGOTES HAVE PRIMARY DYSBETALIPOPROTEINEMIA

Previous studies from different laboratories have demonstrated that
apo E 2 ($Arg_{158} \rightarrow$ Cys) is markedly deficient to bind to LDL receptors
($<$ 2 % of apo E 3, see fig. 2 and ref. 19,20). No significant differences
were noted in these studies in the binding of apo E 3 and apo E 4 to the
receptors. A binding defect of apo E to it's receptor is expected to result in the accumulation of chylomicron-remnants, IDL and apo E containing HDL in plasma thus causing a specific form of hyperlipidemia in
affected individuals (Fig. 1). Surprisingly however subjects with the
E 2/2 phenotype in general have subnormal rather than elevated plasma
cholesterol[21] and apo B levels. In fact the mean plasma cholesterol in
E 2 homozygotes is about 40 mg/dl below that of E 3 homozygotes[21,22],
Table 2). A careful analysis of the plasma lipoproteins of E 2 homozygotes
revealed that the reduction in total cholesterol is almost exclusively
due to a reduction of LDL-cholesterol. Expectedly the lipoprotein fraction
of density $<$ 1.006 g/ml from fasting plasma of E 2 homozygotes contains
remnant lipoproteins that may be demonstrated as a beta-VLDL subfraction
by agarose gel electrophoresis. Upon analysis of the distribution of apo
E among the lipoproteins of apo E 2 homozygotes by crossed immunoelectrophoresis accumulation of apo E may be demonstrated in lipoproteins with
beta-mobility and with slow alpha mobility representing the apo E containing HDL subfraction (Fig. 3). Therefore we have to conclude that the
binding defect of apo E 2 does result in the accumulation of remnants and

Table 2. Cholesterol and Apo B Concentrations (mg/dl) in
German and Finnish Population Samples according
to Apo E Phenotypes[a]

Phenotype	Germans (N=551)			Finns (N=207)	
	Cholesterol	Apo B	Apo E	Cholesterol	Apo B
4/4	197	89	1.9	239	127
4/3	189	83	2.3	223	105
3/3	184	82	2.4	214	100
3/2	166	69	3.4	202	86
2/2	140	47	5.1	173	71
Mean	183	81	2.6	217	101

[a]Covariance Analysis
Global P   < 0.0001   < 0.0001 < 0.0001   n.d.   n.d.

of apo E containing HDL in fating plasma but at the same time in hypocholesterolemia due to a gross reduction in LDL concentration. This hypocholesterolemia is the most frequent monogenic form of hypolipidemia exhibi-
ting a frequency of about 1 % in most caucasian populations.

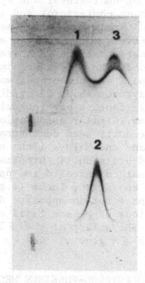

Fig. 3. Crossed immunoelectrophoresis of sera from an E 2/2 homozygote (upper part) and from a subject with type IV hyperlipoproteinemia and apo E 3/3 phenotype (lower part). The
second dimension gel contains monospecific anti-apo E antiserum. Numbers indicate Beta-(1) and Pre-Beta-(2) positions and position of apo E containing HDL (3).

CONTRIBUTION OF APO E ALLELES TO THE NORMAL VARIANCE OF PLASMA LIPOPRO-
TEIN CONCENTRATIONS

The significant effect of the $\mathcal{E}$ 2 allele in homozygous form on plasma
cholesterol, apo B and apo E concentrations (Table 2) prompted investigations on the possible effects of a single copy of the gene $\mathcal{E}$ 2 on lipid
metabolism. About 50 % of apo E molecules in apo E 2 heterozygotes are defective in binding for LDL-receptors. This effect may be demonstrated in
in vitro binding studies were about double the concentration of apo E 3/2
or 4/2 compared to apo E 3/3 is required to compete with [125]J-LDL for
binding to the LDL-receptor (Fig.2). In view of the significant proportion
of most caucasian populations that is heterozygous for apo E 2 (about 14 %)
it is of high interest to know whether E 2 heterozygosity is expressed
in vivo. The first evidence that this is in fact the case was published
by our laboratory in 1979[22]. Since then we and others have confirmed and
extended this finding [23,24], Utermann, in preparation. The results from a
study in which apo E phenotypes, total cholesterol, apo B and apo E concentrations were determined in 551 German blood donors not only confirmed
the effect of the $\mathcal{E}$ 2 allele on lipoprotein concentrations but moreover
suggested an effect of the $\mathcal{E}$ 4 allele on plasma cholesterol, apo B and
apo E (Table 2). When compared to the parent homozygous E 3/3 phenotype
mean cholesterol and apo B concentration were as much as 18 mg/dl and
13 mg/dl lower in E 2 heterozygotes (Table 2). Cholesterol and apo B were
however both higher in groups with the $\mathcal{E}$ 4 allele (4/4 homozygotes and to
a lesser degree 4/3 heterozygotes). The effect of the $\mathcal{E}$ 4 allele if real
seems rather small compared to that of the $\mathcal{E}$ 2 allele and confirmation of
the results by further studies was clearly essential. The Finnish population with its comparetively high frequency of the $\mathcal{E}$ 4 allele (Table 1).
provided an excellent opportunity to confirm the association of the $\mathcal{E}$ 4
allele with high cholesterol levels that had also been observed in studies
of hyperlipidemics[25]. In a collaborative study with Drs. Ehnholm, Lukka,
Kuusi and Nikkilä from Helsinki we could demonstrate that apo E 4/3 heterozygotes and especially apo E 4/4 homozygotes indeed have significantly
higher mean total cholesterol, LDL-cholesterol and apo B concentrations
than E 3/3 subjects (Table 2). Mean cholesterol and apo B concentrations
differed between the Finnish and German population samples studied here
by 34 mg/dl and 20 mg/dl respectively. Nevertheless essentially the same
tendency of apo E gene effects were observed in these two populations that
are of different ethnic background. The difference in mean total cholesterol between apo E 2/2 and apo E 4/4 phenotypes was about 60 mg/dl and
that between 4/3 and 3/2 heterozygotes about 22 mg/dl in both populations,
emphazising the impact of genetic factors particularly apo E genes on
plasma lipoprotein concentrations. Similar associations of apo E genes
with plasma lipoproteins have recently been demonstrated by us for other
ethnic groups e.g.Chinese, Indians and Malays (Saha and Utermann, in preparation) and Japanese (Sata and Utermann, in preparation). In conclusion
it seems well established now that compared to the parent allele $\mathcal{E}$ 3 the
two other common alleles at the apo E gene locus (e.g. $\mathcal{E}$ 2 and $\mathcal{E}$ 4) both
effect plasma cholesterol and apo B but in opposite directions thus determining overlapping distributions of plasma lipid and apolipoprotein
levels in populations (Fig. 4). The effect of the $\mathcal{E}$ 2 allele is however
more pronounced than that of the $\mathcal{E}$ 4 gene.

APO E GENE EFFECTS ON PLASMA CHOLESTEROL-POSSIBLE MECHANISM

The cholesterol delivered to liver cells by receptor mediated endocytosis of lipoproteins (Fig. 1) is believed to regulate several metabolic events involved in cholesterol homeostasis namely the activity of
hydroxymethylglutaryl-CoA-reductase - the rate limiting enzyme in cholesterol biosynthesis - and the number of LDL-receptors on the cell sur-

face. Another class of receptors, the apo E or remnant receptors, that also deliver cholesterol to the liver are not subject to regulation by cholesterol influx[6]. This situation provides a basis for our understanding of the metabolic consequences of apo E 2 homozygosity. As described apo E 2 is defective in its binding to both types of lipoprotein receptors. Lipoproteins depending on apo E for specific uptake by the liver (chylomicron remnants, apo E-HDL) therefore accumulate in the plasma of E 2 homozygotes and less cholesterol of exogenous origin and from peripheral tissues enters the liver. For compensation LDL-receptors may be upregulated with the result of an enhanced uptake of LDL and consequently a lowering of LDL in plasma (Fig. 5). A second mechanism responsible for the low LDL in plasma of E 2 homozygotes may be a delay in the interconversion of IDL to LDL (Fig. 1) a pathway that in part may also depend on the presence of functional apo E on the IDL particles[26]. Even though the model presented is still speculative in some aspects, it does adequately explain the dyslipoproteinemia in E 2 homozygotes and moreover may explain the moderate reduction in LDL-cholesterol and apo B in the plasma of E 2 heterozygotes by the same reasoning. A hyopothesis explaining the mechanism underlying the association of the Є4 allele with elevated plasma cholesterol is still lacking. This association may however result from a similar mechanism than that described above assuming that apo E 4 containing lipoproteins are catabolized more rapidly by the liver than apo E 3 containing lipoproteins. Even though in vitro binding studies have failed to demonstrate any significant differences in the binding of apo E 4 vs apo E 3 to LDL-receptors [19,20], Fig. 2) recent in vivo turnover studies by Gregg et al[27] have provided direct experimental evidence for such a hypothesis. Secondly we have recently observed a peculiar relationship between apo B and apo E concentrations in plasma. Apo B and apo E levels do not correlate upon analysis of a population sample (N = 373; r = -0.083, G.Utermann, unpublished result). However when plotted on the basis of apo E phenotypes an obviously inverse relationship was noted. Whereas apo B concentrations increased in the order E 2/2, 3/2, 3/3, 4/3, 4/4; apo E concentrations decreased in the same order (Table 2). The lower plasma concentration of apo E in individuals with the Є4 allele is consistent with a faster catabolism of apo E 4 in vivo. Apo E genes presumably will primarily effect apo E and not apo B concentrations. We therefore conclude that apo B concentrations in

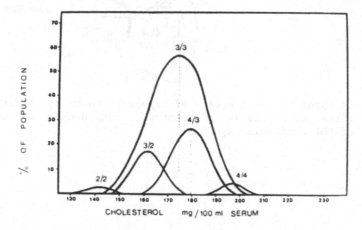

Fig. 4. Schematic representation of the relationship between apolipoprotein E phenotypes and plasma cholesterol concentrations in the population.

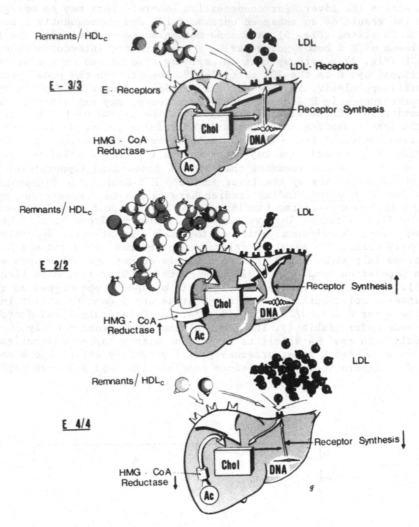

Fig. 5. Receptor mediated uptake of lipoproteins by the liver in the three homozygous apo E phenotypes. HDL$_c$ denotes the subfraction of HDL containing apo E.

plasma in part are regulated by the metabolism of apo E containing lipoproteins. One reason for the faster catabolism and lower concentration in
plasma of apo E 4 may be that this isoform in contrast to apo E 3 and apo
E 2 is unable to form mixed disulfides with apo A-II due to absence of cysteine in the E 4 polypeptide[1]. As shown by Mahley and coworkers[28] the covalent binding of apo A-II to apo E containing lipoproteins does prevent
their recognition by LDL-receptors. In conclusion the association of the
ε 4 allele with high cholesterol may be explained by a mechanism opposite
to that described above for the effect of the ε 2 allele: More cholesterol
is delivered to liver cells by apo E mediated uptake of lipoproteins containing apo E 4, resulting in a more effective downregulation of LDL-receptors in apo E 4/4 subjects and in turn in an elevation of LDL in their
plasma. The key assumption of the model presented in fig. 5 is that apo E
mediated uptake of lipoproteins by the liver effectively regulates LDL-receptor activity on liver cells and thereby LDL concentrations in plasma.
One testible hypothesis derived from this model is that the response of
LDL plasma levels to diatary cholesterol should be more pronounced in E
4/4 than in E 3/3 subjects.

APO E PHENOTYPES AND HYPERLIPIDEMIA

About 10 years ago we noted that patients with type III hyperlipoproteinemia are homozygous for an apo E abnormality that is nowadays
recognized as the apo E 2/2 phenotype[29]. More than 90 % of type III
patients reported to date in the literature were apo E 2/2 homozygotes,
the remainder being E 2 heterozygotes or homozygotes or compounds for
rare apo E mutants that all have in common a defect in binding to the LDL-
receptor[10-12,14,15]. In one family type III hyperlipidemia has resulted
from apo E deficiency[30]. Hence the major defect in the type III disorder
is a functional abnormality or deficiency of the E apoprotein. This abnormality does however not fully explain the pathogenesis of the disease.
The combined data from extended genetic, epidemiologic, biochemical and
nutritional studies have cumulated in a model that explains type III hyperlipidemia as a multifactorial disorder (Fig. 6) and summarizes our present
understanding of the factors involved in the pathogenesis of the type III
disease. In addition to the functional abnormality of apo E other genes
that by themselfes may result in hyperlipidemia (but not type III) as
well as other endogenous and exogenous factors like age, sex, hormones,
nutrition and alcohol consumption will act in concert to produce the final
phenotype. This model explains many features of type III disease, including its association with diabetes mellitus and obesity, the lability
of the phenotypic expression of the disease and the frequent occurance of
none-type III forms of hyperlipidemia in first degree relatives. It should
be borne in mind that the additional factors that may precipitate type
III disease in an E 2 homozygote may be different in any given patient.
However one of the most frequent factors in severe type III disease seems
to be the coexistence of the gene for familial combined hyperlipdidemia[31,
32]. An analyisis of the distribution of apo E phenotypes in patients with
various forms of primary hyperlipidemia demonstrated an association of
apo E genes with hyperlipidemia that reaches far beyond the one with type
III hyperlipidemia described above[25]. Isoform E 2 is significantly more
frequent in patients with hypertriglyceridemia even when E 2 homozygote
type III subjects are excluded from the analysis. Even more intriguing
is the observation that isoform E 4 is significantly more frequent in
patients with hypercholesterolemia. Both isoforms, E 2 and E 4 are more
frequent in patients with mixed hyperlipidemia where as many as 20 % of
patients had one of the rare phenotypes E 4/4, -4/2, or -2/2. From this
result we have concluded that alleles ε 2 and ε 4 both contribute to the
suspectibility for and/or phenotypic expression of hyperlipidemia[25]. The
mechanism underlying the association of the ε 4 gene with hyper-

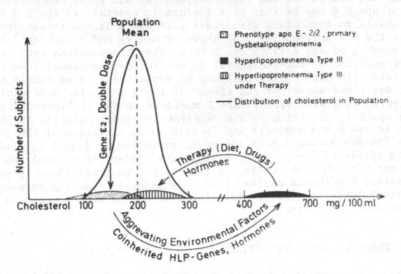

Fig. 6. Factors involved in the pathogenesis and modulation of phenotypic expression of hyperlipidemia type III.

cholesterolemia is presently unclear but the model presented in a previous section (Fig. 5) to explain the association of $\varepsilon 4$ with elevated plasma cholesterol in the population may also give a framework for our understanding of the association with hypercholesterolemia. The assumption that both genes $\varepsilon 2$ and $\varepsilon 4$ have direct effects on lipoprotein pathways that are different from $\varepsilon 3$ does best explain the distorted frequency distribution of apo E phenotypes in patients with hyperlipidemia.

The associations of apo E genes with lipid levels in the population and with hyperlipidemic states suggest that subjects of different apo E phenotypes may also have different risks for the development of atherosclerotic vascular disease. The relationship between apo E phenotypes, plasma lipid levels and the occurance of atherosclerotic vascular disease however predictedly is complex in nature. Recent results on an association of apo E phenotypes with the occurance of coronary heart disease are conflicting[32-34]. Certainly more studies are needed to clarify this important issue.

REFERENCES

1. S. C. Rall, K. H. Weisgraber, and R. W. Mahley, Human apolipoprotein E. The complete amino acid sequence, J Biol Chem. 257:4171 (1982).
2. S. K. Basu, M. S. Brown, Y. K. Ho, R. J. Havel, and J. L. Goldstein, Mouse macrophages synthesize and secrete a protein resembling apolipoprotein E, Proc Natl Acad Sci USA 78:7545 (1981).

3. M. L. Blue, D. L. Williams, S. Zucker, S. A. Khan, and C. B. Blum, Apolipoprotein E synthesis in human kidney, adrenal gland, and liver, Proc Natl Acad Sci, USA. 80:283 (1983).

4. N. A. Elshourbagy, W. S. Liao, R. W. Mahley, and J. M. Taylor, Apolipoprotein E mRNA is abundant in the brain and adrenals, as well as in the liver, and is present in other peripheral tissues of rats and marmosets, Proc Natl Acad Sci, USA. 82:203 (1985).

5. R. W. Mahley, and T. L. Innerarity, Lipoprotein Receptors And Cholesterol Homeostasis, Biochem Biophys acta. 737:197 (1983).

6. M. S. Brown, and J. L. Goldstein, Lipoprotein Receptors in the Liver. Control Signals for Plasma Cholesterol Traffic, J Clin Invest. 72:743 (1983).

7. G. Utermann, A. Steinmetz, and W. Weber, Genetic Control of Human Apolipoprotein E Polymorphism: Comparison of One- and Two-Dimensional Techniques of Isoprotein Analysis, Hum Genet. 60:344 (1982).

8. V. I. Zannis, P. W. Just, and J. L. Breslow, Human apolipoprotein E isoprotein subclasses are genetically determined, Am J Hum Genet. 33:11 (1981).

9. D. Bouthillier, C. F. Sing, and J. Davignon, Apolipoprotein E phenotyping with a single gel method-application to the study of informative matings, J Lipid Res. 24:1060 (1983).

10. K. H. Weisgraber, S. C. Rall, T. L. Innerarity, R. W. Mahley, T. Kuusi, and C. Ehnholm, A Novel Electrophoretic Variant of Human Apolipoprotein E: Identification and Characterization of Apolipoprotein E1, J Clin Invest. 73:1024 (1984).

11. S. C. Rall, K. H. Weisgraber, T. L. Innerarity, and R. W. Mahley, Structural basis for receptor binding heterogeneity of apolipoprotein E from type III hyperlipoproteinemic subjects, Proc Natl Acad Sci, USA. 79:4696 (1982).

12. S. C. Rall, K. H. Weisgraber, T. L. Innerarity, T. P. Bersot, R. W. Mahley, and C. B. Blum, Identification of a new structural variant of human apolipoprotein E, E2(Lys$_{146}$ → Gln), in a type III hyperlipoproteinemic subject with the E3/2 phenotype, J Clin Invest. 72:1288 (1983).

13. J. W. Mc Lean, N. A. Elshourbagy, D. J. Chang, R. W. Mahley, and J. M. Taylor, Human apolipoprotein E mRNA. cDNA cloning and nucleotide sequencing of a new variant, J Biol Chem. 259:6498 (1984).

14. R. E. Gregg, G. Ghiselli, and H. B. Brewer, Apolipoprotein E$_{Bethesda}$: A New Variant of Apolipoprotein E Associated with Type III Hyperlipoproteinemia, J Clin Endocrinol Metabol. 57:969 (1983).

15. L. M. Havekes, J. A. Gevers Leuven, E. Van Corven, E. De Wit, and J. J. Emeis, Apolipoprotein E3-Leiden. Isolation, Partial Characterization And Mode Of Inheritance Of a Variant Of Human Apolipoprotein E3 Associated With Familial Type III Hyperlipoproteinemia, Eur Clin Invest. 14:7 (1984).

16. T. Yamamura, A. Yamamoto, K. Hiramori, and S. Nambu, A New Isoform of Apolipoprotein E - Apo E-5 - Associated with Hyperlipidemia and Atherosclerosis, Atherosclerosis. 50:159 (1984).

17. T. Yamamura, A. Yamamoto, T. Sumiyoshi, K. Hiramori, Y. Nishioeda, and S. Nambu, New Mutants of Apolipoprotein E Associated with Atherosclerotic Diseases But not to Type III Hyperlipoproteinemia, J Clin Invest. 74:1229 (1984).

18. R. J. Havel, L. Kotite, J. P. Kane, P. Tun, and T. Bersot, Atypical Familial Dysbetalipoproteinemia Associated with Apolipoprotein Phenotype E3/3, J Clin Invest. 72:379 (1983).

19. W. J. Schneider, P. T. Kovanen, M. S. Brown, G. Utermann, W. Weber, R. J. Havel, L. Kotite, J. P. Kane, T. L. Innerarity, and R. W. Mahley, Familial Dysbetalipoproteinemia. Abnormal Binding of Mutant Apoprotein E to Low Density Lipoprotein Receptors of Human Fibroblasts and Membranes from Liver and Adrenal of Rats, Rabbits, and Cows, J Clin Invest. 68:1075 (1981).

20. K. H. Weisgraber, T. L. Innerarity, and R. W. Mahley, Abnormal lipo-protein receptor-binding activity of the human E apoprotein due to cysteine-arginine interchange at a single site, J Biol Chem. 257: 2518 (1982).

21. G. Utermann, M. Hees, and A. Steinmetz, Polymorphism of apolipoprotein E and occurance of dysbetalipoproteinemia in man, Nature. 269:604 (1977b).

22. G. Utermann, N. Pruin, and A. Steinmetz, Polymorphism of apolipoprotein E. III. Effect of a single polymorphic gene locus on plasma lipid levels in man, Clin Genet. 15:63 (1979a).

23. C. F. Sing, and J. Davignon, Role of the Apolipoprotein E Polymorphism in Determining Normal Plasma Lipid and Lipoprotein Variation, Am J Hum Genet. 37:268 (1985).

24. F. W. Robertson, and A. M. Cummung, Effects of Apoprotein E Polymorphism on Serum Lipoprotein Concentration, Atherosclerosis. 5:283 (1985).

25. G. Utermann, I. Kindermann, H. Kaffarnik, and A. Steinmetz, Apolipo-protein E phenotypes and hyperlipidemia, Hum Genet. 65:232 (1984).

26. C. Ehnholm, R. W. Mahley, D. A. Chappell, K. H. Weisgraber, E. Ludwig, and J. L. Witztum, Role of apolipoprotein E in the lipolytic con-version of ß-very low density lipoproteins to low density lipo-proteins in type III hyperlipoproteinemia, Proc Natl Acad Sci, USA. 81:5566 (1984).

27. R. E. Gregg, R. Ronan, L. A. Zech, G. Ghiselli, E. J. Schaefer, and H. B. Brewer, Abnormal metabolism of apolipoprotein E4, Circulation. 66:II:160 (1982).

28. T. L. Innerarity, R. W. Mahley, K. H. Weisgraber, and T. P. Bersot, Apoprotein (E—A-II) Complex of Human Plasma Lipoproteins, J Biol Chem. 253(Nr.17):6289 (1978).

29. G. Utermann, M. Jaeschke, and J. Menzel, Familial hyperlipoprotein-emia type III. Deficiency of a specific apolipoprotein (Apo E-III) in the very low-density lipoproteins, FEBS Lett. 56:352 (1975).

30. G. Ghiselli, E. J. Schaefer, P. Gascon, and H. B. Brewer, Type III hyperlipoproteinemia associated with apolipoprotein E deficiency, Science. 214:1239 (1981).

31. G. Utermann, K. H. Vogelberg, A. Steinmetz, W. Schoenborn, N. Pruin, M. Jaeschke, M. Hees, and H. Canzler, Polymorphism of apolipo-protein E. II. Genetics of hyperlipoproteinemia type III, Clin Genet. 15:37 (1979b).

32. W. R. Hazzard, G. R. Warnick, G. Utermann, and J. J. Albers, Genetic transmission of isoapolipoprotein E phenotypes in a large kindred: relationship to dysbetalipoproteinemia and hyperlipidemia, Meta-bolism. 30:79 (1981).

33. J. Menzel, K. G. Kladetzky, and G. Assmann, Apoprotein E polymorphism and coronary artery disease, Arteriosclerosis. 4:310 (1983).

34. A. M. Cumming, and F. W. Robertson, Polymorphism at the Apoprotein-E locus in relation to risk of coronary disease, Clin Genet. 25:310 (1984).

35. G. Utermann, A. Hardewig, and F. Zimmer, Apolipoprotein E phenotype in patients with myocardial infarction, Hum Genet. 65:237 (1984).

TYPE III HYPERLIPOPROTEINEMIA:

A FOCUS ON LIPOPROTEIN RECEPTOR-APOLIPOPROTEIN E2 INTERACTIONS

Thomas L. Innerarity, David Y. Hui, Thomas P. Bersot,
and Robert W. Mahley

Gladstone Foundation Laboratories for Cardiovascular Disease
Cardiovascular Research Institute, Departments of Pathology
and Medicine, University of California, San Francisco, CA

INTRODUCTION

Type III hyperlipoproteinemia is a genetic abnormality characterized by elevated levels of plasma cholesterol and triglycerides and by the presence of cholesterol-enriched β-migrating very low density lipoproteins (β-VLDL). Patients suffering from this abnormality develop tuberous xanthomas and premature atherosclerosis. The expression of the disease depends upon the interaction of a known apoprotein genotype with other genetic and/or environmental factors; thus, type III hyperlipoproteinemia is an especially good model for furthering our understanding of the interplay of genetic and environmental factors in the regulation of lipoprotein metabolism and the development of atherosclerosis.

LIPID ABNORMALITIES IN TYPE III HYPERLIPOPROTEINEMIA

The most serious consequence of type III hyperlipoproteinemia is the development of premature atherosclerosis involving both the coronary and peripheral arteries. (See References 1, 2, and 3 for a more comprehensive discussion of the clinical features of type III hyperlipoproteinemia.) The premature atherosclerosis is associated with elevated plasma lipid levels, and specifically with the occurrence of the β-VLDL.[4] In untreated familial type III hyperlipoproteinemic individuals, cholesterol levels of 300 mg/dl are common, and in some patients, they reach 1000 mg/dl. The plasma triglyceride levels are typically equal to or higher than the plasma cholesterol levels and are usually in the range of 200 to 800 mg/dl. The cholesterol and triglyceride concentrations in a given patient can vary greatly and are especially sensitive to caloric intake (see below). The elevated cholesterol and triglycerides are not due to increased concentrations of normal VLDL and low density lipoproteins (LDL) but, for the most part, to an elevation of cholesterol-rich lipoproteins not usually detected in the plasma of normal individuals.

These cholesterol-rich lipoproteins are referred to as β-VLDL and were initially described as "floating beta lipoproteins." By analytical ultracentrifugation the Schlieren pattern of the plasma from type III hyperlipoproteinemic subjects appears as massive amounts of S_f = 20–400 (d < 1.006 g/ml). Levels of LDL (d = 1.02–1.063 g/ml) are either normal or lower than in normal subjects. When the d < 1.006 g/ml lipoproteins are subjected to paper electrophoresis, a broad band with beta migration is observed. By Pevikon block or starch block electrophoresis, the d < 1.006 g/ml fraction can be separated into two lipoprotein species, the pre-β-VLDL and the β-VLDL. The pre-β-VLDL (or α_2-migrating VLDL) are similar to VLDL found in normal individuals.

The β-VLDL have a unique lipid and apolipoprotein composition. Compared to normal VLDL, the β-VLDL have increased percentages of cholesterol and cholesteryl esters and decreased percentages of triglycerides. The apolipoprotein composition of the β-VLDL is characterized by increased amounts of apolipoprotein (apo-) E and decreased amounts of the C apolipoproteins compared to normal VLDL. The β-VLDL can be subfractionated by gel filtration chromatography into two distinct populations of lipoproteins. One subclass, designated Fraction I, is larger in diameter (90–200 nm) and is of intestinal origin, most likely representing chylomicron remnants. The smaller subfraction (20–60 nm in diameter), designated Fraction II, is postulated to be of hepatic origin, and probably represents cholesterol-rich VLDL remnants. Both subclasses contain apo-E; however, Fraction I contains primarily apo-B48, whereas Fraction II possesses apo-B100.

THE MOLECULAR BASIS FOR TYPE III HYPERLIPOPROTEINEMIA

To understand the molecular basis of this disease, it is necessary to review the salient features of the normal catabolism of chylomicron and VLDL remnants by the liver. In normal subjects, chylomicrons (synthesized by the intestine) and nascent VLDL (synthesized by the liver) are hydro-

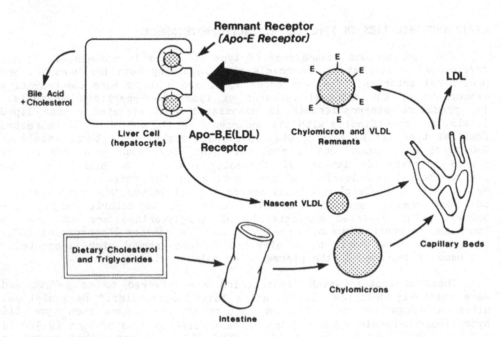

Fig. 1. Schematic representation of the sites of synthesis and catabolism of chylomicron remnants and VLDL remnants in man.

274

lyzed in the capillary beds by the action of lipoprotein lipase. The resulting triglyceride-depleted chylomicron remnant particles and part of the VLDL remnants are rapidly removed from the plasma by the liver (see Figure 1). The lipoprotein clearance process of the liver has now been defined as receptor-mediated endocytosis. The remnant particles bind to receptors on the surface of the hepatocytes and are internalized and degraded. The cholesterol from the lipoprotein particles is either reused for nascent lipoprotein production or is secreted from the body in the form of bile acids or free cholesterol[1-3] (Figure 1).

There are two hepatic lipoprotein receptors responsible for the binding of VLDL and chylomicron remnants.[4] One is the LDL or apo-B,E(LDL) receptor; this receptor appears to be identical to the classic LDL receptor. The other is the remnant receptor. Recently, we have isolated a protein from canine and human liver membranes that has all of the characteristics of the remnant receptor. This receptor, referred to as the apo-E receptor, has an apparent molecular weight of 56,000 on SDS-polyacrylamide gels and is distinct from the apo-B,E(LDL) receptor.[5]

Both the hepatic apo-E receptor and the apo-B,E(LDL) receptor recognize and bind to apo-E-containing lipoproteins. One of the main physiologic roles of apo-E is to bind to hepatic lipoprotein receptors, thus facilitating the uptake of remnants that are enriched in apo-E. The remnants bind to the receptors with very high affinity and normally are cleared from the plasma very rapidly.[1-3]

The underlying genetic defect of type III hyperlipoproteinemia is known. The abnormality involves the defective clearance of chylomicron and VLDL remnants by the liver. Type III hyperlipoproteinemic patients possess an abnormal form of apo-E that binds poorly to both the hepatic apo-E and apo-B,E(LDL) receptors. As a result, the intestinal- and liver-derived remnants are not cleared efficiently from the circulation and the remnant β-VLDL accumulate to high levels in the plasma.

Utermann and associates first discovered that apo-E is polymorphic and that type III hyperlipoproteinemia is associated with one particular isoform of apo-E.[6] Zannis, Breslow, and associates demonstrated that the apo-E polymorphism is due to three common alleles at a single genetic locus.[7] The three homozygous apo-E phenotypes, as determined by isoelectric focusing, are E4/4, E3/3, and E2/2. Almost all clinical subjects with type III hyperlipoproteinemia are homozygous for the apo-E2 allele. The apo-E phenotypes are due to structural differences in the apo-E protein. The apo-E3 is the most common form present in the human population and the structure of the variants are compared to this isoform. The apo-E4 differs from apo-E3 by the substitution of arginine for the normally occurring cysteine at residue 112. The most common form of apo-E2, which has less than 2% of the normal receptor binding activity, differs from apo-E3 by a substitution of cysteine for arginine at residue 158[8] (Table 1).

In addition to the three common alelles, a number of other variant forms of apo-E, which are defective in receptor binding, have been isolated from type III hyperlipoproteinemic subjects.[9] The sites at which the variants differ from apo-E3, are listed in Table 1. All of the variants have neutral amino acids substituted for either arginine or lysine in the center portion of the apo-E molecule, and all of the variants have defective receptor binding.[9] Recently, allelic forms of apo-E that migrate on isoelectric focusing gels in the E5 and E7 position have been identified. These isoforms were isolated from patients with hypertriglyceridemia.[10,11] Their structure and receptor binding capacity is unknown.

Table 1. Human Apolipoprotein E Variants

Isoelectric Focusing Position	Substitution(s)	Receptor Binding Activity Relative to Apo-E3[a] (%)
E3	Parent form	100
E4	$Cys_{112} \to Arg$	100
E2	$Arg_{158} \to Cys$	<2
E2	$Arg_{145} \to Cys$	45
E2	$Lys_{146} \to Gln$	40
E3	$Cys_{112} \to Arg$, $Arg_{142} \to Cys$	<20
E3	$Ala_{99} \to Thr$, $Ala_{152} \to Pro$	Unknown
E1	$Gly_{127} \to Asp$, $Arg_{158} \to Cys$	4

[a]Binding to the apo-B,E(LDL) receptor on human fibroblasts.

INTERACTION OF APOLIPOPROTEIN E WITH LIPOPROTEIN RECEPTORS

Because the primary genetic defect in type III hyperlipoproteine m i a involves defective apo-E binding to hepatic lipoprotein receptors,[12] the molecular nature of the defective binding is of interest. The interaction of apo-E with the receptor is now known in some detail, and it is possible to visualize how variant forms of apo-E with single amino acid substitutions interact abnormally with the apo-B,E(LDL) receptor.[9] Based upon the information from receptor binding studies using the mutant forms of apo-E, fragments of apo-E, and monoclonal antibodies against apo-E, the receptor binding domain of apo-E has been elucidated.[13,14] As shown in Figure 2, the binding domain is centered in the region of residues 140 to 150. Seven of the 11 amino acid residues are basic amino acids. The importance of basic amino acids in receptor binding has been demonstrated previously by chemical modification studies. The selective modification of the lysine or arginine residues of apo-E abolishes receptor binding to the apo-B,E(LDL) and apo-E receptors.[15] The amino acid substitutions that disrupt normal receptor binding (residues 142, 145, 146, and 158) are shown in Figure 2 as residues with two shared amino acids. As will be discussed later, we have shown that the substitution of cysteine for arginine at residue 158 appears to disrupt the conformation of the binding domain and is itself not part of the binding domain.[16]

The derived amino acid sequence of the apo-B,E(LDL) receptor has been determined from the nucleotide sequence of the cDNA.[17] The apo-B,E(LDL) receptor is a $M_r = 97,000$ protein with a number of functional domains, including a membrane spanning domain near the carboxyl-terminal region of the protein. The receptor binding domain has been located to the amino-terminal portion of the molecule. It has been shown that the $M_r = 60,000$ amino-terminal fragment of the receptor will bind to LDL.[18] Within this $M_r = 60,000$ fragment are 7 repeated sequences of about 40 amino acids each. Each of these repeated sequences possesses a cluster of acidic amino acids that have been postulated to represent 7 binding sites (Figure 3). This region is enriched in the acidic amino acids aspartic and glutamic acid.[17] Also shown in Figure 3 is the receptor binding domain of human apo-E. A strong ionic interaction between apo-E and the apo-B,E(LDL) receptor is suggested by studies involving chemical modification of lysine and arginine residues, which abolish binding[9,15] and also by the observation that

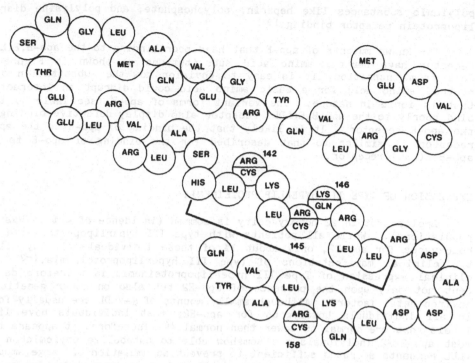

Fig. 2. The portion of the apo-E molecule that includes the receptor
binding domain. The binding site is centered between residues
140 and 150. The amino acid substitutions known to affect
receptor binding are shown (residues 142, 145, 146, and 158).
The Chou-Fasmann logarithm predicts that the secondary structure of the 140 to 150 region is α-helical as shown.

<u>Consensors Sequence for the Postulated Ligand Binding
Domain of the Apo-B,E(LDL) Receptor</u>

(-Cys-ASP-XXX-ASP-Cys-X-Asp-Gly-Ser-ASP-GLU-)

<u>Receptor Binding Domain of Apo-E</u>

HIS-Leu-ARG/Cys-LYS-Leu-ARG/Cys-LYS/Gln-ARG-Leu-Leu-ARG
140   150

Fig. 3. Comparison of the amino acid sequence of the
postulated ligand binding domain of the apo-
B,E(LDL) receptor and the receptor binding
domain of apo-E. Positively charged amino
acids in the apo-E sequence and negatively
charged amino acids in the apo-B,E(LDL) receptor are designated by capital letters.
The amino acid substitutions of apo-E that
disrupt receptor binding are also shown below the underlined amino acids.

polyionic substances like heparin, polyphosphate, and polylysine disrupt lipoprotein receptor binding.[19]

The known mutants of apo-E that have poor binding to the apo-B,E(LDL) receptor have neutral amino acid substitutions as shown in Figure 3. Just from inspection, it is easy to envision how the substitution of a neutral amino acid for a basic amino acid could disrupt an interaction that is ionic in nature. The variant forms of apo-E (see Table 1) that bind poorly to the apo-B,E(LDL) receptor also display defective binding to the apo-E receptor. It is likely that the binding of apo-E to the apo-E receptor is similar to that described for the binding of apo-E to the apo-B,E(LDL) receptor.

EXPRESSION OF TYPE III HYPERLIPOPROTEINEMIA

Apolipoprotein E2 homozygosity is common (incidence of ~2% in Western populations). While all patients with type III hyperlipoproteinemia are homozygous for apo-E2, only about 1% of those individuals display all of the clinical manifestations of type III hyperlipoproteinemia.[1-3] The clinical expression of Type III hyperlipoproteinemia is therefore dependent not only upon the presence of apo-E2 but also on other genetic or environmental factors. Although small amounts of β-VLDL are usually found in all individuals homozygous for apo-E2, most individuals have lipid levels that are normal or lower than normal.[20] Therefore, it appears that most apo-E2/2 individuals are somehow able to catabolize chylomicron and VLDL remnants at rates sufficient to prevent accumulation of large amounts of β-VLDL in the plasma. Thus, the presence of receptor defective apo-E2 is a necessary but not a sufficient condition in many cases for the development of type III hyperlipoproteinemia. Other necessary factors that may be present include age, sex, obesity, hypothyroidism, diabetes, and familial combined hyperlipoproteinemia. The mechanism(s) of how these factors modulate the expression of type III hyperlipoproteinemia is not known. However, a number of possibilities have been suggested and are summarized below.[1,21]

First, the development of type III hyperlipoproteinemia may result, in some cases, from an overproduction of hepatic cholesterol and hepatic VLDL. Various factors known to cause the expression of type III hyperlipoproteinemia, e.g., obesity, obesity with hypothyroidism, diabetes, and age, have been shown to stimulate hepatic synthesis of VLDL and cholesterol.[22]

A second feasible mechanism centers on an apparent defect in the lipase-mediated processing of LDL from d < 1.006 lipoproteins. Normal and hypocholesterolemic apo-E2 subjects possess β-VLDL and low levels of LDL, which may reflect this impaired processing. Normal apo-E3 may play a role in the conversion of the d < 1.006 lipoproteins to LDL while the abnormal apo-E2 found in β-VLDL from type III hyperlipoproteinemic subjects appears to impede this conversion.[1,23]

Third, the level of expression of the hepatic apo-B,E(LDL) receptor may play an important role in the modulation of type III hyperlipoproteinemia. Hepatic apo-B,E(LDL) receptors can be regulated by a number of factors including diet, drugs, hormones, and the plasma concentration of certain lipoproteins and bile acids.[24,25] It is reasonable to believe that a greater number of hepatic apo-B,E(LDL) receptors would facilitate the clearance of chylomicron remnants, even those containing apo-E2. A low number of receptors would exacerbate the already perilous clearance of the apo-E2 remnants. For example, the expression of hepatic apo-B,E(LDL) receptors appears to be decreased in hypothyroidism; however, after thyroxin treatment, which appears to increase receptor activity, there is an alleviation of the type III hyperlipoproteinemia in the apo-E2/2 individuals.[26] Moreover, the beneficial effects of estrogen treatment on type III hyperlipoproteinemia may be due in part to an increased expression of hepatic apo-B,E(LDL) receptors.[27,28]

Finally, the expression of type III familial hyperlipoproteinemia may depend on the variable ability of the chylomicron remnants (β-VLDL) to bind to the hepatic lipoprotein receptors. Several possibilities were considered to explain how remnants from hypo- and normolipidemic apo-E2 homozygous individuals could be cleared satisfactorily, while β-VLDL from other type III patients could bind poorly to hepatic receptors and as a result accumulate in the plasma. The first mechanism tested centers on a possible difference in the structure of lipoproteins from normolipidemic and hyperlipidemic apo-E2 homozygous individuals. Could chylomicron remnants (β-VLDL) bind to the hepatic apo-B,E(LDL) receptor through apo-B instead of through apo-E2? This possibility was tested in vitro using cultured fibroblasts to measure binding to the apo-B,E(LDL) receptor and by an in vitro membrane assay to measure binding to the apo-E receptor on adult dog liver membranes. Monoclonal antibodies against apo-E and apo-B, which inhibit receptor binding, were used to determine which apolipoprotein ligands were recognized by the lipoprotein receptor. These studies demonstrated that chylomicron remnants and apo-E2 β-VLDL interact with lipoprotein receptors almost exclusively through apo-E even though the apo-E2 β-VLDL bind poorly because of the presence of the receptor-defective apo-E2.[29,30]

Consideration was also given to the possibility that the apo-E2 from hypo- and normolipidemic apo-E2 homozygous individuals might be structurally different from the apo-E2 of type III hyperlipoproteinemic subjects and differ in its ability to interact with hepatic lipoprotein receptors. However, apo-E2 isolated from hypo- or normolipidemic subjects was shown to be structurally identical to the apo-E2 from type III hyperlipoproteinemic subjects.[31] Moreover, the apo-E2 from hypo- and normolipidemic individuals displayed defective binding when combined with the phospholipid dimyristoylphosphatidylcholine (DMPC) and tested for binding to the apo-B,E(LDL) receptor on cultured human fibroblasts.[31] These results indicated no detectable differences between the structure of the apo-E2 from type III hyperlipoproteinemic subjects and from apo-E2 homozygous individuals with normal or low levels of plasma cholesterol and triglycerides. Although the protein structure of apo-E2 from normolipidemic and hyperlipidemic subjects is the same, the lipid composition and structure of the β-VLDL are not the same. Thus, apo-E2 in normolipidemic and hyperlipidemic β-VLDL are in different microenvironments on the surface of the peptide, and this raises the possibility that the conformation or configuration of the apo-E2 in the two types of lipoprotein particles could be different and could modulate receptor binding activity. As presented below, this later possibility appears feasible.

VARIATION OF THE RECEPTOR BINDING OF APOLIPOPROTEIN E2

We have found that the ability of apo-E2 to bind to the apo-B,E(LDL) receptor can fluctuate dramatically under certain conditions. Two examples of this variable binding have been observed. The first series of experiments examines the receptor binding of apo-E2 that has been isolated, modified, and recombined with phospholipid. The second series examines the receptor binding of apo-E2 β-VLDL isolated from a patient during different metabolic states. These studies suggest that the binding activity of apo-E2 can vary, secondarily to alterations in the conformation of the binding domain of the apo-E molecule.

Recently, we have obtained evidence that apo-E2·DMPC, which binds poorly to the apo-B,E(LDL) receptor, can under certain conditions be modified to bind normally to the apo-B,E(LDL) receptor.[16] (Apolipoprotein E2-(Arg$_{158}$→Cys) was used in all of the studies reviewed in this section of the paper.) As summarized in Figure 4, apo-E2 was combined with the phospholipid DMPC, isolated by density gradient centrifugation, and tested for its ability to compete with ^{125}I-LDL for binding to the apo-B,E(LDL) receptor on cultured human fibroblasts. The apo-E2·DMPC had only ~0.5% of the binding activity of normal apo-E3·DMPC. However, when apo-E2 was treated with cysteamine, which converted the cysteine in the apo-E2 to a lysine analog and added a positive charge, its receptor binding activity was increased 20-fold (Figure 4). In a second treatment, the abnormal

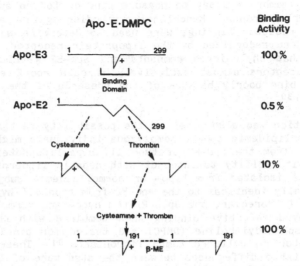

Fig. 4. Diagram summarizing the changes in receptor binding activity of apo-E2 after treatment with cysteamine and/or thrombin. β–ME, β-mercaptoethanol.

apo-E2 was cleaved with thrombin and the amino terminal M_r = 22,000 fragment was combined with DMPC and tested for receptor binding. The M_r = 22,000 thrombolytic fragment of apo-E2 also demonstrated ~20-fold enhancement in receptor binding activity. Thus, the binding activity of this fragment was similar to intact apo-E2 treated with cysteamine. The combination of the two treatments (cysteamine and thrombolytic removal of the carboxyl-terminal one-third of the molecule) further enhanced the binding activity of the M_r = 22,000 fragment of apo-E2. In fact, the cysteamine modified M_r = 22,000 fragment had a receptor binding activity similar to that obtained with normal apo-E3·DMPC. When the cysteamine was removed by treatment of the M_r= 22,000 fragment of apo-E2 with β-mercaptoethanol and the apo-E2·DMPC fragment was tested immediately for receptor binding, most of its enhanced binding activity was still retained. Therefore, the

Table 2. Clinical Data on Subject D.R.

Date	Body Weight (lb.)	Plasma Cholesterol (mg/dl)	Plasma TG[a] (mg/dl)	d < 1.006 Cholesterol (mg/dl)	d < 1.006 Ch:TG	Plasma Apo-E (mg/dl)
1981	270	725	670	465	0.91	----
04/26/82	----	248	246	113	0.46	25.75
07/08/82	265	215	193	85	0.44	----
08/23/82	247	204	202	63	0.31	----
10/13/82	240	108	72	10	0.14	13.4
11/15/82	237	92	77	23	0.30	15.4
02/16/83	257	316	325	112	0.34	----

[a]Abbreviations used: TG, triglyceride; Ch, cholesterol.

enhanced binding activity appears to be due to conformational changes induced in the binding domain by the combination of thrombin removal of the carboxyl-terminal portion of the apo-E molecule and cysteamine treatments. The receptor active conformation of the M_r = 22,000 fragment of apo-E2 is maintained and stabilized by the phospholipid and is not due merely to the presence of a positive charge at residue 158. These studies not only highlight the potential importance of conformation around the binding domain as modulating binding but also indicate that the positive charge at residue 158 is not directly involved in mediating the binding of the apo-E to the lipoprotein receptors.

These results reinforced the possibility that the binding activity of apo-E2 may be modulated by its environment within the lipoprotein particle. For example, the interaction of apo-E2 with other nearby apolipoproteins or with different lipid environments might alter the conformation of the binding domain and affect binding activity. This possibility was tested by examining the binding activity of β-VLDL isolated from a single type III hyperlipoproteinemic patient before and after dietary intervention. At the beginning of the study, subject D.R. [E2(Arg$_{158}$→Cys)] displayed the clinical characteristics of a typical familial type III hyperlipoproteinemic subject. He had a plasma cholesterol level of 725 mg/dl and a triglyceride level of 670 mg/dl (Table 2). Almost two-thirds of the plasma cholesterol (64%) was in the d < 1.006 g/ml fraction, which contained mostly β-VLDL. As shown in Figure 5, untreated β-VLDL isolated from patient D.R. displayed low affinity binding to the LDL receptors on human fibroblasts. After treatment of the β-VLDL with cysteamine, the lipoproteins demonstrated enhanced binding activity, as was previously observed for cysteamine treatment of apo-E2·DMPC (Figure 4). When the results were plotted by the method of Scatchard, the higher affinity binding of the cysteamine-treated β-VLDL (as indicated by the more vertical line) is apparent. The untreated β-VLDL also bound poorly to the apo-E receptor on hepatic membranes from adult dogs (data not shown). Monoclonal antibodies that blocked either apo-B or apo-E binding to the apo-B,E(LDL) receptors were used to demonstrate that the β-VLDL, both before

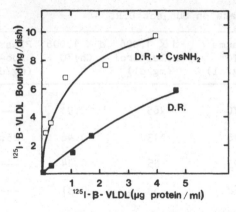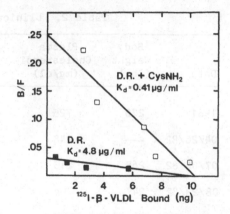

Fig. 5. Effect of cysteamine modification on β-VLDL binding to the apo-B,E(LDL) receptor on human fibroblasts. (Left panel) Concentration-dependent binding at 4°C of ^{125}I-β-VLDL from subject D.R. before (■) and after (□) cysteamine treatment. (Right panel) Binding data plotted by the method of Scatchard. The cysteine residue in the β-VLDL was modified by incubation of β-VLDL (150 μg of protein in 250 μl) in 0.1 M NH_4HCO_3 with 20 μl of a cysteamine solution (100 mg/ml) for 4 hr at 37°C. The control sample was treated in a similar manner without cysteamine. The samples were dialyzed in 0.15 M NaCl, 0.01% EDTA before use. Cultured human fibroblasts on 35-mm dishes were incubated with 1 ml of medium containing 10% lipoprotein-deficient serum and increasing concentrations of D.R.'s ^{125}I-β-VLDL (■) or D.R.'s ^{125}I-β-VLDL modified by cysteamine (□). Specific binding was calculated by comparing the amount of ^{125}I-labeled lipoprotein bound to the cells with the amount bound in the presence of normal human LDL (500 μg of LDL protein). The average amount of cell protein per dish was 141 μg.

and after cysteamine treatment, bound to the apo-B,E(LDL) receptors on cultured fibroblasts through apo-E2 but not apo-B (data not shown).

After the initial experiments, subject D.R. undertook a low fat, low calorie diet, which resulted in a reduction of body weight from 270 lb to 237 lb. His plasma triglyceride levels decreased dramatically from 600-700 mg/dl to more normal levels, and his plasma cholesterol concentration decreased from 725 mg/dl to a hypolipidemic level of 92 mg/dl (11/15/82, Table 2). In the d < 1.006 g/ml fraction from D.R., the cholesterol to triglyceride ratio also decreased substantially after weight loss; however, this fraction still contained β-VLDL. Note that the plasma apo-E level also decreased from 25 mg/dl to 13 and 15 mg/dl, a level which is still about threefold greater than the normal apo-E level (~3-5 mg/dl).[9] The huge diminution of plasma cholesterol and triglyceride levels after dietary intervention is characteristic of type III hyperlipoprotein-emia.[32]

The β-VLDL were isolated from subject D.R. while he was hypolipidemic and tested for binding to apo-B,E(LDL) and apo-E receptors. As shown in Figure 6, the β-VLDL were very active in binding both to the apo-B,E(LDL) receptors on cultured human fibroblasts and to the apo-E receptors on canine hepatic membranes. The binding could not be activated by cysteamine treatment. The β-VLDL used in the initial experiments (Figure 5) will be referred to as inactive β-VLDL and the β-VLDL used in the experiment after weight loss as active β-VLDL. Monoclonal antibody studies

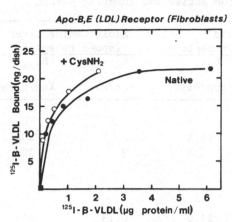
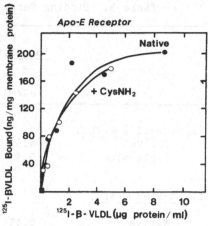

Fig. 6. Binding of receptor-active ^{125}I-β-VLDL to the apo-B,E(LDL) receptors on fibroblasts (left panel) and the apo-E receptor on canine hepatic membranes (right panel) before and after cysteamine treatment. The receptor-active ^{125}I-β-VLDL were treated with cysteamine according to the method described in the legend to Figure 5 (fibroblast binding). (Left panel) The binding of ^{125}I-β-VLDL to the cultured human fibroblasts was performed as described in the legend to Figure 5. (Right panel) Binding to the apo-E receptor was performed by incubation of the ^{125}I-β-VLDL with adult dog liver membranes (60 µg of protein in 0.1 ml) for 1 hr at 4°C. Specific binding of the ^{125}I-β-VLDL was determined by the differences in the amount of binding observed in the presence or absence of 20 mM EDTA.

demonstrated that both the active, as well as the inactive β-VLDL, were binding through apo-E2 and not apo-B. The differences in the receptor binding of active and inactive β-VLDL are summarized in Table 3. There was a tremendous difference in the binding affinity of active β-VLDL to cultured fibroblasts as compared to the inactive β-VLDL. The active β-VLDL bound with over a 30-fold higher affinity, as indicated by the equilibrium dissociation constant (K_d).

Subsequently, the patient D.R. failed to adhere to his diet and regained weight. A blood plasma sample taken 3 months later (2/16/83) indicated that his plasma cholesterol level had risen from 92 mg/dl to 316 mg/dl and his plasma triglyceride level had increased from 77 mg/dl to 325 mg/dl (Table 2). The β-VLDL isolated from subject D.R. at this time again bound poorly to the hepatic apo-B,E(LDL) and apo-E receptors (data not shown).

We have demonstrated that apo-E2 in the active and inactive β-VLDL lipoproteins has the same structure and function when combined with DMPC. The apo-E2 from subject D.R., which has been isolated more than 15 times during the last 5 years, has demonstrated poor binding activity to the apo-B,E(LDL) and the apo-E receptor in every experiment. Moreover, there is no evidence that subject D.R. possesses more than one isoform of apo-E2 (the apo-E2 of D.R. has been sequenced).[33] These data suggest that the altered receptor binding activity before and after dietary intervention (active vs. inactive β-VLDL) is due to differences in the conformation of the binding domain of the apo-E2.

Table 3. Binding Parameters for Active and Inactive β-VLDL

| | Fibroblasts (apo-B,E(LDL) receptor) | | Adult Canine Liver (apo-E receptor) | |
| | K_d | B_{max} | K_d | B_{max} |
	(μg/ml)	(ng/dish)	(μg/ml)	(ng/dish)
I. Inactive β-VLDL				
Native	4.8[a]	10.4	---	----
Cysteamine	0.4	10.4	3	104
II. Active β-VLDL				
Native	0.15	21.0	0.73	214
Cysteamine	0.13	23.2	0.68	221

[a]The data was calculated from the experiments shown in Figures 5 and 6.

The conclusion that the binding of the active and inactive β-VLDL is modulated by different conformations of the apo-E2 was strengthened by studies using thrombin cleavage of the apo-E2 on the β-VLDL. The precedent for the use of thrombin as a probe for apo-E conformation was established by studies of Gianturco et al.[34] Treatment of hypertriglyceridemic VLDL with thrombin abolished the ability of the VLDL to bind to the apo-B,E(LDL) receptor. Hypertriglyceridemic VLDL, like β-VLDL, bind to lipoprotein receptors through apo-E. These investigators showed that thrombin cleaved a portion of the apo-E and that neither the intact apo-E that remained nor apo-B would bind to apo-B,E(LDL) receptors on cultured human fibroblasts. They concluded that thrombin cleaved the "active" apo-E (the apo-E mediating receptor binding) and, furthermore, that the remaining apo-E on the lipoprotein particle was in an alternate protein conformation, which precluded both thrombin cleavage and receptor binding.

To determine the ability of thrombin to cleave apo-E2 on the lipoprotein particles and to determine the effects of thrombin on the receptor binding activity of active and inactive β-VLDL, we treated the active and inactive β-VLDL with thrombin. As shown in Figure 7, all of the apo-E2 inthe active β-VLDL was cleaved, while less than half of the apo-E2 in the inactive β-VLDL was hydrolyzed. The receptor binding activity of both the active and inactive β-VLDL were abolished by the thrombin treatment (data not shown). These results agree with those of Gianturco et al.[34] and suggest that the receptor active apo-E2 is susceptible to thrombin cleavage, while the receptor inactive apo-E2 is resistant. These results, taken as a whole, suggest that differences in the lipid and protein composition, structure, or size of the lipoprotein particle alter the conformation of apo-E2.

CONCLUSION

Familial type III hyperlipoproteinemia is a genetic abnormality in which the primary dysfunction is known. A variant form of apo-E (apo-E2) interacts poorly with hepatic lipoprotein receptors, and the clearance of remnant particles by the liver is retarded. Recently, the structure of

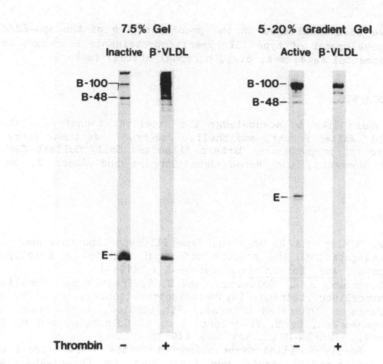

Fig. 7. SDS-polyacrylamide gels of thrombin-digested inactive (left) and active (right) β-VLDL. The β-VLDL were incubated at room temperature for 18 hr in the presence (+) or absence (−) of thrombin at a protein ratio of 20:1. The β-VLDL were reisolated by centrifugation at 55,000 rpm for 18 hr in a SW55 rotor. The β-VLDL, which floated at the top of the tube, were isolated and applied to a 5-20% gradient SDS-polyacrylamide gel for analysis. The gel was stained with silver nitrate for identification of the proteins.

both the E apolipoprotein and the apo-B,E(LDL) receptor has been elucidated, and it is possible to visualize how the variant forms of apo-E could disrupt normal interaction of apo-E with the apo-B,E(LDL) receptor. Several possibilities as to why the vast majority of the individuals homozygous for apo-E2 never express the clinical manifestations of the disease have been considered. Evidence has been presented indicating that the composition and/or structure of the lipoproteins synthesized by these patients could alter the conformation of the apo-E2 on the surface of the particles and modulate receptor binding activity of the apo-E2-containing lipoproteins. It is possible that predisposing factors, such as obesity and high fat, high caloric diets could alter the composition of the remnant lipoproteins, modifying the conformation of the apo-E2 and precipitating the development of hyperlipoproteinemia by interfering with hepatic clearance of these lipoproteins. Numerous other mechanisms, including overproduction of specific lipoproteins, impaired lipolytic processing, and regulation of expression of the level of hepatic apo-B,E(LDL) receptors, have been considered in the predisposition of the apo-E2/2 individual to development of type III hyperlipoproteinemia secondary to genetic or environmental (age, sex, diet, hormonal status) factors.

ACKNOWLEDGMENTS

We would like to acknowledge the excellent technical assistance of Kay Arnold, Walter Brecht, and Shellie Jacobson. We thank Kerry Humphrey for manuscript preparation, Barbara Allen and Sally Gullatt Seehafer for editorial comments, and Norma Jean Gargasz and James X. Warger for graphics.

REFERENCES

1. R. W. Mahley and B. Angelin, Type III hyperlipoproteinemia. Recent insights into the genetic defect of familial dysbetalipoproteinemia, Adv. Intern. Med. 29:385-411, (1984).
2. M. S. Brown, J. L. Goldstein, and D. S. Fredrickson, Familial type 3 hyperlipoproteinemia (dysbetalipoproteinemia), in: "The Metabolic Basis of Inherited Disease," 5th edition, J. B. Stanbury, J. B. Wyngaarden, D. S. Fredrickson, J. L. Goldstein, and M. S. Brown, eds., McGraw-Hill, New York (1983).
3. R. J. Havel, Familial dysbetalipoproteinemia. New aspects of pathogenesis and diagnosis, Med. Clin. North Am. 66:441-454 (1982).
4. R. W. Mahley, Atherogenic lipoproteins and coronary artery disease: concepts derived from recent advances in cellular and molecular biology, Circulation 72:in press (1985).
5. D. Y. Hui, W. J. Brecht, E. A. Hall, G. Friedman, T. L. Innerarity, and R. W. Mahley, Purification of the Hepatic Apolipoprotein E Receptor, Submitted for publication.
6. G. Utermann, M. Hees, and A. Steinmetz, Polymorphism of apolipoprotein E and occurrence of dysbetalipoproteinaemia in man, Nature 269:604-607 (1977).
7. V. I. Zannis and J. L. Breslow, Human very low density lipoprotein apolipoprotein E isoprotein polymorphism is explained by genetic variation and post-translational modification, Biochemistry 20: 1033-1041 (1981).
8. K. H. Weisgraber, T. L. Innerarity, and R. W. Mahley, Abnormal lipoprotein receptor-binding activity of the human E apoprotein due to cysteine-arginine interchange at a single site, J. Biol. Chem. 257:2518-2521 (1982).
9. R. W. Mahley, T. L. Innerarity, S. C. Rall, Jr., and K. H. Weisgraber, Plasma lipoproteins: apolipoprotein structure and function, J. Lipid Res. 25:1277-1294 (1984).
10. T. Yamamura, A. Yamamoto, S. Nambu, and K. Hiramori, Plasma apolipoprotein E mutants associated with atherosclerotic diseases, Arteriosclerosis 4:549a (1984).
11. T. Yamamura, A. Yamamoto, T. Sumiyoshi, K. Hiramori, Y. Nishioeda, and S. Nambu, New mutants of apolipoprotein E associated with atherosclerotic diseases but not to type III hyperlipoproteinemia, J. Clin. Invest. 74:1229-1237 (1984).
12. R. J. Havel, Y.-S. Chao, E. E. Windler, L. Kotite, and L. S. S. Guo, Isoprotein specificity in the hepatic uptake of apolipoprotein E and the pathogenesis of familial dysbetalipoproteinemia, Proc. Natl. Acad. Sci. USA 77:4349-4353 (1980).
13. T. L. Innerarity, E. J. Friedlander, S. C. Rall, Jr., K. W. Weisgraber, and R. W. Mahley, The receptor binding domain of human apolipoprotein E: binding of apolipoprotein E fragments, J. Biol. Chem. 258:12341-12347 (1983).

14. K. H. Weisgraber, T. L. Innerarity, K. J. Harder, R. W. Mahley, R. W. Milne, Y. L. Marcel, and J. T. Sparrow, The receptor binding domain of human apolipoprotein E: monoclonal antibody inhibition of binding, J. Biol. Chem. 258:12348-12354 (1983).

15. R. W. Mahley and T. L. Innerarity, Lipoprotein receptors and cholesterol homeostasis, Biochim. Biophys. Acta 737:197-222 (1983).

16. T. L. Innerarity, K. H. Weisgraber, K. S. Arnold, S. C. Rall, Jr., and R. W. Mahley, Normalization of receptor binding of apolipoprotein E2. Evidence for modulation of the binding site conformation, J. Biol. Chem. 259:7261-7267 (1984).

17. T. Yamamoto, C. G. Davis, M. S. Brown, W. J. Schneider, M. L. Casey, J. L. Goldstein, and D. W. Russell, The human LDL receptor: a cysteine-rich protein with multiple Alu sequences in its mRNA, Cell 39:27-38 (1984).

18. J. L. Goldstein, M. S. Brown, R. G. W. Anderson, D. W. Russell, and W. J. Schneider, Receptor-mediated endocytosis: concepts emerging from the LDL receptor system, Annu. Rev. Cell Biol., in press (1985).

19. M. S. Brown, T. F. Deuel, S. K. Basu, J. L. Goldstein, Inhibition of the binding of low density lipoprotein to its cell surface receptor in human fibroblasts by positively charged proteins, J. Supramol. Struct. 8:223-234 (1978).

20. G. Utermann, N. Pruin, and A. Steinmetz, Polymorphism of apolipoprotein E. III. Effect of a single polymorphic gene locus on plasma lipid levels in man, Clin. Genet. 15:63-72 (1979).

21. R. W. Mahley, T. L. Innerarity, S. C. Rall, Jr., and K. H. Weisgraber, Lipoproteins of special significance in atherosclerosis: insights provided by studies of type III hyperlipoproteinemia, Ann. NY Acad. Sci. USA, 454:209-221 (1985).

22. S. M. Grundy, Cholesterol metabolism in man, West. J. Med. 128:13-25 (1978).

23. C. Ehnholm, R. W. Mahley, D. A. Chappell, K. H. Weisgraber, E. Ludwig, and J. L. Witztum, The role of apolipoprotein E in the lipolytic conversion of β-very low density lipoproteins to low density lipoproteins in type III hyperlipoproteinemia, Proc. Natl. Acad. Sci. USA 81:5566-5570 (1984).

24. M. S. Brown and J. L. Goldstein, Lipoprotein receptors in the liver. Control signals for plasma cholesterol traffic, J. Clin. Invest. 72:743-747 (1983).

25. B. Angelin, C. A. Raviola, T. L. Innerarity, and R. W. Mahley, Regulation of hepatic lipoprotein receptors in the dog. Rapid regulation of apolipoprotein B,E receptors, but not of apolipoprotein E receptors, by intestinal lipoproteins and bile acids, J. Clin. Invest. 71:816-831 (1983).

26. W. R. Hazzard, Primary type III hyperlipoproteinemia, in: "Hyperlipidemia Diagnosis and Therapy," B.M. Rifkind and R.I. Levy, eds., Grune & Stratton, New York (1977).

27. R. S. Kushwaha, W. R. Hazzard, C. Gagne, A. Chait, J. J. Albers, Type III hyperlipoproteinemia: paradoxical hypolipidemic response to estrogen, Ann. Intern. Med. 87:517-525 (1977).

28. E. E. Windler, P. T. Kovanen, Y. S. Chao, M. S. Brown, R. J. Havel, and J. L. Goldstein, The estradiol-stimulated lipoprotein receptor of rat liver. A binding site that mediates the uptake of rat lipoproteins containing apoproteins B and E, J. Biol. Chem. 255:10464-10471 (1980).

29. D. Y. Hui, T. L. Innerarity, and R. W. Mahley, Defective hepatic lipoprotein receptor binding of β-very low density lipoproteins from type III hyperlipoproteinemic patients. Importance of apolipoprotein E, J. Biol. Chem. 259:860-869 (1984).

30. D. Y. Hui, T. L. Innerarity, R. W. Milne, Y. L. Marcel, and R. W. Mahley, Binding of chylomicron remnants and β-very low density

lipoproteins to hepatic and extrahepatic lipoprotein receptors. Aprocess independent of apolipoprotein B48, <u>J. Biol. Chem</u>. 259: 15060-15068 (1984).

31. S. C. Rall, Jr., K. H. Weisgraber, T. L. Innerarity, R. W. Mahley, and G. Assmann, Identical structural and receptor binding defects in apolipoprotein E2 in hypo-, normo-, and hypercholesterolemic dysbetalipoproteinemia, <u>J. Clin. Invest</u>. 71:1023-1031 (1983).

32. R. I. Levy, D. S. Fredrickson, R. Shulman, D. W. Bilheimer, J. L. Breslow, N. J. Stone, S. E. Lux, H. R. Sloan, R. M. Kraus, and P. N. Herbert, Dietary and drug treatment of primary hyperlipoproteinemia, <u>Ann. Intern. Med</u>. 77:267-294 (1972).

33. S. C. Rall, Jr., K. H. Weisgraber, and R. W. Mahley, Human apolipoprotein E. The complete amino acid sequence, <u>J. Biol. Chem</u>. 257: 4171-4178 (1982).

34. S. H. Gianturco, A. M. Gotto, Jr., S.-L. C. Hwang, J. B. Karlin, A. H. Y. Lin, S. C. Prasad, and W. A. Bradley, Apolipoprotein E mediates uptake of S_f 100-400 hypertriglyceridemic very low density lipoproteins by the low density lipoprotein receptor pathway in normal human fibroblasts, <u>J. Biol. Chem</u>. 258:4526-4533 (1983).

THE ROLE OF APOLIPOPROTEIN E IN MODULATING THE METABOLISM OF APOLIPO-PROTEIN B-48 AND APOLIPOPROTEIN B-100 CONTAINING LIPOPROTEINS IN HUMANS

Richard E. Gregg and H. Bryan Brewer, Jr.

Molecular Disease Branch
National Heart, Lung, and Blood Institute
National Institutes of Health
Building 10, Room 7N117
Bethesda, Maryland 20892

ApoE[*] is a glycoprotein of 299 amino acids which is associated primarily with VLDL and HDL in human plasma (1,2). It is a polymorphic protein with three common alleles inherited in a co-dominant fashion (3,4), the most common allele coding for apoE$_3$. Homozygosity for this allele is present in approximately 60% of normal subjects, and this is considered to be the normal phenotype (5). Two other common alleles, which code for apoE$_2$ and apoE$_4$, differ from apoE$_3$ by only one amino acid each (6). The allelic frequency of each of these two alleles is approximately 10-15%. Homozygosity for the apoE$_2$ allele is associated with type III hyperlipoproteinemia (3,4) while apoE$_4$ is associated with type V hyperlipoproteinemia (7). Previous investigations have demonstrated that apoE$_2$ is catabolized abnormally slowly in vivo (8) and has a reduced binding affinity for cellular receptors (9). ApoE$_4$ has a normal cellular receptor binding activity when compared to apoE$_3$ (9). The results of these and other studies have led to the proposal that apoE has a major role in modulating the catabolism of remnants of triglyceride-rich lipoprotein particles from the intestine.

---

[*]Abbreviations: apo, apolipoprotein; VLDL, very low density lipoproteins; IDL, intermediate density lipoproteins; LDL, low density lipoproteins; HDL, high density lipoproteins; d, density; FCR, fractional catabolic rate

The plasma cholesterol and LDL cholesterol concentrations are regulated by multiple factors. These include genetic, hormonal, and environmental factors. The most extensively investigated modulator of the LDL cholesterol concentration is the LDL receptor (10). More recently, epidemiologic studies have indicated that the different apoE phenotypes are of major importance in determining plasma and LDL cholesterol concentrations (11-15). Compared to the normal homozygous apoE$_3$ phenotype, results from studies have demonstrated that apoE$_2$ heterozygosity and homozygosity are associated with a decrease in the total plasma and LDL cholesterol concentration (11,12) while apoE$_4$ heterozygosity and homozygosity are associated with an increase in these levels (13-15). The different apoE phenotypes are of more significance in the modulation of the cholesterol level in the population, as a whole, than are mutations in the LDL receptor (16). The mechanism by which apoE modulates LDL metabolism has not been extensively investigated. In order to investigate this mechanism, we performed a series of _in vivo_ kinetic investigations.

IN VIVO KINETICS OF APOLIPOPROTEIN E

The kinetics of the three apoE isoforms were investigated utilizing radioiodinated apoE (8,17-19). ApoE was isolated from subjects homozygous for each of the three common apoE alleles. VLDL from each phenotype were isolated by ultracentrifugation, delipidated with chloroform/methanol, and the apoE isolated by heparin affinity and gel permeation chromatography. Each apoE isoform was iodinated by the iodine monochloride procedure with either ^{125}I or ^{131}I, reassociated with lipoproteins in plasma, and the lipoprotein associated radioiodinated apoE isolated by ultracentrifugation at a density of 1.21 g/ml. Radiolabeled apoE containing lipoproteins were injected into study subjects eating a controlled diet, plasma samples obtained at periodic time intervals for up to seven days, and the radioactivity quantitated in the plasma and lipoprotein subfractions. The mass of apoE was quantitated by radioimmunoassay, the residence times calculated by computer assisted multiexponential curve fitting (SAAM 27), and the production rates calculated.

The kinetics of apoE$_2$ and apoE$_4$ were compared to apoE$_3$ in subjects with a homozygous E$_3$ phenotype. All three isoforms were analyzed in subjects with the same phenotype so that the only variable being investigated was the isoform of the injected radioiodinated apoE. This approach has proven to be the most effective way to compare the kinetics of the different isoforms of apoE.

The catabolism of apoE$_3$ from plasma was multiexponential and the residence time was 0.7 days (19). The decay was most rapid within VLDL and was the slowest from HDL. The residence time of apoE$_2$ was slightly prolonged in these subjects at 0.8 days (17) which was significant at p < 0.01. In the lipoprotein subfractions, approximately twice as much apoE$_2$ as apoE$_3$ was present on HDL, and only one-half as much apoE$_2$ was on VLDL (20). ApoE$_2$, like apoE$_3$ was catabolized most rapidly from VLDL and the slowest from HDL.

When the kinetics of apoE$_4$ were compared to apoE$_3$, the opposite situation was observed. ApoE$_4$ had a shortened residence time of 0.4 days (p < 0.001) (18), there was twice as much radioiodinated apoE$_4$ on VLDL as apoE$_3$, and there was one-half the concentration of apoE$_4$ on HDL (20). As was the case for apoE$_2$ and apoE$_3$, apoE$_4$ was catabolized the most rapidly from VLDL and slowest from HDL. In summary, both apoE$_2$ and apoE$_4$ were metabolically abnormal, when compared to apoE$_3$. They both have abnormal affinities for the different lipoprotein density fractions, and they have abnormal rates of catabolism _in vivo_ in humans, but the abnormalities are in opposite directions for apoE$_2$ and apoE$_4$. The results of the _in vivo_ metabolism of apoE$_2$ are consistent with the _in vitro_ receptor binding data for apoE$_2$ (9). The _in vitro_ binding data for apoE$_4$ (9), however, is at variance with the _in vivo_ metabolic studies since the binding studies reported for apoE$_4$ suggests normal receptor binding properties. The large population studies and the _in vivo_ results are consistent with apoE$_4$ being metabolically abnormal.

LOW DENSITY LIPOPROTEIN METABOLISM IN SUBJECTS WITH A HOMOZYGOUS E$_2$ PHENOTYPE

In order to investigate the mechanism of the decreased LDL concentrations in subjects with a homozygous E$_2$ phenotype, _in vivo_ LDL apoB kinetic studies were performed (21). LDL (d 1.019-1.063 g/ml) were isolated by ultracentrifugation from normolipidemic subjects with either a homozygous E$_2$ or E$_3$ phenotype. The isolated LDL were iodinated by the iodine monochloride method with either ^{125}I or ^{131}I, and both types of LDL were injected into normolipidemic subjects homozygous for each apoE isoform. Blood samples were obtained periodically for 14 days, the radioactivity quantitated in plasma, and the rate of removal of LDL apoB from plasma determined by computer assisted curve fitting. The LDL apoB mass was quantitated by radioimmunoassay, and the production rates of apoB were calculated.

In both the normolipidemic E_2 or E_3 subjects, the E_2 LDL (LDL isolated from the subject with a homozygous E_2 phenotype) had a slower FCR than the E_3 LDL (21). In addition, the FCR of the LDL from either type of subject was faster in the subjects with the homozygous E_2 phenotype than in the homozygous E_3 subjects. When the metabolic parameters for the LDL from the E_2 subject injected into the E_2 subject were compared to the metabolic parameters for the LDL from the E_3 subject injected into the E_3 subject, the following results were obtained. The LDL apoB mass was 16 mg/dl vs 63 mg/dl, the LDL apoB fractional catabolic rate was 0.70 day^{-1} vs 0.46 day^{-1}, and the LDL apoB production rate was 4.5 mg/kg–day vs 11.8 mg/kg–day for E_2 and the E_3 subjects, respectively. The following conclusions were drawn: 1) the LDL from E_2 subjects is metabolically abnormal and is catabolized more slowly than LDL from E_3 subjects, 2) the subjects homozygous for the E_2 phenotype have an upregulated fractional catabolic rate for LDL, 3) the E_2 subjects have a decreased LDL synthetic rate, and 4) the low concentrations of LDL in the E_2 subjects are due to both the increased fractional catabolic rate and the decreased synthetic rate in these individuals. These results are consistent with apoE having an important modulatory affect on LDL apoB metabolism. First, the data indicate that apoE modulates, either directly or indirectly, the rate of synthesis and/or conversion of lipoproteins to LDL. Secondly, these results are consistent with $apoE_3$ being necessary for the formation of metabolically normal LDL. Thirdly, apoE regulates the FCR of LDL. We propose that this regulation is achieved by modulation of the number of LDL receptors which is influenced by the delivery of cholesterol to cells via an apoE mediated lipoprotein remnant catabolic pathway. $ApoE_2$ and $apoE_4$ have opposite _in vivo_ metabolic abnormalities, and therefore, it is also very possible that they may also have opposite effects on the modulation of LDL metabolism. If this is, in fact, the case, it would provide a metabolic explanation for the elevated LDL cholesterol and apoB concentrations in subjects with an $apoE_4$ phenotype.

METABOLISM OF ApoB–48 AND ApoB–100 IN SUBJECTS WITH HOMOZYGOUS FAMILIAL HYPERCHOLESTEROLEMIA AND WITH ApoE DEFICIENCY

In humans triglyceride-rich lipoprotein particles containing apoB–48 or apoB–100 are secreted by the intestine and the liver, respectively (22). These lipoprotein particles and their remnants contain three apolipoproteins that are potential ligands for the LDL (or apoB/E) receptor, apoB-48, apoB-100, and apoE. In order to further delineate the role of apoE in modulating the metabolism of remnants containing either apoB-48 or apoB-100, and the role of the LDL receptor in modulating the metabolism of these same remnants, the following studies were performed. Lipoproteins of d < 1.006 g/ml from a subject with apoE deficiency, which contain remnants of both intestinal and hepatic triglyceride-rich lipoprotein particles (23), were isolated by ultracentrifugation, radioiodinated with ^{125}I by the iodine monochloride method, and injected into normal subjects, subjects with homozygous familial hypercholesterol-emia, and a subject with apoE deficiency. Blood was obtained periodically over a 14 day period, and VLDL, IDL, and LDL density fractions were isolated by sequential ultracentrifugation. In each of these fractions the apoB-48 and apoB-100 were separated by agarose-acrylamide gel electrophoresis (24), the respective bands were cut from the gels, and the radioactivity in each band at each time point quantitated. From these data the resident times (1/fractional catabolic rate) were determined for apoB-48 and apoB-100 in VLDL for each subject. In addition, the fraction of apoB-100 and apoB-48 that was directly catabolized from VLDL, and the fraction that was converted to IDL and LDL was estimated.

In the normal subjects the apoB-48 containing remnant particles were catabolized very rapidly from VLDL with a residence time of approximately 0.5 hours (25,26). In addition, there was virtually no conversion of the apoB-48 to IDL or LDL. The metabolism of apoB-48 containing remnants in the homozygous familial hypercholesterolic subjects was very similar to the normal subjects with very rapid catabolism of apoB-48 from VLDL and virtually no conversion to IDL or LDL (26). These results are consistent with the LDL receptor not playing an important role in the catabolism of apoB-48 containing remnant particles. A very different situation was observed in the subject with apoE deficiency (25). The residence time of apoB-48 containing remnant particles was greatly prolonged in VLDL, and there was partial conversion of apoB-48 to IDL in this subject which occurred at a slow rate. These results indicate that apoE is necessary for the rapid catabolism of apoB-48 containing lipoprotein remnant particles, and that apoE mediated apoB-48 catabolism is not through the LDL receptor. In addition, in the absence of rapid catabolism, there can be additional remodeling of these particles resulting in the slow formation of apoB-48 containing particles with a hydrated density greater than 1.006 g/ml.

The metabolism of remnants of triglyceride-rich lipoprotein particles containing apoB-100 was strikingly different than for the apoB-48 containing remnants. In the normal subjects the apoB-100 containing remnants were metabolized from the VLDL density range with a residence time of approximately one hour with 85% of this metabolism involving the direct degradation of apoB-100 from VLDL and the remaining 15% being converted onto IDL and LDL (26). In the subjects with homozygous familial hypercholesterolemia, the residence time of these apoB-100 containing remnant particles was more than three times longer in normal subjects. In addition, in contrast to the normal subjects, approximately 50 percent of the apoB-100 containing remnants were metabolized to IDL and LDL while the other one-half were directly degraded from VLDL, even in the absence of functional LDL receptors. These results are interpreted as indicating that the LDL receptor is important in modulating the metabolism of apoB-100 containing remnant particles since in the absence of functional LDL receptors, there is a slower catabolic rate for apoB-100 remnants and a greater fraction of these remnants are metabolized to IDL and LDL. This last result could explain, at least in part, the over production of LDL apoB-100 in subjects with homozygous familial hypercholesterolemia. These results, also, indicate that there are at least two mechanisms for the direct catabolism of apoB-100 containing remnants from VLDL. The first involves direct catabolism mediated by the LDL receptor. In the homozygous familial hypercholesterolemic subjects, though, one-half of the apoB-100 remnants were still rapidly catabolized directly from VLDL by a second mechanism that is independent of the LDL receptor. The metabolism of apoB-100 remnants from VLDL in the apoE deficient subject was extremely slow (25). Both the rate of the direct degradation of apoB-100 from VLDL and the rate of the metabolism on to IDL and LDL were decreased in this subject compared to normals. This is consistent with apoE having an important modulatory role in the direct degradation of apoB-100 containing remnants from VLDL by both the LDL receptor dependent and the LDL receptor independent mechanisms, and also being important in modulating the metabolism of these remnants on to IDL and LDL. Therefore, apoE plays a central role in the regulation of metabolism of both apoB-48 and apoB-100 containing lipoprotein remnant particles.

METABOLIC SCHEME FOR THE ROLE OF ApoE IN MODULATING THE METABOLISM OF ApoB-48 AND ApoB-100 CONTAINING LIPOPROTEINS

Figure 1 schematically represents the proposed role for apoE in regulating the metabolism of apoB containing lipoproteins. Dietary lipids

294

are processed in the gastrointestinal tract and secreted as constituents of chylomicrons into the blood stream. The triglyceride on the newly secreted chylomicron particles undergoes hydrolysis by lipoprotein lipase with the particles being converted to chylomicron remnants. The chylomicron remnants are then directly and rapidly catabolized from the plasma by an apoE mediated process that is independent of the LDL receptor. The internalized cholesterol can modulate the intracellular regulatory mechanisms for cholesterol metabolism. As an example, the

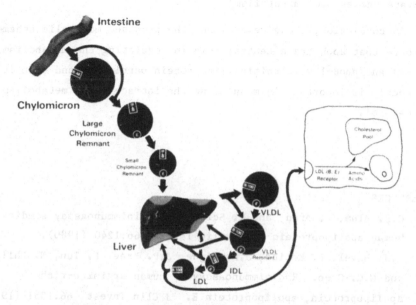

Figure 1. Schematic representation of the role of apoE in modulating the metabolism of apoB containing lipoproteins.

defective internalization of lipoprotein cholesterol associated with an E_2 phenotype would lead to decreased intracellular cholesterol and an up regulation of the LDL receptor. Conversely, if apoE$_4$ promotes the rapid catabolism of chylomicron and VLDL remnants, there would be an increased intracellular cholesterol pool which would result in a down regulation of the LDL receptor. Endogenous lipid is processed in the liver and secreted as constituents of VLDL. Similarly to chylomicrons, the triglyceride in the VLDL is hydrolyzed by lipoprotein lipase as the particles are converted into VLDL remnants. These remnants have three possible metabolic fates. They may be further metabolized by an apoE mediated process to IDL, directly catabolized by an apoE:apoE receptor mediated process, or directly catabolized by an LDL receptor mediated process with either apoB or apoE being the ligand for the receptor. IDL has the same metabolic potential as the VLDL remnants. They may be processed to LDL by an apoE mediated process or directly catabolized by receptor mediated endocytosis by either the LDL or the apoE receptor. LDL is directly catabolized by the LDL receptor. At each step of the catabolic pathway, the internalized cholesterol can be utilized to modulate the various intracellular factors including the number of LDL receptors needed to regulate cholesterol metabolism.

In conclusion, these results and the proposed metabolic scheme indicate that apoE has a central role in regulating the metabolism of apoB-48 and apoB-100 containing lipoprotein particles, and that it indirectly is important in modulating the intracellular metabolism of cholesterol.

REFERENCES

1. C.B. Blum, L. Aron, and R. Sciacca. Radioimmunoassay studies on human apolipoprotein E. J Clin Invest 66:1240 (1980).

2. R.J. Havel, L. Kotite, J.L. Vigne, J.P. Kane, P. Tun, N. Phillips, and G.C. Chen. Radioimmunoassay of human arginine-rich apolipoprotein, apolipoprotein E. J Clin Invest 66:1351 (1980).

3. G. Utermann, M. Jaeschke, and J. Menzel. Familial hyperlipo-proteinemia type III: deficiency of a specific apolipoprotein (apoE-III) in the very low density lipoproteins. FEBS Lett 56:352 (1975).

4. V.I. Zannis, and J.L. Breslow. Human very low density lipoprotein apolipoprotein E isoprotein polymorphism is explained by genetic variation and post-translational modification. Biochemistry 20:1033 (1981).

5. G. Ghiselli, R.E. Gregg, L.A. Zech, E.J. Schaefer, and H.B. Brewer, Jr. Phenotype study of apolipoprotein E isoforms in hyperlipoproteinemia patients. Lancet ii:405 (1982).

6. S.C. Rall, Jr., K.H. Weisgraber, and R.W. Mahley. Human apolipoprotein E. The complete amino acid sequence. J Biol Chem 257:4171 (1982).

7. G. Ghiselli, E.J. Schaefer, L.A. Zech, R.E. Gregg, and H.B. Brewer, Jr. Increased prevalence of apolipoprotein E_4 in type V hyperlipoproteinemia. J Clin Invest 70:474 (1982).

8. R.E. Gregg, L.A. Zech, E.J. Schaefer, and H.B. Brewer, Jr. Type III hyperlipoproteinemia: defective metabolism of an abnormal apolipoprotein E. Science 211:584 (1981).

9. K. H. Weisgraber, T.L. Innerarity, and R.W. Mahley. Abnormal lipoprotein receptor-binding activity of the human E apoprotein due to cysteine-arginine interchange at a single site. J Biol Chem 257:2518 (1982).

10. J.L. Goldstein, M.S. Brown, in The Metabolic Basis of Inherited Disease, J.B. Stanbury, J.B. Wyngaarden, D.S. Frederickson, J.L. Goldstein, and M.S. Brown, Eds., Mc Graw-Hill, New York, N.Y. pp. 672-712 (1983).

11. G. Utermann, K.H. Vogelberg, A. Steinmetz, W. Schoenborn, N. Pruin, M. Jaeschke, M. Hees. Polymorphism of apolipoprotein E. II. Genetics of hyperlipoproteinemia type III. Clin Genet 15:37 (1979).

12. H.-J. Menzel, R.-G. Kladetzky, G. Assmann. Apolipoprotein E polymorphism and coronary artery disease. Arteriosclerosis 3:310 (1983).

13. D. Bouthillier, C.F. Sing, and J. Davignon. Apolipoprotein E phenotyping with a single gel method: Application to the study of informative matings. J Lipid Res 24:1060 (1983).

14. G. Utermann, I. Kindermann, H. Kaffarnik, and A. Steinmetz. Apolipoprotein E phenotypes and hyperlipidemia. Hum Genet 65:232 (1984).

15. G. Assmann, G. Schmitz, H.-J. Menzel, and H. Schulte. Apolipoprotein E polymorphism and hyperlipidemia. Clin Chem 30:641 (1984).

16. C.F. Sing, and J. Davignon. Role of the apolipoprotein E polymorphism in determining normal plasma lipid and lipoprotein variation. Am J Hum Genet 37:268 (1985).

17. H.B. Brewer, Jr., L.A. Zech, R.E. Gregg, D.S. Schwartz, and E.J. Schaefer. Type III hyperlipoproteinemia: diagnosis, molecular defects, pathology, and treatment. Ann Intern Med 98:623 (1983).

18. R.E. Gregg, R. Ronan, L.A. Zech, G. Ghiselli, E.J. Schaefer, and H.B. Brewer, Jr. Abnormal metabolism of apolipoprotein E_4. Arteriosclerosis 2:420a (1982).

19. R.E. Gregg, L.A. Zech, E.J. Schaefer, and H.B. Brewer, Jr. Apolipoprotein E metabolism in normolipoproteinemic human subjects. J Lipid Res 25:1167 (1984).

20. R.E. Gregg, E.J. Schaefer, L.A. Zech, H.B. Brewer, Jr. Lipoprotein distribution of radioiodinated type III and type V apolipoprotein E (apoE). Arteriosclerosis 1:85 (1981).

21. R.E. Gregg, L.A. Zech, C. Gabelli, D. Stark, D. Wilson, H.B. Brewer, Jr. LDL catabolism in normolipidemic apoE$_2$ homozygotes. Circulation 70, Supp II:312 (1984).

22. J.P. Kane. Apolipoprotein B: structure and metabolic heterogeneity. Ann Rev Physiol 45:632 (1983).

23. G. Ghiselli, E.J. Schaefer, P. Gascon, and H.B. Brewer, Jr. Type III hyperlipoproteinemia associated with plasma apolipoprotein E deficiency. Science 214:1239 (1981).

24. C. Gabelli, D.G. Stark, R.E. Gregg, H.B. Brewer, Jr. Separation of apolipoprotein B species by agarose-acrylamide gel electrophoresis. J Lipid Res (in press).

25. E.J. Schaefer, L.A. Zech, G. Ghiselli, H.B. Brewer, Jr. ApoB-48 metabolism: retarded catabolism due to apoE deficiency. Arteriosclerosis 2:424a (1982).

26. C. Gabelli, R.E. Gregg, L.A. Zech, J.M. Hoeg, H.B. Brewer, Jr. ApoB-48 catabolism in normal and homozygous familial hypercholesterolemic individuals. Clin Res 32:116a (1984).

Printed in the United States
by Baker & Taylor Publisher Services